Nonparametric Statistics with Applications to Science and Engineering

THE WILEY BICENTENNIAL–KNOWLEDGE FOR GENERATIONS

Each generation has its unique needs and aspirations. When Charles Wiley first opened his small printing shop in lower Manhattan in 1807, it was a generation of boundless potential searching for an identity. And we were there, helping to define a new American literary tradition. Over half a century later, in the midst of the Second Industrial Revolution, it was a generation focused on building the future. Once again, we were there, supplying the critical scientific, technical, and engineering knowledge that helped frame the world. Throughout the 20th Century, and into the new millennium, nations began to reach out beyond their own borders and a new international community was born. Wiley was there, expanding its operations around the world to enable a global exchange of ideas, opinions, and know-how.

For 200 years, Wiley has been an integral part of each generation's journey, enabling the flow of information and understanding necessary to meet their needs and fulfill their aspirations. Today, bold new technologies are changing the way we live and learn. Wiley will be there, providing you the must-have knowledge you need to imagine new worlds, new possibilities, and new opportunities.

Generations come and go, but you can always count on Wiley to provide you the knowledge you need, when and where you need it!

WILLIAM J. PESCE
PRESIDENT AND CHIEF EXECUTIVE OFFICER

PETER BOOTH WILEY
CHAIRMAN OF THE BOARD

Nonparametric Statistics with Applications to Science and Engineering

Paul H. Kvam

Georgia Institute of Technology
The H. Milton Stewart School of Industrial and Systems Engineering
Atlanta, GA

Brani Vidakovic

Georgia Institute of Technology and Emory University School of Medicine
The Wallace H. Coulter Department of Biomedical Engineering
Atlanta, GA

WILEY-INTERSCIENCE
A John Wiley & Sons, Inc., Publication

Published by John Wiley & Sons, Inc., Hoboken, New Jersey.
Published simultaneously in Canada.

For general information on our other products and services or for technical support, please contact our Customer Care Department within the United States at (800) 762-2974, outside the United States at (317) 572-3993 or fax (317) 572-4002.

Wiley also publishes its books in a variety of electronic formats. Some content that appears in print may not be available in electronic format. For information about Wiley products, visit our web site at www.wiley.com.

Wiley Bicentennial Logo: Richard J. Pacifico

Library of Congress Cataloging-in-Publication Data is available.

ISBN 978-0-470-08147-1

10 9 8 7 6 5 4 3 2 1

Contents

Preface

> Danger lies not in what we don't know, but in what we think we know that just ain't so.
>
> Mark Twain (1835 – 1910)

As Prefaces usually start, the author(s) explain why they wrote the book in the first place – and we will follow this tradition. Both of us taught the graduate course on nonparametric statistics at the School of Industrial and Systems Engineering at Georgia Tech (ISyE 6404) several times. The audience was always versatile: PhD students in Engineering Statistics, Electrical Engineering, Management, Logistics, Physics, to list a few. While comprising a non homogeneous group, all of the students had solid mathematical, programming and statistical training needed to benefit from the course. Given such a nonstandard class, the text selection was all but easy.

There are plenty of excellent monographs/texts dealing with nonparametric statistics, such as the encyclopedic book by Hollander and Wolfe, *Nonparametric Statistical Methods*, or the excellent evergreen book by Conover, *Practical Nonparametric Statistics*, for example. We used as a text the 3rd edition of Conover's book, which is mainly concerned with what most of us think of as traditional nonparametric statistics: proportions, ranks, categorical data, goodness of fit, and so on, with the understanding that the text would be supplemented by the instructor's handouts. Both of us ended up supplying an increasing number of handouts every year, for units such as density and function estimation, wavelets, Bayesian approaches to nonparametric problems, the EM algorithm, splines, machine learning, and other arguably modern nonparametric topics. About a year ago, we decided to merge the handouts and fill the gaps.

There are several novelties this book provides. We decided to intertwine informal comments that might be amusing, but tried to have a good balance. One could easily get carried away and produce a preface similar to that of celebrated Barlow and Proschan's, *Statistical Theory of Reliability and Life Testing: Probability Models*, who acknowledge greedy spouses and obnoxious children as an impetus to their book writing. In this spirit, we featured photos and sometimes biographic details of statisticians who made fundamental contributions to the field of nonparametric statistics, such as Karl Pearson, Nathan Mantel, Brad Efron, and Baron Von Munchausen.

Computing. Another specificity is the choice of computing support. The book is integrated with MATLAB© and for many procedures covered in this book, MATLAB's m-files or their core parts are featured. The choice of software was natural: engineers, scientists, and increasingly statisticians are communicating in the "MATLAB language." This language is, for example, taught at Georgia Tech in a core computing course that every freshman engineering student takes, and almost everybody around us "speaks MATLAB." The book's website:

`http://www2.isye.gatech.edu/NPbook`

contains most of the m-files and programming supplements easy to trace and download. For Bayesian calculation we used WinBUGS, a free software from Cambridge's Biostatistics Research Unit. Both MATLAB and WinBUGS are briefly covered in two appendices for readers less familiar with them.

Outline of Chapters. For a typical graduate student to cover the full breadth of this textbook, two semesters would be required. For a one-semester course, the instructor should necessarily cover Chapters 1–3, 5–9 to start. Depending on the scope of the class, the last part of the course can include different chapter selections.

Chapters 2–4 contain important background material the student needs to understand in order to effectively learn and apply the methods taught in a nonparametric analysis course. Because the ranks of observations have special importance in a nonparametric analysis, Chapter 5 presents basic results for order statistics and includes statistical methods to create tolerance intervals.

Traditional topics in estimation and testing are presented in Chapters 7–10 and should receive emphasis even to students who are most curious about advanced topics such as density estimation (Chapter 11), curve-fitting (Chapter 13) and wavelets (Chapter 14). These topics include a core of rank tests that are analogous to common parametric procedures (*e.g.*, t-tests, analysis of variance).

Basic methods of categorical data analysis are contained in Chapter 9. Although most students in the biological sciences are exposed to a wide variety of statistical methods for categorical data, engineering students and other students in the physical sciences typically receive less schooling in this quintessential branch of statistics. Topics include methods based on tabled data, chi-square tests and the introduction of general linear models. Also included in the first part of the book is the topic of "goodness of fit" (Chapter 6), which refers to testing data not in terms of some unknown parameters, but the unknown distribution that generated it. In a way, goodness of fit represents an interface between distribution-free methods and traditional parametric methods of inference, and both analytical and graphical procedures are presented. Chapter 10 presents the nonparametric alternative to maximum likelihood estimation and likelihood ratio based confidence intervals.

The term "regression" is familiar from your previous course that introduced you to statistical methods. Nonparametric regression provides an alternative method of analysis that requires fewer assumptions of the response variable. In Chapter 12 we use the regression platform to introduce other important topics that build on linear regression, including isotonic (constrained) regression, robust regression and generalized linear models. In Chapter 13, we introduce more general curve fitting methods. Regression models based on wavelets (Chapter 14) are presented in a separate chapter.

In the latter part of the book, emphasis is placed on nonparametric procedures that are becoming more relevant to engineering researchers and practitioners. Beyond the conspicuous rank tests, this text includes many of the newest nonparametric tools available to experimenters for data analysis. Chapter 17 introduces fundamental topics of statistical learning as a basis for data mining and pattern recognition, and includes discriminant analysis, nearest-neighbor classifiers, neural networks and binary classification trees. Computational tools needed for nonparametric analysis include bootstrap resampling (Chapter 15) and the EM Algorithm (Chapter 16). Bootstrap methods, in particular, have become indispensable for uncertainty analysis with large data sets and elaborate stochastic models.

The textbook also unabashedly includes a review of Bayesian statistics and an overview of nonparametric Bayesian estimation. If you are familiar with Bayesian methods, you might wonder what role they play in nonparametric statistics. Admittedly, the connection is not obvious, but in fact nonparametric Bayesian methods (Chapter 18) represent an important set of tools for complicated problems in statistical modeling and learning, where many of the models are nonparametric in nature.

The book is intended both as a reference text and a text for a graduate course. We hope the reader will find this book useful. All comments, suggestions, updates, and critiques will be appreciated.

Acknowledgments. Before anyone else we would like to thank our wives, Lori Kvam and Draga Vidakovic, and our families. Reasons they tolerated our disorderly conduct during the writing of this book are beyond us, but we love them for it.

We are especially grateful to Bin Shi, who supported our use of MATLAB and wrote helpful coding and text for the Appendix A. We are grateful to MathWorks Statistics team, especially to Tom Lane who suggested numerous improvements and updates in that appendix. Several individuals have helped to improve on the primitive drafts of this book, including Saroch Boonsiripant, Lulu Kang, Hee Young Kim, Jongphil Kim, Seoung Bum Kim, Kichun Lee, and Andrew Smith.

Finally, we thank Wiley's team, Melissa Yanuzzi, Jacqueline Palmieri and Steve Quigley, for their kind assistance.

PAUL H. KVAM
School of Industrial and System Engineering
Georgia Institute of Technology

BRANI VIDAKOVIC
School of Biomedical Engineering
Georgia Institute of Technology

1

Introduction

For every complex question, there is a simple answer.... and it is wrong.

H. L. Mencken

Jacob Wolfowitz (Figure 1.1a) first coined the term *nonparametric*, saying "We shall refer to this situation [*where a distribution is completely determined by the knowledge of its finite parameter set*] as the parametric case, and denote the opposite case, where the functional forms of the distributions are unknown, as the non-parametric case" (Wolfowitz, 1942). From that point on, nonparametric statistics was defined by what it is not: traditional statistics based on known distributions with unknown parameters. Randles, Hettmansperger, and Casella (2004) extended this notion by stating "nonparametric statistics can and should be broadly defined to include all methodology that does not use a model based on a single parametric family."

Traditional statistical methods are based on parametric assumptions; that is, that the data can be assumed to be generated by some well-known family of distributions, such as normal, exponential, Poisson, and so on. Each of these distributions has one or more parameters (e.g., the normal distribution has μ and σ^2), at least one of which is presumed unknown and must be inferred. The emphasis on the normal distribution in linear model theory is often justified by the central limit theorem, which guarantees *approximate normality* of sample means provided the sample sizes are large enough. Other distributions also play an important role in science and engineering. Physical failure mechanisms often characterize the lifetime distribution of industrial compo-

(a) (b)

Fig. 1.1 (a) Jacob Wolfowitz (1910–1981) and (b) Wassily Hoeffding (1914–1991), pioneers in nonparametric statistics.

nents (e.g., Weibull or lognormal), so parametric methods are important in reliability engineering.

However, with complex experiments and messy sampling plans, the generated data might not be attributed to any well-known distribution. Analysts limited to basic statistical methods can be trapped into making parametric assumptions about the data that are not apparent in the experiment or the data. In the case where the experimenter is not sure about the underlying distribution of the data, statistical techniques are needed which can be applied regardless of the true distribution of the data. These techniques are called *nonparametric methods*, or *distribution-free methods*.

> The terms nonparametric and distribution-free are not synonymous... Popular usage, however, has equated the terms ... Roughly speaking, a nonparametric test is one which makes no hypothesis about the value of a parameter in a statistical density function, whereas a distribution-free test is one which makes no assumptions about the precise form of the sampled population.
>
> J. V. Bradley (1968)

It can be confusing to understand what is implied by the word "nonparametric". What is termed *modern nonparametrics* includes statistical models that are quite refined, except the distribution for error is left unspecified. Wasserman's recent book *All Things Nonparametric* (Wasserman, 2005) emphasizes only modern topics in nonparametric statistics, such as curve fitting, density estimation, and wavelets. Conover's *Practical Nonparametric Statistics* (Conover, 1999), on the other hand, is a classic nonparametrics textbook, but mostly limited to traditional binomial and rank tests, contingency tables, and tests for goodness of fit. Topics that are not really under the distribution-free umbrella, such as robust analysis, Bayesian analysis, and statistical learning also have important connections to nonparametric statistics, and are all featured in this book. Perhaps this text could have been titled *A Bit Less of Parametric Statistics with Applications in Science and Engineering*, but it surely would have sold fewer copies. On the other hand, if sales were the primary objective, we would have titled this *Nonparametric Statistics for Dummies* or maybe *Nonparametric Statistics with Pictures of Naked People.*

1.1 EFFICIENCY OF NONPARAMETRIC METHODS

It would be a mistake to think that nonparametric procedures are simpler than their parametric counterparts. On the contrary, a primary criticism of using parametric methods in statistical analysis is that they oversimplify the population or process we are observing. Indeed, parametric families are not more useful because they are perfectly appropriate, rather because they are perfectly convenient.

Nonparametric methods are inherently less powerful than parametric methods. This must be true because the parametric methods are assuming more information to construct inferences about the data. In these cases the estimators are inefficient, where the efficiencies of two estimators are assessed by comparing their variances for the same sample size. This inefficiency of one method relative to another is measured in power in hypothesis testing, for example.

However, even when the parametric assumptions hold perfectly true, we will see that nonparametric methods are only slightly less powerful than the more presumptuous statistical methods. Furthermore, if the parametric assumptions about the data fail to hold, only the nonparametric method is valid. A t-test between the means of two normal populations can be dangerously misleading if the underlying data are not actually normally distributed. Some examples of the relative efficiency of nonparametric tests are listed in Table 1.1, where asymptotic relative efficiency (A.R.E.) is used to compare parametric procedures (2nd column) with their nonparametric counterparts (3rd column). Asymptotic relative efficiency describes the relative efficiency of two estimators of a parameter as the sample size approaches infinity. The A.R.E. is listed for the normal distribution, where parametric assumptions are justified, and the double-exponential distribution. For example, if the underlying data are normally distributed, the t-test requires 955 observations in order to have the same power of the Wilcoxon signed-rank test based on 1000 observations.

Parametric assumptions allow us to extrapolate away from the data. For example, it is hardly uncommon for an experimenter to make inferences about a population's extreme upper percentile (say 99th percentile) with a sample so small that none of the observations would be expected to exceed that percentile. If the assumptions are not justified, this is grossly unscientific.

Nonparametric methods are seldom used to extrapolate outside the range

Table 1.1 Asymptotic relative efficiency (A.R.E.) of some nonparametric tests

	Parametric Test	Nonparametric Test	A.R.E. (normal)	A.R.E. (double exp.)
2-Sample Test	t-test	Mann-Whitney	0.955	1.50
3-Sample Test	one-way layout	Kruskal-Wallis	0.864	1.50
Variances Test	F-test	Conover	0.760	1.08

of observed data. In a typical nonparametric analysis, little or nothing can be said about the probability of obtaining future data beyond the largest sampled observation or less than the smallest one. For this reason, the actual measurements of a sample item means less compared to its rank within the sample. In fact, nonparametric methods are typically based on *ranks* of the data, and properties of the population are deduced using *order statistics* (Chapter 5). The measurement scales for typical data are

Nominal Scale: Numbers used only to categorize outcomes (e.g., we might define a random variable to equal one in the event a coin flips heads, and zero if it flips tails).

Ordinal Scale: Numbers can be used to order outcomes (e.g., the event X is greater than the event Y if X = *medium* and Y = *small*).

Interval Scale: Order between numbers as well as distances between numbers are used to compare outcomes.

Only interval scale measurements can be used by parametric methods. Nonparametric methods based on ranks can use ordinal scale measurements, and simpler nonparametric techniques can be used with nominal scale measurements.

The binomial distribution is characterized by counting the number of independent observations that are classified into a particular category. Binomial data can be formed from measurements based on a *nominal scale* of measurements, thus binomial models are most encountered models in nonparametric analysis. For this reason, Chapter 3 includes a special emphasis on statistical estimation and testing associated with binomial samples.

1.2 OVERCONFIDENCE BIAS

Be slow to believe what you worst want to be true

Samual Pepys

Confirmation Bias or *Overconfidence Bias* describes our tendency to search for or interpret information in a way that confirms our preconceptions. Business and finance has shown interest in this psychological phenomenon (Tversky and Kahneman, 1974) because it has proven to have a significant effect on personal and corporate financial decisions where the decision maker will actively seek out and give extra weight to evidence that confirms a hypothesis they already favor. At the same time, the decision maker tends to ignore evidence that contradicts or disconfirms their hypothesis.

Overconfidence bias has a natural tendency to effect an experimenter's data analysis for the same reasons. While the dictates of the experiment and the data sampling should reduce the possibility of this problem, one of the clear pathways open to such bias is the infusion of parametric assumptions into the data analysis. After all, if the assumptions seem plausible, the researcher has much to gain from the extra certainty that comes from the assumptions in terms of narrower confidence intervals and more powerful statistical tests.

Nonparametric procedures serve as a buffer against this human tendency of looking for the evidence that best supports the researcher's underlying hypothesis. Given the subjective interests behind many corporate research findings, nonparametric methods can help alleviate doubt to their validity in cases when these procedures give statistical significance to the corporations's claims.

1.3 COMPUTING WITH MATLAB

Because a typical nonparametric analysis can be computationally intensive, computer support is essential to understand both theory and applications. Numerous software products can be used to complete exercises and run nonparametric analysis in this textbook, including SAS, **R**, S-Plus, MINITAB, StatXact and JMP (to name a few). A student familiar with one of these platforms can incorporate it with the lessons provided here, and without too much extra work.

It must be stressed, however, that demonstrations in this book rely entirely on a single software tool called MATLAB© (by MathWorks Inc.) that is used widely in engineering and the physical sciences. MATLAB (short for *MATrix LABoratory*) is a flexible programming tool that is widely popular in engineering practice and research. The program environment features user-friendly front-end and includes menus for easy implementation of program commands. MATLAB is available on Unix systems, Microsoft Windows and

Apple Macintosh. If you are unfamiliar with MATLAB, in the first appendix we present a brief tutorial along with a short description of some MATLAB procedures that are used to solve analytical problems and demonstrate nonparametric methods in this book. For a more comprehensive guide, we recommend the handy little book *MATLAB Primer* (Sigmon and Davis, 2002).

We hope that many students of statistics will find this book useful, but it was written primarily with the scientist and engineer in mind. With nothing against statisticians (some of our best friends know statisticians) our approach emphasizes the application of the method over its mathematical theory. We have intentionally made the text less heavy with theory and instead emphasized applications and examples. If you come into this course thinking the history of nonparametric statistics is dry and unexciting, you are probably right, at least compared to the history of ancient Rome, the British monarchy or maybe even Wayne Newton[1]. Nonetheless, we made efforts to convince you otherwise by noting the interesting historical context of the research and the personalities behind its development. For example, we will learn more about Karl Pearson (1857–1936) and R. A. Fisher (1890–1962), legendary scientists and competitive arch-rivals, who both contributed greatly to the foundation of nonparametric statistics through their separate research directions.

Fig. 1.2 "Doubt is not a pleasant condition, but certainty is absurd" – Francois Marie Voltaire (1694–1778).

In short, this book features techniques of data analysis that rely less on the assumptions of the data's good behavior – the very assumptions that can get researchers in trouble. Science's gravitation toward distribution-free techniques is due to both a deeper awareness of experimental uncertainty and the availability of ever-increasing computational abilities to deal with the implied ambiguities in the experimental outcome. The quote from Voltaire

[1]Strangely popular Las Vegas entertainer.

(Figure 1.2) exemplifies the attitude toward uncertainty; as science progresses, we are able to see some truths more clearly, but at the same time, we uncover more uncertainties and more things become less "black and white".

1.4 EXERCISES

1.1. Describe a potential data analysis in engineering where parametric methods are appropriate. How would you defend this assumption?

1.2. Describe another potential data analysis in engineering where parametric methods may not be appropriate. What might prevent you from using parametric assumptions in this case?

1.3. Describe three ways in which overconfidence bias can affect the statistical analysis of experimental data. How can this problem be overcome?

REFERENCES

Bradley, J. V. (1968), *Distribution Free Statistical Tests*, Englewood Cliffs, NJ: Prentice Hall.

Conover, W. J. (1999), *Practical Nonparametric Statistics*, New York: Wiley.

Randles, R. H., Hettmansperger, T.P., and Casella, G. (2004), Introduction to the Special Issue "Nonparametric Statistics," *Statistical Science*, 19, 561-562.

Sigmon, K., and Davis, T.A. (2002), *MATLAB Primer*, 6th Edition, MathWorks, Inc., Boca Raton, FL: CRC Press.

Tversky, A., and Kahneman, D. (1974), "Judgment Under Uncertainty: Heuristics and Biases," *Science*, 185, 1124-1131.

Wasserman, L. (2006), *All Things Nonparametric*, New York: Springer Verlag.

Wolfowitz, J. (1942), "Additive Partition Functions and a Class of Statistical Hypotheses," *Annals of Statistics*, 13, 247-279.

2

Probability Basics

Probability theory is nothing but common sense reduced to calculation.

Pierre Simon Laplace (1749-1827)

In these next two chapters, we review some fundamental concepts of elementary probability and statistics. If you think you can use these chapters to catch up on all the statistics you forgot since you passed "Introductory Statistics" in your college sophomore year, you are acutely mistaken. What is offered here is an abbreviated reference list of definitions and formulas that have applications to nonparametric statistical theory. Some parametric distributions, useful for models in both parametric and nonparametric procedures, are listed but the discussion is abridged.

2.1 HELPFUL FUNCTIONS

- Permutations. The number of arrangements of n distinct objects is $n! = n(n-1)\dots(2)(1)$. In MATLAB: `factorial(n)`.
- Combinations. The number of distinct ways of choosing k items from a set of n is
$$\binom{n}{k} = \frac{n!}{k!(n-k)!}.$$
In MATLAB: `nchoosek(n,k)`.

- $\Gamma(t) = \int_0^\infty x^{t-1}e^{-x}dx, \quad t > 0$ is called the gamma function. If t is a positive integer, $\Gamma(t) = (t-1)!$. In MATLAB: `gamma(t)`.

- Incomplete Gamma is defined as $\gamma(t,z) = \int_0^z x^{t-1}e^{-x}dx$. In MATLAB: `gammainc(t,z)`. The upper tail Incomplete Gamma is defined as $\Gamma(t,z) = \int_z^\infty x^{t-1}e^{-x}dx$, in MATLAB: `gammainc(t,z,'upper')`. If t is an integer,

$$\Gamma(t,z) = (t-1)!e^{-z}\sum_{i=0}^{t-1} z^i/i!.$$

- Beta Function. $B(a,b) = \int_0^1 t^{a-1}(1-t)^{b-1}dt = \Gamma(a)\Gamma(b)/\Gamma(a+b)$. In MATLAB: `beta(a,b)`.

- Incomplete Beta. $B(x,a,b) = \int_0^x t^{a-1}(1-t)^{b-1}dt, \ 0 \le x \le 1$. In MATLAB: `betainc(x,a,b)` represents normalized Incomplete Beta defined as $I_x(a,b) = B(x,a,b)/B(a,b)$.

- Summations of powers of integers:

$$\sum_{i=1}^n i = \frac{n(n+1)}{2}, \quad \sum_{i=1}^n i^2 = \frac{n(n+1)(2n+1)}{6}, \quad \sum_{i=1}^n i^3 = \left(\frac{n(n+1)}{2}\right)^2.$$

- Floor Function. $\lfloor a \rfloor$ denotes the greatest integer $\le a$. In MATLAB: `floor(a)`.

- Geometric Series.

$$\sum_{j=0}^n p^j = \frac{1-p^{n+1}}{1-p}, \text{ so that for } |p| < 1, \sum_{j=0}^\infty p^j = \frac{1}{1-p}.$$

- Stirling's Formula. To approximate the value of a large factorial,

$$n! \approx \sqrt{2\pi}e^{-n}n^{n+1/2}.$$

- Common Limit for e. For a constant α,

$$\lim_{x \longrightarrow 0}(1+\alpha x)^{1/x} = e^\alpha.$$

This can also be expressed as $(1+\alpha/n)^n \longrightarrow e^\alpha$ as $n \longrightarrow \infty$.

- Newton's Formula. For a positive integer n,

$$(a+b)^n = \sum_{j=0}^{n} \binom{n}{j} a^j b^{n-j}.$$

- Taylor Series Expansion. For a function $f(x)$, its Taylor series expansion about $x = a$ is defined as

$$f(x) = f(a) + f'(a)(x-a) + f''(a)\frac{(x-a)^2}{2!} + \cdots + f^{(k)}(a)\frac{(x-a)^k}{k!} + R_k,$$

where $f^{(m)}(a)$ denotes m^{th} derivative of f evaluated at a and, for some $\bar{a}$ between a and x,

$$R_k = f^{(k+1)}(\bar{a})\frac{(x-a)^{k+1}}{(k+1)!}.$$

- Convex Function. A function h is *convex* if for any $0 \le \alpha \le 1$,

$$h(\alpha x + (1-\alpha)y) \le \alpha h(x) + (1-\alpha)h(y),$$

for all values of x and y. If h is twice differentiable, then h is convex if $h''(x) \ge 0$. Also, if $-h$ is convex, then h is said to be *concave*.

- Bessel Function. $J_n(x)$ is defined as the solution to the equation

$$x^2 \frac{\partial^2 y}{\partial x^2} + x\frac{\partial y}{\partial x} + (x^2 - n^2)y = 0.$$

In MATLAB: `bessel(n,x)`.

2.2 EVENTS, PROBABILITIES AND RANDOM VARIABLES

- The *conditional probability* of an event A occurring given that event B occurs is $P(A|B) = P(AB)/P(B)$, where AB represents the intersection of events A and B, and $P(B) > 0$.

- Events A and B are stochastically *independent* if and only if $P(A|B) = P(B)$ or equivalently, $P(AB) = P(A)P(B)$.

- *Law of Total Probability.* Let $A_1, \ldots, A_k$ be a partition of the sample space Ω, i.e., $A_1 \cup A_2 \cup \cdots \cup A_k = \Omega$ and $A_iA_j = \emptyset$ for $i \ne j$. For event B, $P(B) = \sum_i P(B|A_i)P(A_i)$.

- *Bayes Formula.* For an event B where $P(B) \ne 0$, and partition

$(A_1, \ldots, A_k)$ of Ω,

$$P(A_j|B) = \frac{P(B|A_j)P(A_j)}{\sum_i P(B|A_i)P(A_i)}.$$

- A function that assigns real numbers to points in the sample space of events is called a *random variable*.[1]

- For a random variable X, $F_X(x) = P(X \le x)$ represents its (cumulative) *distribution function*, which is non-decreasing with $F(-\infty) = 0$ and $F(\infty) = 1$. In this book, it will often be denoted simply as CDF. The *survivor function* is defined as $S(x) = 1 - F(x)$.

- If the CDF's derivative exists, $f(x) = \partial F(x)/\partial x$ represents the *probability density function*, or PDF.

- A *discrete random variable* is one which can take on a countable set of values $X \in \{x_1, x_2, x_3, \ldots\}$ so that $F_X(x) = \sum_{t \le x} P(X = t)$. Over the support X, the probability $P(X = x_i)$ is called the probability mass function, or PMF.

- A *continuous random variable* is one which takes on any real value in an interval, so $P(X \in A) = \int_A f(x)dx$, where $f(x)$ is the density function of X.

- For two random variables X and Y, their *joint distribution function* is $F_{X,Y}(x, y) = P(X \le x, Y \le y)$. If the variables are continuous, one can define joint density function $f_{X,Y}(x, y)$ as $\frac{\partial^2}{\partial x \partial y} F_{X,Y}(x, y)$. The conditional density of X, given $Y = y$ is $f(x|y) = f_{X,Y}(x, y)/f_Y(y)$, where $f_Y(y)$ is the density of Y.

- Two random variables X and Y, with distributions F_X and F_Y, are *independent* if the joint distribution $F_{X,Y}$ of (X, Y) is such that $F_{X,Y}(x, y) = F_X(x)F_Y(y)$. For any sequence of random variables $X_1, \ldots, X_n$ that are independent with the same (identical) marginal distribution, we will denote this using *i.i.d.*

2.3 NUMERICAL CHARACTERISTICS OF RANDOM VARIABLES

- For a random variable X with distribution function F_X, the *expected value* of some function $\phi(X)$ is defined as $\mathbb{E}(\phi(X)) = \int \phi(x)dF_X(x)$. If

[1] While writing their early textbooks in Statistics, J. Doob and William Feller debated on whether to use this term. Doob said, "I had an argument with Feller. He asserted that everyone said *random variable* and I asserted that everyone said *chance variable*. We obviously had to use the same name in our books, so we decided the issue by a stochastic procedure. That is, we tossed for it and he won."

F_X is continuous with density $f_X(x)$, then $\mathbb{E}(\phi(X)) = \int \phi(x) f_X(x)dx$. If X is discrete, then $\mathbb{E}(\phi(X)) = \sum_x \phi(x)P(X = x)$.

- The kth *moment* of X is denoted as $\mathbb{E}X^k$. The kth moment about the mean, or k^{th} central moment of X is defined as $\mathbb{E}(X - \mu)^k$, where $\mu = \mathbb{E}X$.

- The *variance* of a random variable X is the second central moment, $\mathbb{V}\text{ar}X = \mathbb{E}(X - \mu)^2 = \mathbb{E}X^2 - (\mathbb{E}X)^2$. Often, the variance is denoted by σ_X^2, or simply by σ^2 when it is clear which random variable is involved. The square root of variance, $\sigma_X = \sqrt{\mathbb{V}\text{ar}X}$, is called the standard deviation of X.

- With $0 \le p \le 1$, the pth *quantile* of F, denoted x_p is the value x such that $P(X \le x) \ge p$ and $P(X \ge x) \ge 1 - p$. If the CDF F is invertible, then $x_p = F^{-1}(p)$. The 0.5^{th} quantile is called the *median* of F.

- For two random variables X and Y, the *covariance* of X and Y is defined as $\mathbb{C}\text{ov}(X, Y) = \mathbb{E}[(X - \mu_X)(Y - \mu_Y)]$, where μ_X and μ_Y are the respective expectations of X and Y.

- For two random variables X and Y with covariance $\mathbb{C}\text{ov}(X, Y)$, the *correlation coefficient* is defined as

$$\mathbb{C}\text{orr}(X, Y) = \frac{\mathbb{C}\text{ov}(X, Y)}{\sigma_X \sigma_Y},$$

where σ_X and σ_Y are the respective standard deviations of X and Y. Note that $-1 \le \rho \le 1$ is a consequence of the Cauchy-Schwartz inequality (Section 2.8).

- The *characteristic function* of a random variable X is defined as

$$\varphi_X(t) = \mathbb{E}e^{itX} = \int e^{itx} dF(x).$$

The *moment generating function* of a random variable X is defined as

$$m_X(t) = \mathbb{E}e^{tX} = \int e^{tx} dF(x),$$

whenever the integral exists. By differentiating r times and letting $t \to 0$ we have that

$$\frac{d^r}{dt^r} m_X(0) = \mathbb{E}X^r.$$

- The *conditional expectation* of a random variable X is given $Y = y$ is defined as

$$\mathbb{E}(X|Y = y) = \int x f(x|y) dx,$$

where $f(x|y)$ is a conditional density of X given Y. If the value of Y is not specified, the conditional expectation $\mathbb{E}(X|Y)$ is a random variable and its expectation is $\mathbb{E}X$, that is, $\mathbb{E}(\mathbb{E}(X|Y)) = \mathbb{E}X$.

2.4 DISCRETE DISTRIBUTIONS

Ironically, parametric distributions have an important role to play in the development of nonparametric methods. Even if we are analyzing data without making assumptions about the distributions that generate the data, these parametric families appear nonetheless. In counting trials, for example, we can generate well-known discrete distributions (e.g., binomial, geometric) assuming only that the counts are independent and probabilities remain the same from trial to trial.

2.4.1 Binomial Distribution

A simple Bernoulli random variable Y is dichotomous with $P(Y = 1) = p$ and $P(Y = 0) = 1-p$ for some $0 \le p \le 1$. It is denoted as $Y \sim \mathcal{B}er(p)$. Suppose an experiment consists of n independent trials $(Y_1, \dots, Y_n)$ in which two outcomes are possible (e.g., success or failure), with $P(\text{success}) = P(Y = 1) = p$ for each trial. If $X = x$ is defined as the number of successes (out of n), then $X = Y_1 + Y_2 + \cdots + Y_n$ and there are $\binom{n}{x}$ arrangements of x successes and $n-x$ failures, each having the same probability $p^x(1-p)^{n-x}$. X is a *binomial* random variable with probability mass function

$$p_X(x) = \binom{n}{x} p^x (1-p)^{n-x}, \quad x = 0, 1, \dots, n.$$

This is denoted by $X \sim \mathcal{B}in(n, p)$. From the moment generating function $m_X(t) = (pe^t + (1-p))^n$, we obtain $\mu = \mathbb{E}X = np$ and $\sigma^2 = \mathbb{V}\text{ar}X = np(1-p)$.

The cumulative distribution for a binomial random variable is not simplified beyond the sum; i.e., $F(x) = \sum_{i \le x} p_X(i)$. However, interval probabilities can be computed in MATLAB using `binocdf(x,n,p)`, which computes the cumulative distribution function at value x. The probability mass function is also computed in MATLAB using `binopdf(x,n,p)`. A "quick-and-dirty" plot of a binomial PDF can be achieved through the MATLAB function `binoplot`.

2.4.2 Poisson Distribution

The probability mass function for the Poisson distribution is

$$p_X(x) = \frac{\lambda^x}{x!} e^{-\lambda}, \qquad x = 0, 1, 2, \ldots$$

This is denoted by $X \sim \mathcal{P}(\lambda)$. From $m_X(t) = \exp\{\lambda(e^t - 1)\}$, we have $\mathbb{E}X = \lambda$ and $\mathbb{V}\text{ar}X = \lambda$; the mean and the variance coincide.

The sum of a finite independent set of Poisson variables is also Poisson. Specifically, if $X_i \sim \mathcal{P}(\lambda_i)$, then $Y = X_1 + \cdots + X_k$ is distributed as $\mathcal{P}(\lambda_1 + \cdots + \lambda_k)$. Furthermore, the Poisson distribution is a limiting form for a binomial model, i.e.,

$$\lim_{n, np \to \infty, \lambda} \binom{n}{x} p^x (1-p)^{n-x} = \frac{1}{x!} \lambda^x e^{-\lambda}. \tag{2.1}$$

MATLAB commands for Poisson CDF, PDF, quantile, and a random number are: `poisscdf`, `poisspdf`, `poissinv`, and `poissrnd`.

2.4.3 Negative Binomial Distribution

Suppose we are dealing with i.i.d. trials again, this time counting the number of successes observed until a fixed number of failures (k) occur. If we observe k consecutive failures at the start of the experiment, for example, the count is $X = 0$ and $P_X(0) = p^k$, where p is the probability of failure. If $X = x$, we have observed x successes and k failures in $x + k$ trials. There are $\binom{x+k}{x}$ different ways of arranging those $x + k$ trials, but we can only be concerned with the arrangements in which the last trial ended in a failure. So there are really only $\binom{x+k-1}{x}$ arrangements, each equal in probability. With this in mind, the probability mass function is

$$p_X(x) = \binom{k + x - 1}{x} p^k (1-p)^x, \qquad x = 0, 1, 2, \ldots$$

This is denoted by $X \sim \mathcal{NB}(k, p)$. From its moment generating function

$$m(t) = \left(\frac{p}{1 - (1-p)e^t} \right)^k,$$

the expectation of a negative binomial random variable is $\mathbb{E}X = k(1-p)/p$ and variance $\mathbb{V}\text{ar}X = k(1-p)/p^2$. MATLAB commands for negative binomial CDF, PDF, quantile, and a random number are: `nbincdf`, `nbinpdf`, `nbininv`, and `nbinrnd`.

2.4.4 Geometric Distribution

The special case of negative binomial for $k = 1$ is called the geometric distribution. Random variable X has geometric $\mathcal{G}(p)$ distribution if its probability mass function is

$$p_X(x) = p(1-p)^x, \qquad x = 0, 1, 2, \ldots$$

If X has geometric $\mathcal{G}(p)$ distribution, its expected value is $\mathbb{E}X = (1-p)/p$ and variance $\mathbb{V}\text{ar}X = (1-p)/p^2$. The geometric random variable can be considered as the discrete analog to the (continuous) exponential random variable because it possesses a "memoryless" property. That is, if we condition on $X \geq m$ for some non-negative integer m, then for $n \geq m$, $P(X \geq n | X \geq m) = P(X \geq n - m)$. MATLAB commands for geometric CDF, PDF, quantile, and a random number are: `geocdf`, `geopdf`, `geoinv`, and `geornd`.

2.4.5 Hypergeometric Distribution

Suppose a box contains m balls, k of which are white and $m - k$ of which are gold. Suppose we randomly select and remove n balls from the box *without replacement*, so that when we finish, there are only $m - n$ balls left. If X is the number of white balls chosen (without replacement) from n, then

$$p_X(x) = \frac{\binom{k}{x}\binom{m-k}{n-x}}{\binom{m}{n}}, \qquad x \in \{0, 1, \ldots, \min\{n, k\}\}.$$

This probability mass function can be deduced with counting rules. There are $\binom{m}{n}$ different ways of selecting the n balls from a box of m. From these (each equally likely), there are $\binom{k}{x}$ ways of selecting x white balls from the k white balls in the box, and similarly $\binom{m-k}{n-x}$ ways of choosing the gold balls.

It can be shown that the mean and variance for the hypergeometric distribution are, respectively,

$$\mathbb{E}(X) = \mu = \frac{nk}{m} \text{ and } \mathbb{V}\text{ar}(X) = \sigma^2 = \left(\frac{nk}{m}\right)\left(\frac{m-k}{m}\right)\left(\frac{m-n}{m-1}\right).$$

MATLAB commands for Hypergeometric CDF, PDF, quantile, and a random number are: `hygecdf`, `hygepdf`, `hygeinv`, and `hygernd`.

2.4.6 Multinomial Distribution

The binomial distribution is based on dichotomizing event outcomes. If the outcomes can be classified into $k \geq 2$ categories, then out of n trials, we have X_i outcomes falling in the category i, $i = 1, \ldots, k$. The probability mass function for the vector $(X_1, \ldots, X_k)$ is

$$p_{X_1,\ldots,X_k}(x_1, \ldots, x_k) = \frac{n!}{x_1! \cdots x_k!} {p_1}^{x_1} \cdots {p_k}^{x_k},$$

where $p_1 + \cdots + p_k = 1$, so there are $k-1$ free probability parameters to characterize the multivariate distribution. This is denoted by $\mathbf{X} = (X_1, \ldots, X_k) \sim \mathcal{M}n(n, p_1, \ldots, p_k)$.

The mean and variance of X_i is the same as a binomial because this is the marginal distribution of X_i, i.e., $\mathbb{E}(X_i) = np_i$, $\mathbb{V}\text{ar}(X_i) = np_i(1-p_i)$. The covariance between X_i and X_j is $\mathbb{C}\text{ov}(X_i, X_j) = -np_ip_j$ because $\mathbb{E}(X_iX_j) = \mathbb{E}(\mathbb{E}(X_iX_j|X_j)) = \mathbb{E}(X_j\mathbb{E}(X_i|X_j))$ and conditional on $X_j = x_j$, X_i is binomial $\mathcal{B}in(n-x_j, p_i/(1-p_j))$. Thus, $\mathbb{E}(X_iX_j) = \mathbb{E}(X_j(n-X_j))p_i/(1-p_j)$, and the covariance follows from this.

2.5 CONTINUOUS DISTRIBUTIONS

Discrete distributions are often associated with nonparametric procedures, but continuous distributions will play a role in how we learn about nonparametric methods. The normal distribution, of course, can be produced in a sample mean when the sample size is large, as long as the underlying distribution of the data has finite mean and variance. Many other distributions will be referenced throughout the text book.

2.5.1 Exponential Distribution

The probability density function for an exponential random variable is

$$f_X(x) = \lambda e^{-\lambda x}, \ x > 0, \lambda > 0.$$

An exponentially distributed random variable X is denoted by $X \sim \mathcal{E}(\lambda)$. Its moment generating function is $m(t) = \lambda/(\lambda - t)$ for $t < \lambda$, and the mean and variance are $1/\lambda$ and $1/\lambda^2$, respectively. This distribution has several interesting features - for example, its *failure rate*, defined as

$$r_X(x) = \frac{f_X(x)}{1 - F_X(x)},$$

is constant and equal to λ.

The exponential distribution has an important connection to the Poisson distribution. Suppose we measure i.i.d. exponential outcomes $(X_1, X_2, \ldots)$, and define $S_n = X_1 + \cdots + X_n$. For any positive value t, it can be shown that $P(S_n < t < S_{n+1}) = p_Y(n)$, where $p_Y(n)$ is the probability mass function for a Poisson random variable Y with parameter λt. Similar to a geometric random variable, an exponential random variable has the *memoryless property* because for $t > x$, $P(X \geq t | X \geq x) = P(X \geq t - x)$.

The median value, representing a typical observation, is roughly 70% of the mean, showing how extreme values can affect the population mean. This is easily shown because of the ease at which the inverse CDF is computed:

$$p \equiv F_X(x;\lambda) = 1 - e^{-\lambda x} \quad \Longleftrightarrow \quad {F_X}^{-1}(p) \equiv x_p = -\frac{1}{\lambda}\log(1-p).$$

MATLAB commands for exponential CDF, PDF, quantile, and a random number are: `expcdf`, `exppdf`, `expinv`, and `exprnd`. MATLAB uses the alternative parametrization with $1/\lambda$ in place of λ. For example, the CDF of random variable $X \sim \mathcal{E}(3)$ distribution evaluated at $x = 2$ is calculated in MATLAB as `expcdf(2, 1/3)`.

2.5.2 Gamma Distribution

The gamma distribution is an extension of the exponential distribution. Random variable X has gamma $\mathcal{G}amma(r, \lambda)$ distribution if its probability density function is given by

$$f_X(x) = \frac{\lambda^r}{\Gamma(r)} x^{r-1} e^{-\lambda x}, \; x > 0, \; r > 0, \; \lambda > 0.$$

The moment generating function is $m(t) = (\lambda/(\lambda - t))^r$, so in the case $r = 1$, gamma is precisely the exponential distribution. From $m(t)$ we have $\mathbb{E}X = r/\lambda$ and $\mathbb{V}\text{ar}X = r/\lambda^2$.

If $X_1, \ldots, X_n$ are generated from an exponential distribution with (rate) parameter λ, it follows from $m(t)$ that $Y = X_1 + \cdots + X_n$ is distributed gamma with parameters λ and n; that is, $Y \sim \mathcal{G}amma(n, \lambda)$. Often, the gamma distribution is parameterized with $1/\lambda$ in place of λ, and this alternative parametrization is used in MATLAB definitions. The CDF in MATLAB is `gamcdf(x, r, 1/lambda)`, and the PDF is `gampdf(x, r, 1/lambda)`. The function `gaminv(p, r, 1/lambda)` computes the pth quantile of the gamma.

2.5.3 Normal Distribution

The probability density function for a normal random variable with mean $\mathbb{E}X = \mu$ and variance $\mathbb{V}\text{ar}X = \sigma^2$ is

$$f_X(x) = \frac{1}{\sqrt{2\pi\sigma^2}} e^{-\frac{1}{2\sigma^2}(x-\mu)^2}, \qquad \infty < x < \infty.$$

The distribution function is computed using integral approximation because no closed form exists for the anti-derivative; this is generally not a problem for practitioners because most software packages will compute interval probabilities numerically. For example, in MATLAB, `normcdf(x, mu, sigma)` and `normpdf(x, mu, sigma)` find the CDF and PDF at x, and `norminv(p, mu, sigma)` computes the inverse CDF with quantile probability p. A random variable X with the normal distribution will be denoted $X \sim \mathcal{N}(\mu, \sigma^2)$.

The central limit theorem (formulated in a later section of this chapter) elevates the status of the normal distribution above other distributions. Despite its difficult formulation, the normal is one of the most important distributions in all science, and it has a critical role to play in nonparametric statistics. Any linear combination of normal random variables (independent or with simple covariance structures) are also normally distributed. In such sums, then, we need only keep track of the mean and variance, because these two parameters completely characterize the distribution. For example, if $X_1, \ldots, X_n$ are i.i.d. $\mathcal{N}(\mu, \sigma^2)$, then the sample mean $\bar{X} = (X_1 + \cdots + X_n)/n \sim \mathcal{N}(\mu, \sigma^2/n)$ distribution.

2.5.4 Chi-square Distribution

The probability density function for an chi-square random variable with the parameter k, called the *degrees of freedom*, is

$$f_X(x) = \frac{2^{-k/2}}{\Gamma(k/2)} x^{k/2-1} e^{-x/2}, \quad -\infty < x < \infty.$$

The chi-square distribution (χ^2) is a special case of the gamma distribution with parameters $r = k/2$ and $\lambda = 1/2$. Its mean and variance are $\mathbb{E}X = \mu = k$ and $\mathbb{V}\text{ar}X = \sigma^2 = 2k$.

If $Z \sim \mathcal{N}(0, 1)$, then $Z^2 \sim \chi_1^2$, that is, a chi-square random variable with one degree-of-freedom. Furthermore, if $U \sim \chi_m^2$ and $V \sim \chi_n^2$ are independent, then $U + V \sim \chi_{m+n}^2$.

From these results, it can be shown that if $X_1, \ldots, X_n \sim \mathcal{N}(\mu, \sigma^2)$ and $\bar{X}$ is the sample mean, then the *sample variance* $S^2 = \sum_i (X_i - \bar{X})^2/(n-1)$ is proportional to a chi-square random variable with $n - 1$ degrees of freedom:

$$\frac{(n-1)S^2}{\sigma^2} \sim \chi_{n-1}^2.$$

In MATLAB, the CDF and PDF for a χ_k^2 is `chi2cdf(x,k)` and `chi2pdf(x,k)`. The pth quantile of the χ_k^2 distribution is `chi2inv(p,k)`.

2.5.5 (Student) t - Distribution

Random variable X has Student's t distribution with k degrees of freedom, $X \sim t_k$, if its probability density function is

$$f_X(x) = \frac{\Gamma\left(\frac{k+1}{2}\right)}{\sqrt{k\pi}\,\Gamma(k/2)} \left(1 + \frac{x^2}{k}\right)^{-\frac{k+1}{2}}, \qquad -\infty < x < \infty.$$

The t-distribution[2] is similar in shape to the standard normal distribution except for the fatter tails. If $X \sim t_k$, $\mathbb{E}X = 0$, $k > 1$ and $\mathbb{V}\text{ar}X = k/(k-2)$, $k > 2$. For $k = 1$, the t distribution coincides with the Cauchy distribution.

The t-distribution has an important role to play in statistical inference. With a set of i.i.d. $X_1, \dots, X_n \sim \mathcal{N}(\mu, \sigma^2)$, we can standardize the sample mean using the simple transformation of $Z = (\bar{X} - \mu)/\sigma_{\bar{X}} = \sqrt{n}(\bar{X} - \mu)/\sigma$. However, if the variance is unknown, by using the same transformation except substituting the sample standard deviation S for σ, we arrive at a t-distribution with $n-1$ degrees of freedom:

$$T = \frac{(\bar{X} - \mu)}{S/\sqrt{n}} \sim t_{n-1}.$$

More technically, if $Z \sim \mathcal{N}(0,1)$ and $Y \sim \chi^2_k$ are independent, then $T = Z/\sqrt{Y/k} \sim t_k$. In MATLAB, the CDF at x for a t-distribution with k degrees of freedom is calculated as `tcdf(x,k)`, and the PDF is computed as `tpdf(x,k)`. The pth percentile is computed with `tinv(p,k)`.

2.5.6 Beta Distribution

The density function for a beta random variable is

$$f_X(x) = \frac{1}{B(a,b)} x^{a-1}(1-x)^{b-1}, \qquad 0 < x < 1, a > 0, b > 0,$$

and B is the beta function. Because X is defined only in (0,1), the beta distribution is useful in describing uncertainty or randomness in proportions or probabilities. A beta-distributed random variable is denoted by $X \sim \mathcal{B}e(a,b)$. The *Uniform distribution* on $(0,1)$, denoted as $\mathcal{U}(0,1)$, serves as a special case

[2] William Sealy Gosset derived the t-distribution in 1908 under the pen name "Student" (Gosset, 1908). He was a researcher for Guinness Brewery, which forbid any of their workers to publish "company secrets".

with $(a,b) = (1,1)$. The beta distribution has moments

$$\mathbb{E}X^k = \frac{\Gamma(a+k)\Gamma(a+b)}{\Gamma(a)\Gamma(a+b+k)} = \frac{a(a+1)\ldots(a+k-1)}{(a+b)(a+b+1)\ldots(a+b+k-1)}$$

so that $\mathbb{E}(X) = a/(a+b)$ and $\mathbb{V}\mathrm{ar}X = ab/[(a+b)^2(a+b+1)]$.

In MATLAB, the CDF for a beta random variable (at $x \in (0,1)$) is computed with `betacdf(x, a, b)` and the PDF is computed with `betapdf(x, a, b)`. The pth percentile is computed `betainv(p,a,b)`. If the mean μ and variance σ^2 for a beta random variable are known, then the basic parameters (a,b) can be determined as

$$a = \mu\left(\frac{\mu(1-\mu)}{\sigma^2} - 1\right), \qquad \text{and} \qquad b = (1-\mu)\left(\frac{\mu(1-\mu)}{\sigma^2} - 1\right). \tag{2.2}$$

2.5.7 Double Exponential Distribution

Random variable X has double exponential $\mathcal{DE}(\mu, \lambda)$ distribution if its density is given by

$$f_X(x) = \frac{\lambda}{2}\, e^{-\lambda|x-\mu|}, \quad -\infty < x < \infty, \lambda > 0.$$

The expectation of X is $\mathbb{E}X = \mu$ and the variance is $\mathbb{V}\mathrm{ar}X = 2/\lambda^2$. The moment generating function for the double exponential distribution is

$$m(t) = \frac{\lambda^2 e^{\mu t}}{\lambda^2 - t^2}, \quad |t| < \lambda.$$

Double exponential is also called *Laplace distribution.* If X_1 and X_2 are independent $\mathcal{E}(\lambda)$, then $X_1 - X_2$ is distributed as $\mathcal{DE}(0, \lambda)$. Also, if $X \sim \mathcal{DE}(0, \lambda)$ then $|X| \sim \mathcal{E}(\lambda)$.

2.5.8 Cauchy Distribution

The Cauchy distribution is symmetric and bell-shaped like the normal distribution, but with much heavier tails. For this reason, it is a popular distribution to use in nonparametric procedures to represent non-normality. Because the distribution is so spread out, it has no mean and variance (none of the Cauchy moments exist). Physicists know this as the *Lorentz distribution.* If $X \sim \mathcal{C}a(a,b)$, then X has density

$$f_X(x) = \frac{1}{\pi}\,\frac{b}{b^2 + (x-a)^2}, \quad -\infty < x < \infty.$$

The moment generating function for Cauchy distribution does not exist but its characteristic function is $\mathbb{E}e^{iX} = \exp\{iat - b|t|\}$. The $\mathcal{C}a(0,1)$ coincides with t-distribution with one degree of freedom.

The Cauchy is also related to the normal distribution. If Z_1 and Z_2 are two independent $\mathcal{N}(0,1)$ random variables, then $C = Z_1/Z_2 \sim \mathcal{C}a(0,1)$. Finally, if $C_i \sim \mathcal{C}a(a_i, b_i)$ for $i = 1,\dots,n$, then $S_n = C_1 + \cdots + C_n$ is distributed Cauchy with parameters $a_S = \sum_i a_i$ and $b_S = \sum_i b_i$.

2.5.9 Inverse Gamma Distribution

Random variable X is said to have an inverse gamma $\mathcal{IG}(r,\lambda)$ distribution with parameters $r > 0$ and $\lambda > 0$ if its density is given by

$$f_X(x) = \frac{\lambda^r}{\Gamma(r)x^{r+1}} e^{-\lambda/x}, \quad x \geq 0.$$

The mean and variance of X are $\mathbb{E}X = \lambda^k/(r-1)$ and $\mathbb{V}\mathrm{ar}X = \lambda^2/((r-1)^2(r-2))$, respectively. If $X \sim \mathcal{G}amma(r,\lambda)$ then its reciprocal X^{-1} is $\mathcal{IG}(r,\lambda)$ distributed.

2.5.10 Dirichlet Distribution

The Dirichlet distribution is a multivariate version of the beta distribution in the same way the Multinomial distribution is a multivariate extension of the Binomial. A random variable $X = (X_1, \dots, X_k)$ with a Dirichlet distribution ($X \sim \mathcal{D}ir(a_1,\dots,a_k)$) has probability density function

$$f(x_1,\dots,x_k) = \frac{\Gamma(A)}{\prod_{i=1}^k \Gamma(a_i)} \prod_{i=1}^k {x_i}^{a_i - 1},$$

where $A = \sum a_i$, and $x = (x_1,\dots,x_k) \geq 0$ is defined on the simplex $x_1 + \cdots + x_k = 1$. Then

$$\mathbb{E}(X_i) = \frac{a_i}{A}, \quad \mathbb{V}\mathrm{ar}(X_i) = \frac{a_i(A-a_i)}{A^2(A+1)}, \quad \text{and } \mathbb{C}\mathrm{ov}(X_i, X_j) = -\frac{a_i a_j}{A^2(A+1)}.$$

The Dirichlet random variable can be generated from gamma random variables $Y_1,\dots,Y_k \sim \mathcal{G}amma(a,b)$ as $X_i = Y_i/S_Y$, $i = 1,\dots,k$ where $S_Y = \sum_i Y_i$. Obviously, the marginal distribution of a component X_i is $\mathcal{B}e(a_i, A - a_i)$.

2.5.11 F Distribution

Random variable X has F distribution with m and n degrees of freedom, denoted as $F_{m,n}$, if its density is given by

$$f_X(x) = \frac{m^{m/2} n^{n/2} \ x^{m/2-1}}{B(m/2, n/2) \ (n+mx)^{-(m+n)/2}}, \ x > 0.$$

The CDF of the F distribution has no closed form, but it can be expressed in terms of an incomplete beta function.

The mean is given by $\mathbb{E}X = n/(n-2), n > 2$, and the variance by $\mathbb{V}\text{ar}X = [2n^2(m+n-2)]/[m(n-2)^2(n-4)], n > 4$. If $X \sim \chi^2_m$ and $Y \sim \chi^2_n$ are independent, then $(X/m)/(Y/n) \sim F_{m,n}$. If $X \sim \mathcal{B}e(a,b)$, then $bX/[a(1-X)] \sim F_{2a,2b}$. Also, if $X \sim F_{m,n}$ then $mX/(n+mX) \sim \mathcal{B}e(m/2, n/2)$.

The F distribution is one of the most important distributions for statistical inference; in introductory statistical courses test of equality of variances and ANOVA are based on the F distribution. For example, if S_1^2 and S_2^2 are sample variances of two independent normal samples with variances σ_1^2 and σ_2^2 and sizes m and n respectively, the ratio $(S_1^2/\sigma_1^2)/(S_2^2/\sigma_2^2)$ is distributed as $F_{m-1,n-1}$.

In MATLAB, the CDF at x for a F distribution with m, n degrees of freedom is calculated as `fcdf(x,m,n)`, and the PDF is computed as `fpdf(x,m,n)`. The pth percentile is computed with `finv(p,m,n)`.

2.5.12 Pareto Distribution

The Pareto distribution is named after the Italian economist Vilfredo Pareto. Some examples in which the Pareto distribution provides a good-fitting model include wealth distribution, sizes of human settlements, visits to encyclopedia pages, and file size distribution of internet traffic. Random variable X has a Pareto $\mathcal{P}a(x_0, \alpha)$ distribution with parameters $0 < x_0 < \infty$ and $\alpha > 0$ if its density is given by

$$f(x) = \frac{\alpha}{x_0} \left(\frac{x_0}{x} \right)^{\alpha+1}, \ x \geq x_0, \ \alpha > 0.$$

The mean and variance of X are $\mathbb{E}X = \alpha x_0/(\alpha - 1)$ and $\mathbb{V}\text{ar}X = \alpha {x^2}_0/((\alpha-1)^2(\alpha-2))$. If $X_1, \dots, X_n \sim \mathcal{P}a(x_0, \alpha)$, then $Y = 2x_0 \sum \ln(X_i) \sim {\chi^2}_{2n}$.

2.6 MIXTURE DISTRIBUTIONS

Mixture distributions occur when the population consists of heterogeneous subgroups, each of which is represented by a different probability distribution. If the sub-distributions cannot be identified with the observation, the observer is left with an unsorted mixture. For example, a finite mixture of k distributions has probability density function

$$f_X(x) = \sum_{i=1}^{k} p_i f_i(x)$$

where f_i is a density and the weights ($p_i \geq 0,\ i = 1, \dots, k$) are such that $\sum_i p_i = 1$. Here, p_i can be interpreted as the probability that an observation will be generated from the subpopulation with PDF f_i.

In addition to applications where different types of random variables are mixed together in the population, mixture distributions can also be used to characterize extra variability (dispersion) in a population. A more general continuous mixture is defined via a *mixing distribution* $g(\theta)$, and the corresponding mixture distribution

$$f_X(x) = \int_\theta f(t;\theta) g(\theta) d\theta.$$

Along with the mixing distribution, $f(t;\theta)$ is called the *kernel distribution.*

Example 2.1 Suppose an observed count is distributed $\mathcal{B}in(n,p)$, and over-dispersion is modeled by treating p as a mixing parameter. In this case, the binomial distribution is the kernel of the mixture. If we allow $g_P(p)$ to follow a beta distribution with parameters (a, b), then the resulting mixture distribution

$$p_X(x) = \int_0^1 p_{X|p}(t;p) g_P(p;a,b) dp = \binom{n}{x} \frac{B(a+x, n+b-x)}{B(a,b)}$$

is the *beta-binomial* distribution with parameters (n, a, b) and B is the beta function.

Example 2.2 In 1 MB dynamic random access memory (DRAM) chips, the distribution of defect frequency is approximately exponential with $\mu = 0.5/cm^2$. The 16 MB chip defect frequency, on the other hand, is exponential with $\mu = 0.1/cm^2$. If a company produces 20 times as many 1 MB chips as they produce 16 MB chips, the overall defect frequency is a mixture of exponentials:

$$f_X(x) = \frac{1}{21} 10 e^{-10x} + \frac{20}{21} 2 e^{-2x}.$$

In MATLAB, we can produce a graph (see Figure 2.1) of this mixture using the following code:

```
>> x = 0:0.01:1;
```

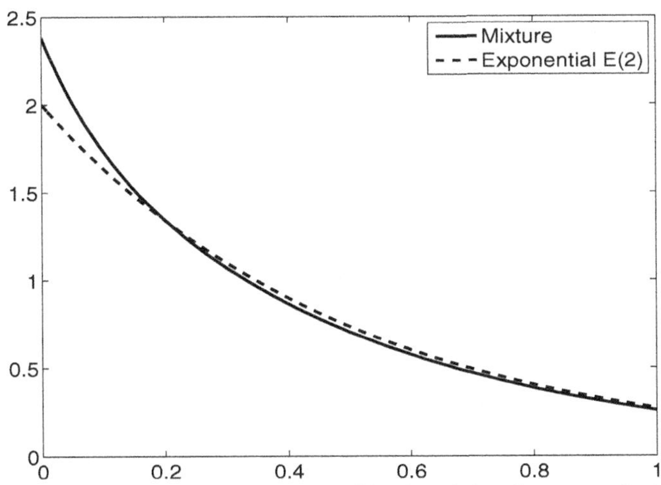

Fig. 2.1 Probability density function for DRAM chip defect frequency (*solid*) against exponential PDF (*dotted*).

```
>> y = (10/21)*exp(-x*10)+(40/21)*exp(-x*2);
>> z = 2*exp(-2*x);
>> plot(x,y,'k-')
>> hold on
>> plot(x,z,'k--')
```

Estimation problems involving mixtures are notoriously difficult, especially if the mixing parameter is unknown. In Section 16.2, the EM Algorithm is used to aid in statistical estimation.

2.7 EXPONENTIAL FAMILY OF DISTRIBUTIONS

We say that y_i is from the exponential family, if its distribution is of form

$$f(y|\theta,\phi) = \exp\left\{\frac{y\theta - b(\theta)}{\phi} + c(y,\phi)\right\}, \tag{2.3}$$

for some given functions b and c. Parameter θ is called *canonical parameter*, and ϕ dispersion parameter.

Example 2.3 We can write the normal density as

$$\frac{1}{\sqrt{2\pi}\sigma}\exp\left\{-\frac{(y-\mu)^2}{2\sigma^2}\right\} = \exp\left\{\frac{y\mu - \mu^2/2}{\sigma^2} - 1/2[y^2/\sigma^2 + \log(2\pi\sigma^2)]\right\},$$

thus it belongs to the exponential family, with $\theta = \mu$, $\phi = \sigma^2$, $b(\theta) = \theta^2/2$ and $c(y, \phi) = -1/2[y^2/\phi + \log(2\pi\phi)]$.

2.8 STOCHASTIC INEQUALITIES

The following four simple inequalities are often used in probability proofs.

1. *Markov Inequality.* If $X \geq 0$ and $\mu = \mathbb{E}(X)$ is finite, then

$$P(X > t) \leq \mu/t.$$

2. *Chebyshev's Inequality.* If $\mu = \mathbb{E}(X)$ and $\sigma^2 = \mathbb{V}\text{ar}(X)$, then

$$P(|X - \mu| \geq t) \leq \frac{\sigma^2}{t^2}.$$

3. *Cauchy-Schwartz Inequality.* For random variables X and Y with finite variances,

$$\mathbb{E}|XY| \leq \sqrt{\mathbb{E}(X^2)\mathbb{E}(Y^2)}.$$

4. *Jensen's Inequality.* Let $h(x)$ be a convex function. Then

$$h(\mathbb{E}(X)) \leq \mathbb{E}(h(X)).$$

   For example, $h(x) = x^2$ is a convex function and Jensen's inequality implies $[\mathbb{E}(X)]^2 \leq \mathbb{E}(X^2)$.

Most comparisons between two populations rely on direct inequalities of specific parameters such as the mean or median. We are more limited if no parameters are specified. If $F_X(x)$ and $G_Y(y)$ represent two distributions (for random variables X and Y, respectively), there are several direct inequalities used to describe how one distribution is larger or smaller than another. They are stochastic ordering, failure rate ordering, uniform stochastic ordering and likelihood ratio ordering.

Stochastic Ordering. X is smaller than Y in stochastic order ($X \leq_{ST} Y$) iff $F_X(t) \geq G_Y(t)\ \forall\ t$. Some texts use stochastic ordering to describe any general ordering of distributions, and this case is referred to as *ordinary stochastic ordering.*

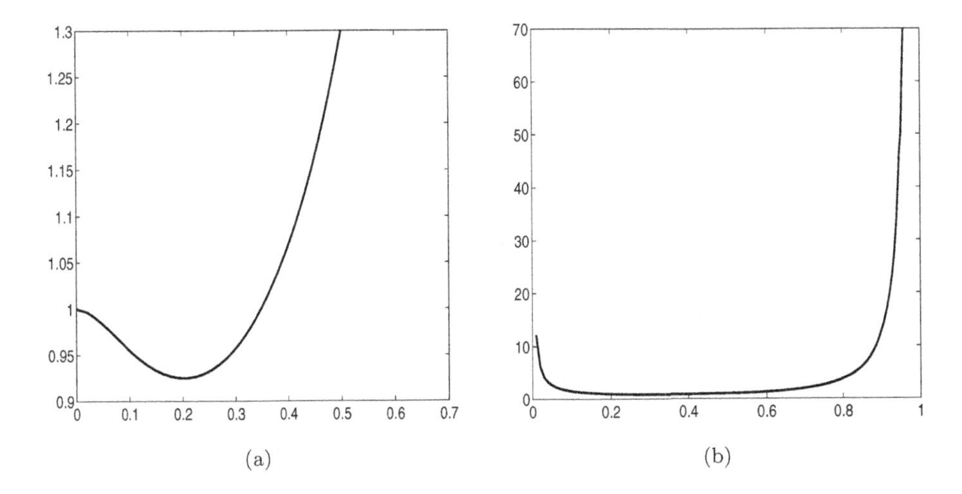

Fig. 2.2 For distribution functions F ($\mathcal{B}e(2,4)$) and G ($\mathcal{B}e(3,6)$): (a) Plot of $(1 - F(x))/(1 - G(x))$ (b) Plot of $f(x)/g(x)$.

Failure Rate Ordering. Suppose F_X and G_Y are differentiable and have probability density functions f_X and g_Y, respectively. Let $r_X(t) = f_X(t)/(1 - F_X(t))$, which is called the *failure rate* or *hazard rate* of X. X is smaller than Y in failure rate order ($X \le_{HR} Y$) iff $r_X(t) \ge r_Y(t)$ $\forall$ t.

Uniform Stochastic Ordering. X is smaller than Y in uniform stochastic order ($X \le_{US} Y$) iff the ratio $(1 - F_X(t))/(1 - G_Y(t))$ is decreasing in t.

Likelihood Ratio Ordering. Suppose F_X and G_Y are differentiable and have probability density functions f_X and g_Y, respectively. X is smaller than Y in likelihood ratio order ($X \le_{LR} Y$) iff the ratio $f_X(t)/g_Y(t)$ is decreasing in t.

It can be shown that uniform stochastic ordering is equivalent to failure rate ordering. Furthermore, there is a natural ordering to the three different inequalities:

$$X \le_{LR} Y \Rightarrow X \le_{HR} Y \Rightarrow X \le_{ST} Y.$$

That is, stochastic ordering is the weakest of the three. Figure 2.2 shows how these orders relate two different beta distributions. The MATLAB code below plots the ratios $(1 - F(x))/(1 - G(x))$ and $f(x)/g(x)$ for two beta random variables that have the same mean but different variances. Figure 2.2(a) shows that they do not have uniform stochastic ordering because $(1 - F(x))/(1 - G(x))$ is not monotone. This also assures us that the distributions do not have likelihood ratio ordering, which is illustrated in Figure 2.2(b).

```
>> x1=0:0.02:0.7;
```

```
>> r1=(1-betacdf(x1,2,4))./(1-betacdf(x1,3,6));
>> plot(x1,r1)
>> x2=0.08:0.02:.99;
>> r2=(betapdf(x2,2,4))./(betapdf(x2,3,6));
>> plot(x2,r2)
```

2.9 CONVERGENCE OF RANDOM VARIABLES

Unlike number sequences for which the convergence has a unique definition, sequences of random variables can converge in many different ways. In statistics, convergence refers to an estimator's tendency to look like what it is estimating as the sample size increases.

For general limits, we will say that $g(n)$ is *small "o" of* n and write $g_n = o(n)$ if and only if $g_n/n \to 0$ when $n \to \infty$. Then if $g_n = o(1)$, $g_n \to 0$. The *"big O" notation* concerns equiconvergence. Define $g_n = O(n)$ if there exist constants $0 < C_1 < C_2$ and integer n_0 so that $C_1 < |g_n/n| < C_2 \quad \forall n > n_0$. By examining how an estimator behaves as the sample size grows to infinity (its *asymptotic limit*), we gain a valuable insight as to whether estimation for small or medium sized samples make sense. Four basic measure of convergence are

Convergence in Distribution. A sequence of random variables $X_1, \ldots, X_n$ converges in distribution to a random variable X if $P(X_n \le x) \to P(X \le x)$. This is also called *weak convergence* and is written $X_n \Longrightarrow X$ or $X_n \to_d X$.

Convergence in Probability. A sequence of random variables $X_1, \cdots, X_n$ converges in probability to a random variable X if, for every $\epsilon > 0$, we have $P(|X_n - X| > \epsilon) \to 0$ as $n \to \infty$. This is symbolized as $X_n \xrightarrow{P} X$.

Almost Sure Convergence. A sequence of random variables $X_1, \ldots, X_n$ converges almost surely (a.s.) to a random variable X (symbolized $X_n \xrightarrow{a.s.} X$) if $P(\lim_{n\to\infty} |X_n - X| = 0) = 1$.

Convergence in Mean Square. A sequence of random variables $X_1, \cdots, X_n$ converges in mean square to a random variable X if $\mathbb{E}|X_n - X|^2 \to 0$ This is also called *convergence in* $\mathbb{L}_2$ and is written $X_n \xrightarrow{\mathbb{L}_2} X$.

Convergence in distribution, probability and almost sure can be ordered; i.e.,

$$X_n \xrightarrow{a.s.} X \;\Rightarrow\; X_n \xrightarrow{P} X \;\Rightarrow\; X_n \Longrightarrow X.$$

The $\mathbb{L}_2$-convergence implies convergence in probability and in distribution but it is not comparable with the almost sure convergence.

If $h(x)$ is a continuous mapping, then the convergence of X_n to X guarantees the same kind of convergence of $h(X_n)$ to $h(X)$. For example, if $X_n \xrightarrow{a.s.} X$ and $h(x)$ is continuous, then $h(X_n) \xrightarrow{a.s.} h(X)$, which further implies that $h(X_n) \xrightarrow{P} h(X)$ and $h(X_n) \Longrightarrow h(X)$.

Laws of Large Numbers (LLN). For i.i.d. random variables $X_1, X_2, \ldots$ with finite expectation $\mathbb{E}X_1 = \mu$, the sample mean converges to μ in the almost-sure sense, that is, $S_n/n \xrightarrow{a.s.} \mu$, for $S_n = X_1 + \cdots + X_n$. This is termed the *strong law of large numbers* (SLLN). Finite variance makes the proof easier, but it is not a necessary condition for the SLLN to hold. If, under more general conditions, $S_n/n = \bar{X}$ converges to μ in probability, we say that the *weak law of large numbers* (WLLN) holds. Laws of large numbers are important in statistics for investigating the consistency of estimators.

Slutsky's Theorem. Let $\{X_n\}$ and $\{Y_n\}$ be two sequences of random variables on some probability space. If $X_n - Y_n \xrightarrow{P} 0$, and $Y_n \Longrightarrow X$, then $X_n \Longrightarrow X$.

Corollary to Slutsky's Theorem. In some texts, this is sometimes called Slutsky's Theorem. If $X_n \Longrightarrow X$, $Y_n \xrightarrow{P} a$, and $Z_n \xrightarrow{P} b$, then $X_nY_n + Z_n \Longrightarrow aX + b$.

Delta Method. If $\mathbb{E}X_i = \mu$ and $\mathbb{V}\mathrm{ar}X_i = \sigma^2$, and if h is a differentiable function in the neighborhood of μ with $h'(\mu) \neq 0$, then $\sqrt{n}\,(h(X_n) - h(\mu)) \Longrightarrow W$, where $W \sim \mathcal{N}(0, [h'(\mu)]^2\sigma^2)$.

Central Limit Theorem (CLT). Let $X_1, X_2, \ldots$ be i.i.d. random variables with $\mathbb{E}X_1 = \mu$ and $\mathbb{V}\mathrm{ar}X_1 = \sigma^2 < \infty$. Let $S_n = X_1 + \cdots + X_n$. Then

$$\frac{S_n - n\mu}{\sqrt{n\sigma^2}} \Longrightarrow Z,$$

where $Z \sim \mathcal{N}(0,1)$. For example, if $X_1, \ldots, X_n$ is a sample from population with the mean μ and finite variance σ^2, by the CLT, the sample mean $\bar{X} = (X_1 + \cdots + X_n)/n$ is approximately normally distributed, $\bar{X} \overset{\text{appr}}{\sim} \mathcal{N}(\mu, \sigma^2/n)$, or equivalently, $(\sqrt{n}(\bar{X} - \mu))/\sigma \overset{\text{appr}}{\sim} \mathcal{N}(0,1)$. In many cases, usable approximations are achieved for n as low as 20 or 30.

Example 2.4 We illustrate the CLT by MATLAB simulations. A single sample of size $n = 300$ from Poisson $\mathcal{P}(1/2)$ distribution is generated as `sample = poissrnd(1/2, [1,300]);` According to the CLT, the sum $S_{300} =$

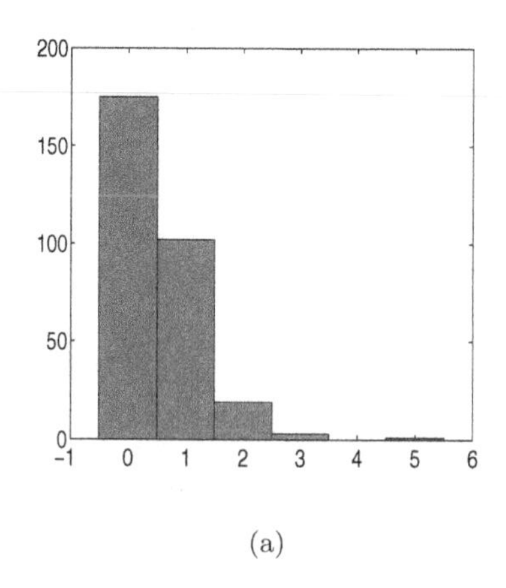

(a)

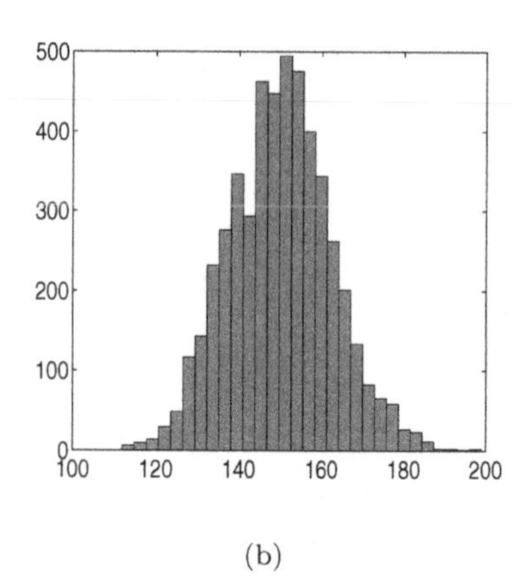

(b)

Fig. 2.3 (a) Histogram of single sample generated from Poisson $\mathcal{P}(1/2)$ distribution. (b) Histogram of S_n calculated from 5,000 independent samples of size $n = 300$ generated from Poisson $\mathcal{P}(1/2)$ distribution.

$X_1 + \cdots + X_{300}$ should be approximately normal $\mathcal{N}(300 \times 1/2, 300 \times 1/2)$. The histogram of the original sample is depicted in Figure 2.3(a). Next, we generated $N = 5000$ similar samples, each of size $n = 300$ from the same distribution and for each we found the sum S_{300}.

```
>> S_300 = [ ];
>> for i = 1:5000
       S_300 = [S_300 sum(poissrnd(0.5,[1,300]))];
   end
>> hist(S_300, 30)
```

The histogram of 5000 realizations of S_{300} is shown in Figure 2.3(b). Notice that the histogram of sums is bell-shaped and normal-like, as predicted by the CLT. It is centered near $300 \times 1/2 = 150$.

A more general central limit theorem can be obtained by relaxing the assumption that the random variables are identically distributed. Let $X_1, X_2, \ldots$ be independent random variables with $\mathbb{E}(X_i) = \mu_i$ and $\mathbb{V}\text{ar}(X_i) = \sigma_i^2 < \infty$. Assume that the following limit (called *Lindeberg's condition*) is satisfied:

For $\epsilon > 0$,

$$(D_n^2)^{-1} \sum_{i=1}^{n} \mathbb{E}[(X_i - \mu_i)^2] \mathbf{1}_{\{|X_i - \mu_i| \geq \epsilon D_n\}} \to 0, \text{ as } n \to \infty, \tag{2.4}$$

where

$$D_n^2 = \sum_{i=1}^{n} \sigma_i^2.$$

Extended CLT. Let $X_1, X_2, \ldots$ be independent (not necessarily identically distributed) random variables with $\mathbb{E}X_i = \mu_i$ and $\mathbb{V}\text{ar}X_i = \sigma_i^2 < \infty$. If condition (2.4) holds, then

$$\frac{S_n - \mathbb{E}S_n}{D_n} \Longrightarrow Z,$$

where $Z \sim \mathcal{N}(0,1)$ and $S_n = X_1 + \cdots + X_n$.

Continuity Theorem. Let $F_n(x)$ and $F(x)$ be distribution functions which have characteristic functions $\varphi_n(t)$ and $\varphi(t)$, respectively. If $F_n(x) \Longrightarrow F(x)$, then $\varphi_n(t) \longrightarrow \varphi(t)$. Furthermore, let $F_n(x)$ and $F(x)$ have characteristic functions $\varphi_n(t)$ and $\varphi(t)$, respectively. If $\varphi_n(t) \longrightarrow \varphi(t)$ and $\varphi(t)$ is continuous at 0, then $F_n(x) \Longrightarrow F(x)$.

Example 2.5 Consider the following array of independent random variables

$$\begin{array}{llll}
X_{11} & & & \\
X_{21} & X_{22} & & \\
X_{31} & X_{32} & X_{33} & \\
\vdots & \vdots & \vdots & \ddots
\end{array}$$

where $X_{nk} \sim \mathcal{B}er(p_n)$ for $k = 1, \ldots, n$. The X_{nk} have characteristic functions

$$\varphi_{X_{nk}}(t) = p_n e^{it} + q_n$$

where $q_n = 1 - p_n$. Suppose $p_n \to 0$ in such a way that $np_n \to \lambda$, and let $S_n = \sum_{k=1}^n X_{nk}$. Then

$$\begin{array}{rclcl}
\varphi_{S_n}(t) & = & \prod_{k=1}^n \varphi_{X_{nk}}(t) & = & (p_n e^{it} + q_n)^n \\
& = & (1 + p_n e^{it} - p_n)^n & = & [1 + p_n(e^{it} - 1)]^n \\
& \approx & [1 + \frac{\lambda}{n}(e^{it} - 1)]^n & \to & \exp[\lambda(e^{it} - 1)],
\end{array}$$

which is the characteristic function of a Poisson random variable. So, by the Continuity Theorem, $S_n \Longrightarrow \mathcal{P}(\lambda)$.

2.10 EXERCISES

2.1. For the characteristic function of a random variable X, prove the three following properties:

(i) $\varphi_{aX+b}(t) = e^{ib}\varphi_X(at)$.

(ii) If $X = c$, then $\varphi_X(t) = e^{ict}$.

(iii) If $X_1, X_2, \cdot X_n$ are independent, then $S_n = X_1 + X_2 + \cdot + X_n$ has characteristic function $\varphi_{S_m}(t) = \prod_{i=1}^{n} \varphi_{X_i}(t)$.

2.2. Let $U_1, U_2, \ldots$ be independent uniform $\mathcal{U}(0,1)$ random variables. Let $M_n = \min\{U_1, \ldots, U_n\}$. Prove $nM_n \Longrightarrow X \sim \mathcal{E}(1)$, the exponential distribution with rate parameter λ=1.

2.3. Let $X_1, X_2, \ldots$ be independent geometric random variables with parameters $p_1, p_2, \ldots$. Prove, if $p_n \to 0$, then $p_n X_n \Longrightarrow \mathcal{E}(1)$.

2.4. Show that for continuous distributions that have continuous density functions, failure rate ordering is equivalent to uniform stochastic ordering. Then show that it is weaker than likelihood ratio ordering and stronger than stochastic ordering.

2.5. Derive the mean and variance for a Poisson distribution using (a) just the probability mass function and (b) the moment generating function.

2.6. Show that the Poisson distribution is a limiting form for a binomial model, as given in equation (2.1) on page 15.

2.7. Show that, for the exponential distribution, the median is less than 70% of the mean.

2.8. Use a Taylor series expansion to show the following:

(i) $e^{-ax} = 1 - ax + (ax)^2/2! - (ax)^3/3! + \cdots$

(ii) $\log(1 + x) = x - x^2/2 + x^3/3 - \cdots$

2.9. Use MATLAB to plot a mixture density of two normal distributions with mean and variance parameters (3,6) and (10,5). Plot using weight function $(p_1, p_2) = (0.5, 0.5)$.

2.10. Write a MATLAB function to compute, in table form, the following quantiles for a χ^2 distribution with ν degrees of freedom, where ν is a function (user) input:

$$\{0.005, 0.01, 0.025, 0.05, 0.10, 0.90, 0.95, 0.975, 0.99, 0.995\}.$$

REFERENCES

Gosset, W. S. (1908), "The Probable Error of a Mean," *Biometrika*, 6, 1-25.

3

Statistics Basics

Daddy's rifle in my hand felt reassurin',
he told me "Red means run, son. Numbers add up to nothin'."
But when the first shot hit the dog, I saw it comin'...

Neil Young (from the song *Powderfinger*)

In this chapter, we review fundamental methods of statistics. We emphasize some statistical methods that are important for nonparametric inference. Specifically, tests and confidence intervals for the binomial parameter p are described in detail, and serve as building blocks to many nonparametric procedures. The empirical distribution function, a nonparametric estimator for the underlying cumulative distribution, is introduced in the first part of the chapter.

3.1 ESTIMATION

For distributions with unknown parameters (say θ), we form a point estimate $\hat{\theta}_n$ as a function of the sample $X_1, \dots, X_n$. Because $\hat{\theta}_n$ is a function of random variables, it has a distribution itself, called the *sampling distribution*. If we sample randomly from the same population, then the sample is said to be independently and identically distributed, or i.i.d.

An *unbiased estimator* is a statistic $\hat{\theta}_n = \hat{\theta}_n(X_1, \dots, X_n)$ whose expected value is the parameter it is meant to estimate; i.e., $\mathbb{E}(\hat{\theta}_n) = \theta$. An estimator is weakly *consistent* if, for any $\epsilon > 0$, $P(|\hat{\theta}_n - \theta| > \epsilon) \to 0$ as $n \to \infty$ (i.e., $\hat{\theta}_n$ converges to θ in probability). In compact notation: $\hat{\theta}_n \xrightarrow{P} \theta$.

Unbiasedness and consistency are desirable qualities in an estimator, but there are other ways to judge an estimate's efficacy. To compare estimators, one might seek the one with smaller mean squared error (MSE), defined as

$$\text{MSE}(\hat{\theta}_n) = \mathbb{E}(\hat{\theta}_n - \theta)^2 = \mathbb{V}\text{ar}(\hat{\theta}_n) + [\text{Bias}(\hat{\theta}_n)]^2,$$

where $\text{Bias}(\hat{\theta}_n) = \mathbb{E}(\hat{\theta}_n - \theta)$. If the bias and variance of the estimator have limit 0 as $n \to \infty$, (or equivalently, $\text{MSE}(\hat{\theta}_n) \to 0$) the estimator is consistent. An estimator is defined as *strongly* consistent if, as $n \to \infty$, $\hat{\theta}_n \xrightarrow{a.s.} \theta$.

Example 3.1 Suppose $X \sim \mathcal{B}in(n,p)$. If p is an unknown parameter, $\hat{p} = X/n$ is unbiased and strongly consistent for p. This is because the SLLN holds for i.i.d. $\mathcal{B}er(p)$ random variables, and X coincides with S_n for the Bernoulli case; see Laws of Large Numbers on p. 29.

3.2 EMPIRICAL DISTRIBUTION FUNCTION

Let $X_1, X_2, \dots, X_n$ be a sample from a population with continuous CDF F. An *empirical (cumulative) distribution function* (EDF) based on a random sample is defined as

$$F_n(x) = \frac{1}{n} \sum_{i=1}^{n} \mathbf{1}(X_i \le x), \tag{3.1}$$

where $\mathbf{1}(\rho)$ is called the *indicator function* of ρ, and is equal to 1 if the relation ρ is true, and 0 if it is false. In terms of ordered observations $X_{1:n} \le X_{2:n} \le \cdots \le X_{n:n}$, the empirical distribution function can be expressed as

$$F_n(x) = \begin{cases} 0 & \text{if } x < X_{1:n} \\ k/n & \text{if } X_{k:n} \le x < X_{k+1:n} \\ 1 & \text{if } x \ge X_{n:n} \end{cases}$$

We can treat the empirical distribution function as a random variable with a sampling distribution, because it is a function of the sample. Depending on the argument x, it equals one of $n+1$ discrete values, $\{0/n, 1/n, \dots, (n-1)/n, 1\}$. It is easy to see that, for any fixed x, $nF_n(x) \sim \mathcal{B}in(n, F(x))$, where $F(x)$ is the true CDF of the sample items.

Indeed, for $F_n(x)$ to take value k/n, $k = 0, 1, \dots, n$, k observations from $X_1, \dots, X_n$ should be less than or equal to x, and $n-k$ observations larger than x. The probability of an observation being less than or equal to x is $F(x)$. Also, the k observations less than or equal to x can be selected from the sample in $\binom{n}{k}$ different ways. Thus,

$$P\left(F_n(x) = \frac{k}{n}\right) = \binom{n}{k}(F(x))^k\ (1 - F(x))^{n-k},\ k = 0, 1, \ldots, n.$$

From this it follows that $\mathbb{E}F_n(x) = F(x)$ and $\mathbb{V}\mathrm{ar}F_n(x) = F(x)(1 - F(x))/n$.

A simple graph of the EDF is available in MATLAB with the `plotedf(x)` function. For example, the code below creates Figure 3.1 that shows how the EDF becomes more refined as the sample size increases.

```
>> y1 = randn(20,1);
>> y2 = randn(200,1);
>> x = -3:0.05:3;
>> y = normcdf(x,0,1);
>> plot(x,y);
>> hold on;
>> plotedf(y1);
>> plotedf(y2);
```

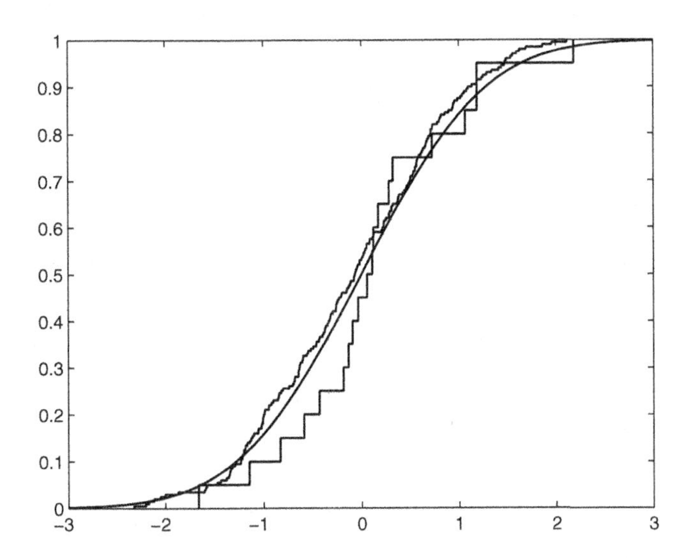

Fig. 3.1 EDF of normal samples (sizes 20 and 200) plotted along with the true CDF.

3.2.1 Convergence for EDF

The mean squared error (MSE) is defined for F_n as $\mathbb{E}(F_n(x)-F(x))^2$. Because $F_n(x)$ is unbiased for $F(x)$, the MSE reduces to $\mathbb{V}\text{ar}F_n(x) = F(x)(1-F(x))/n$, and as $n \to \infty$, $\text{MSE}(F_n(x)) \to 0$, so that $F_n(x) \xrightarrow{P} F(x)$.

There are a number of convergence properties for F_n that are of limited use in this book and will not be discussed. However, one fundamental limit theorem in probability theory, the Glivenko-Cantelli Theorem, is worthy of mention.

Theorem 3.1 *(Glivenko-Cantelli) If $F_n(x)$ is the empirical distribution function based on an i.i.d. sample $X_1, \ldots, X_n$ generated from $F(x)$,*

$$\sup_x |F_n(x) - F(x)| \xrightarrow{a.s.} 0.$$

3.3 STATISTICAL TESTS

> I shall not require of a scientific system that it shall be capable of being singled out, once and for all, in a positive sense; but I shall require that its logical form shall be such that it can be singled out, by means of empirical tests, in a negative sense: it must be possible for an empirical scientific system to be refuted by experience.
>
> Karl Popper, Philosopher (1902–1994)

Uncertainty associated with the estimator is a key focus of statistics, especially *tests of hypothesis* and *confidence intervals.* There are a variety of methods to construct tests and confidence intervals from the data, including Bayesian (see Chapter 4) and frequentist methods, which are discussed in Section 3.3.3. Of the two general methods adopted in research today, methods based on the *Likelihood Ratio* are generally superior to those based on *Fisher Information.*

In a traditional set-up for testing data, we consider two hypotheses regarding an unknown parameter in the underlying distribution of the data. Experimenters usually plan to show new or alternative results, which are typically conjectured in the *alternative hypothesis* (H_1 or H_a). The *null hypothesis*, designated H_0, usually consists of the parts of the parameter space not considered in H_1.

When a test is conducted and a claim is made about the hypotheses, two distinct errors are possible:

Type I error. The type I error is the action of rejecting H_0 when H_0 was actually true. The probability of such error is usually labeled by α, and referred to as *significance level* of the test.

Type II error. The type II error is an action of failing to reject H_0 when H_1 was actually true. The probability of the type II error is denoted by β. *Power* is defined as $1-\beta$. In simple terms, the power is propensity of a test to reject wrong alternative hypothesis.

3.3.1 Test Properties

A test is *unbiased* if the power is always as high or higher in the region of H_1 than anywhere in H_0. A test is *consistent* if, over all of H_1, $\beta \to 0$ as the sample sizes goes to infinity.

Suppose we have a hypothesis test of $H_0 : \theta = \theta_0$ versus $H_1 : \theta \neq \theta_0$. The *Wald* test of hypothesis is based on using a normal approximation for the test statistic. If we estimate the variance of the estimator θ_n by plugging in θ_n for θ in the variance term $\sigma^2_{\theta_n}$ (denote this $\hat{\sigma}^2_{\theta_n}$), we have the z-test statistic

$$z_0 = \frac{\theta_n - \theta_0}{\hat{\sigma}_{\theta_n}}.$$

The critical region (or rejection region) for the test is determined by the quantiles z_q of the normal distribution, where q is set to match the type I error.

p-values: The p-value is a popular but controversial statistic for describing the significance of a hypothesis given the observed data. Technically, it is the probability of observing a result as "rejectable" (according to H_0) as the observed statistic that actually occurred but from a new sample. So a p-value of 0.02 means that if H_0 is true, we would expect to see results more reflective of that hypothesis 98% of the time in repeated experiments. Note that if the p-value is less than the set α level of significance for the test, the null hypothesis should be rejected (and otherwise should not be rejected).

In the construct of classical hypothesis testing, the p-value has potential to be misleading with large samples. Consider an example in which $H_0 : \mu = 20.3$ versus $H_1 : \mu \neq 20.3$. As far as the experimenter is concerned, the null hypothesis might be conjectured only to three significant digits. But if the sample is large enough, $\bar{x} = 20.30001$ will eventually be rejected as being too far away from H_0 (granted, the sample size will have to be *awfully* large, but you get our point?). This problem will be revisited when we learn about goodness-of-fit tests for distributions.

Binomial Distribution. For binomial data, consider the test of hypothesis

$$H_0 : p \leq p_0 \qquad \text{vs} \qquad H_1 : p > p_0.$$

If we fix the type I error to α, we would have a critical region (or *rejection region*) of $\{x : x > x_0\}$, where x_0 is chosen so that $\alpha = P(X > x_0 \mid p = p_0)$. For instance, if $n = 10$, an $\alpha = 0.0547$ level test for $H_0 : p \leq 0.5$ vs $H_1 : p > 0.5$ is to reject H_0 if $X \geq 8$. The test's power is plotted in Figure 3.2 based on the following MATLAB code. The figure illustrates how our chance at rejecting the null hypothesis in favor of specific alternative $H_1 : p = p_1$ increases as p_1 increases past 0.5.

```
>> p1=0.5:0.01:0.99;
>> pow=1-binocdf(7,10,p1);
>> plot(p1,pow)
```

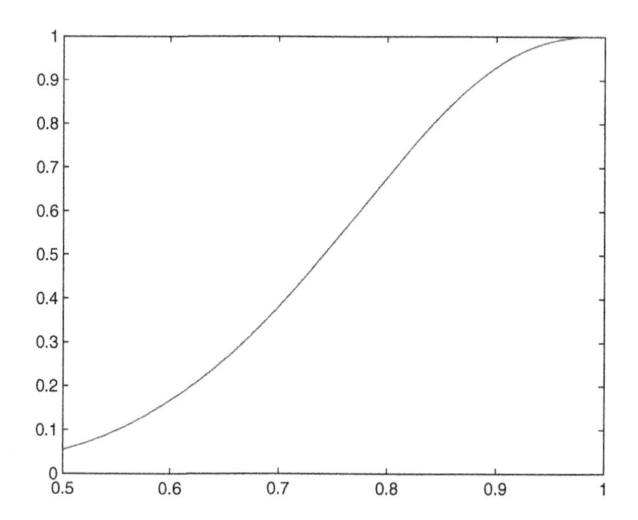

Fig. 3.2 Graph of statistical test power for binomial test for specific alternative $H_1 : p = p_1$. Values of p_1 are given on the horizontal axis.

Example 3.2 A semiconductor manufacturer produces an unknown proportion p of defective integrative circuit (IC) chips, so that chip *yield* is defined as $1 - p$. The manufacturer's reliability target is 0.9. With a sample of 25 randomly selected microchips, the Wald test will reject $H_0 : p \leq 0.10$ in favor of $H_1 : p > 0.10$ if

$$\frac{\hat{p} - 0.1}{\sqrt{(0.1)(0.9)/100}} > z_\alpha,$$

or for the case $\alpha = 0.05$, if the number of defective chips $X > 14.935$.

3.3.2 Confidence Intervals

A $1-\alpha$ level *confidence interval* is a statistic, in the form of a region or interval, that contains an unknown parameter θ with probability $1-\alpha$. For communicating uncertainty in layman's terms, confidence intervals are typically more suitable than tests of hypothesis, as the uncertainty is illustrated by the length of the interval constructed, along with the adjoining confidence statement.

A two-sided confidence interval has the form $(L(X), U(X))$, where X is the observed outcome, and $P(L(X) \leq \theta \leq U(X)) = 1-\alpha$. These are the most commonly used intervals, but there are cases in which one-sided intervals are more appropriate. If one is concerned with how large a parameter might be, we would construct an *upper bound* $U(X)$ such that $P(\theta \leq U(X)) = 1-\alpha$. If small values of the parameter are of concern to the experimenter, a *lower bound* $L(X)$ can be used where $P(L(X) \leq \theta) = 1-\alpha$.

Example 3.3 Binomial Distribution. To construct a two-sided $1-\alpha$ confidence interval for p, we solve the equation

$$\sum_{k=0}^{x} \binom{n}{k} p^k (1-p)^{n-k} = \alpha/2$$

for p to obtain the upper $1-\alpha$ limit for p, and solve

$$\sum_{k=x}^{n} \binom{n}{k} p^k (1-p)^{n-k} = \alpha/2$$

to obtain the lower limit. One sided $1-\alpha$ intervals can be constructed by solving just one of the equations using α in place of $\alpha/2$. Use MATLAB functions `binup(n,x,α)` and `binlow(n,x,α)`. This is named the Clopper-Pearson interval (Clopper and Pearson, 1934), where Pearson refers to Egon Pearson, Karl Pearson's son.

This exact interval is typically *conservative*, but not conservative like a G.O.P. senator from Mississippi. In this case, conservative means the *coverage probability* of the confidence interval is at least as high as the *nominal* coverage probability $1-\alpha$, and can be much higher. In general, "conservative" is synonymous with risk averse. The nominal and actual coverage probabilities disagree frequently with discrete data, where an interval with the exact coverage probability of $1-\alpha$ may not exist. While the guaranteed confidence in a conservative interval is reassuring, it is potentially inefficient and misleading.

Example 3.4 If $n = 10, x = 3$, then $\hat{p} = 0.3$ and a 95% (two-sided) confidence interval for p is computed by finding the upper limit p_1 for which $F_X(3; p_1) = 0.025$ and lower limit p_2 for which $1 - F_X(2; p_2) = 0.025$, where F_X is the CDF for the binomial distribution with $n = 10$. The resulting interval, $(0.06774, 0.65245)$ is not symmetric in p.

Intervals Based on Normal Approximation. The interval in Example 3.4 is "exact", in contrast to more commonly used intervals based on a normal approximation. Recall that $\bar{x} \pm z_{\alpha/2}\sigma/\sqrt{n}$ serves as a $1-\alpha$ level confidence interval for μ with data generated from a normal distribution. Here z_α represents the α quantile of the standard normal distribution. With the normal approximation (see *Central Limit Theorem* in Chapter 2), $\hat{p}$ has an approximate normal distribution if n is large, so if we estimate $\sigma^2_{\hat{p}} = p(1-p)/n$ with $\hat{\sigma}^2_{\hat{p}} = \hat{p}(1-\hat{p})/n$, an approximate $1-\alpha$ interval for p is

$$\hat{p} \pm z_{\alpha/2}\sqrt{x(n-x)/n^3}.$$

This is called the Wald interval because it is based on inverting the (Wald) z-test statistic for $H_0 : p = p_0$ versus $H_1 : p \neq p_0$. Agresti (1998) points out that both the exact and Wald intervals perform poorly compared to the *score interval* which is based on the Wald z-test of hypothesis, but instead of using $\hat{p}$ in the error term, it uses the value p_0 for which $(\hat{p}-p_0)/\sqrt{p_0(1-p_0)/n} = \pm z_{\alpha/2}$. The solution, first stated by Wilson (1927), is the interval

$$\frac{\hat{p} + \frac{z^2_{\alpha/2}}{2n} \pm z_{\alpha/2}\sqrt{\frac{\hat{p}(1-\hat{p}) + z^2_{\alpha/2}/4n}{n}}}{1 + z^2_{\alpha/2}/n}.$$

This actually serves as an example of *shrinkage*, which is a statistical phenomenon where better estimators are sometimes produced by "shrinking" or adjusting treatment means toward an overall (sample) mean. In this case, one can show that the middle of the confidence interval shrinks a little from $\hat{p}$ toward 1/2, although the shrinking becomes negligible as n gets larger. Use MATLAB function `binomial_shrink_ci(n,x,alpha)` to generate a two-sided Wilson's confidence interval.

Example 3.5 In the previous example, with $n = 10$ and $x = 3$, the exact 2-sided 95% confidence interval (0.06774, 0.65245) has length 0.5847, so the inference is rather vague. Using the normal approximation, the interval computes to (0.0616, 0.5384) and has length 0.4786. The shrinkage interval is (0.1078, 0.6032) and has length 0.4954. Is this accurate? In general, the exact interval will have coverage probability exceeding $1-\alpha$, and the Wald interval sometimes has coverage probability below $1-\alpha$. Overall, the shrinkage interval has coverage probability closer to $1-\alpha$. In the case of the binomial, the word "exact" does not imply a confidence interval is better.

```
>> x=0:10;
>> y=binopdf(x,10,0.3);
>> bar(x,y)
>> barh([1 2 3],[0.067 0.652; 0.061, 0.538; 0.213 0.405],'stacked')
```

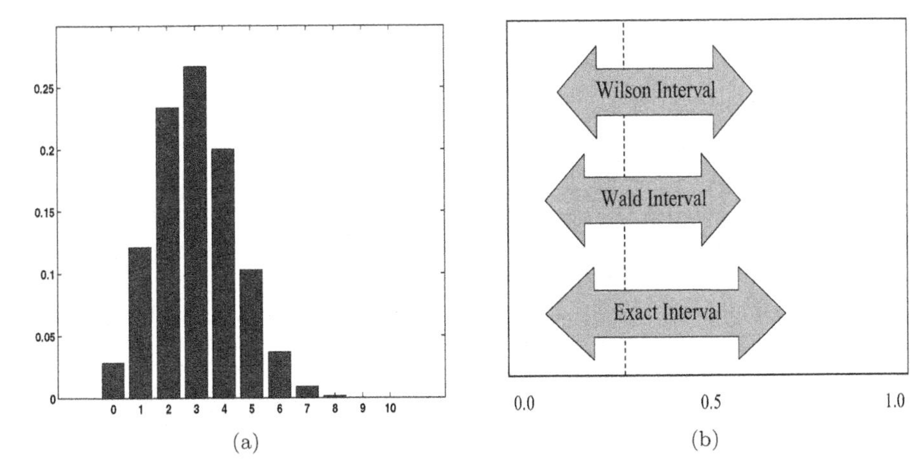

Fig. 3.3 (a) The binomial $\mathcal{B}in(10, 0.3)$ PMF; (b) 95% confidence intervals based on exact, Wald and Wilson method.

3.3.3 Likelihood

Sir Ronald Fisher, perhaps the greatest innovator of statistical methodology, developed the concepts of likelihood and sufficiency for statistical inference. With a set of random variables $X_1, \ldots, X_n$, suppose the joint distribution is a function of an unknown parameter θ: $f_n(x_1, \ldots, x_n; \theta)$. The *likelihood function* pertaining to the observed data $L(\theta) = f_n(x_1, \ldots, x_n; \theta)$ is associated with the probability of observing the data at each possible value θ of an unknown parameter. In the case the sample consists of i.i.d. measurements with density function $f(x; \theta)$, the likelihood simplifies to

$$L(\theta) = \prod_{i=1}^{n} f(x_i; \theta).$$

The likelihood function has the same numerical value as the PDF of a random variable, but it is regarded as a function of the parameters θ, and treats the data as fixed. The PDF, on the other hand, treats the parameters as fixed and is a function of the data points. The *likelihood principle* states that after x is observed, all relevant experimental information is contained in the likelihood function for the observed x, and that θ_1 supports the data more than θ_2 if $L(\theta_1) \geq L(\theta_2)$. The *maximum likelihood estimate* (MLE) of θ is that value of θ in the parameter space that maximizes $L(\theta)$. Although the MLE is based strongly on the parametric assumptions of the underlying density function $f(x; \theta)$, there is a sensible nonparametric version of the likelihood introduced in Chapter 10.

MLEs are known to have optimal performance if the sample size is sufficient and the densities are "regular"; for one, the support of $f(x;\theta)$ should not depend on θ. For example, if $\hat{\theta}$ is the MLE, then

$$\sqrt{n}(\hat{\theta}-\theta) \Longrightarrow \mathcal{N}(0, i^{-1}(\theta)),$$

where $i(\theta) = \mathbb{E}([\partial \log f/\partial\theta]^2)$ is the *Fisher Information* of θ. The regularity conditions also demand that $i(\theta) \geq 0$ is bounded and $\int f(x;\theta)dx$ is thrice differentiable. For a comprehensive discussion about regularity conditions for maximum likelihood, see Lehmann and Casella (1998).

The optimality of the MLE is guaranteed by the following result:

Cramer-Rao Lower Bound. From an i.i.d. sample $X_1, \ldots, X_n$ where X_i has density function $f_X(x)$, let $\hat{\theta}_n$ be an unbiased estimator for θ. Then

$$\mathbb{V}\text{ar}(\hat{\theta}_n) \geq (i(\theta)n)^{-1}.$$

Delta Method for MLE. The *invariance property* of MLEs states that if g is a one-to-one function of the parameter θ, then the MLE of $g(\theta)$ is $g(\hat{\theta})$. Assuming the first derivative of g (denoted g') exists, then

$$\sqrt{n}(g(\hat{\theta}) - g(\theta)) \Longrightarrow \mathcal{N}(0, g'(\theta)^2/i(\theta)).$$

Example 3.6 After waiting for the kth success in a repeated process with constant probabilities of success and failure, we recognize the probability distribution of X = no. of failures is *negative binomial.* To estimate the unknown success probability p, we can maximize

$$L(p) = p_X(x;p) \propto p^k(1-p)^x, \qquad 0 < p < 1.$$

Note the combinatoric part of p_X was left off the likelihood function because it plays no part in maximizing L. From $\log L(p) = k\log(p) + x\log(1-p)$, $\partial L/\partial p = 0$ leads to $\hat{p} = k/(k+x)$, and $i(p) = k/(p^2(1-p))$, thus for large n, $\hat{p}$ has an approximate normal distribution, i.e.,

$$\hat{p} \overset{\text{appr}}{\sim} \mathcal{N}(p, p^2(1-p)/k).$$

Example 3.7 In Example 3.6, suppose that $k = 1$, so X has a geometric $\mathcal{G}(p)$ distribution. If we are interested in estimating θ = probability that m

or more failures occur before a success occurs, then

$$\theta = g(p) = \sum_{j=m}^{\infty} p(1-p)^j = (1-p)^m,$$

and from the invariance principle, the MLE of θ is $\hat{\theta} = (1-\hat{p})^m$. Furthermore,

$$\sqrt{n}(\hat{\theta} - \theta) \Longrightarrow \mathcal{N}(0, \sigma_{\hat{\theta}}^2),$$

where $\sigma_{\hat{\theta}}^2 = g'(p)^2/i(p) = p^2(1-p)^{2m-1}m^2$.

3.3.4 Likelihood Ratio

The likelihood ratio function is defined for a parameter set θ as

$$R(\theta_0) = \frac{L(\theta_0)}{\sup_\theta L(\theta)}, \tag{3.2}$$

where $\sup_\theta L(\theta) = L(\hat{\theta})$ and $\hat{\theta}$ is the MLE of θ. Wilks (1938) showed that under the previously mentioned regularity conditions, $-2\log R(\theta)$ is approximately distributed χ^2 with k degrees of freedom (when θ is a correctly specified vector of length k).

The likelihood ratio is useful in setting up tests and intervals via the parameter set defined by $\mathcal{C}(\theta) = \{\theta : R(\theta) \geq r_0\}$ where r_0 is determined so that if $\theta = \theta_0$, $P(\hat{\theta} \in \mathcal{C}) = 1 - \alpha$. Given the chi-square result above, we have the following $1-\alpha$ confidence interval for θ based on the likelihood ratio:

$$\left\{\theta : -2\log R \leq {\chi_p}^2(1-\alpha)\right\}, \tag{3.3}$$

where ${\chi_p}^2(1-\alpha)$ is the $1-\alpha$ quantile of the ${\chi_p}^2$ distribution. Along with the nonparametric MLE discussed in Chapter 10, there is also a nonparametric version of the likelihood ratio, called the *empirical likelihood* which we will introduce also in Chapter 10.

Example 3.8 If $X_1, \ldots, X_n \sim \mathcal{N}(\mu, 1)$, then

$$L(\mu) = \prod_{i=1}^{n} (2\pi)^{-n/2} e^{-\frac{1}{2}\sum_{i=1}^{n}(x_i-\mu)^2}.$$

Because $\hat{\mu} = \bar{x}$ is the MLE, $R(\mu) = L(\mu)/L(\bar{x})$ and the interval defined in (3.3) simplifies to

$$\left\{\mu : \sum_{i=1}^{n}(x_i-\mu)^2 - \sum_{i=1}^{n}(x_i-\bar{x})^2 \leq {\chi_1}^2(1-\alpha)\right\}.$$

By expanding the sums of squares, one can show (see Exercise 3.6) that this interval is equivalent to the Fisher interval $\bar{x} \pm z_{\alpha/2}/\sqrt{n}$,

3.3.5 Efficiency

Let ϕ_1 and ϕ_2 be two different statistical tests (i.e., specified critical regions) based on the same underlying hypotheses. Let n_1 be the sample size for ϕ_1. Let n_2 be the sample size needed for ϕ_2 in order to make the type I and type II errors identical. The *relative efficiency* of ϕ_1 with respect to ϕ_2 is $RE = n_2/n_1$. The *asymptotic relative efficiency ARE* is the limiting value of RE as $n_1 \to \infty$. Nonparametric procedures are often compared to their parametric counterparts by computing the ARE for the two tests.

If a test or confidence interval is based on assumptions but tends to come up with valid answers even when some of the assumptions are not, the method is called *robust*. Most nonparametric procedures are more robust than their parametric counterparts, but also less efficient. Robust methods are discussed in more detail in Chapter 12.

3.3.6 Exponential Family of Distributions

Let $f(y|\theta)$ be a member of the *exponential family* with natural parameter θ. Assume that θ is univariate. Then the log likelihood $\ell(\theta) = \sum_{i=1}^n \log(f(y_i|\theta) = \sum_{i=1}^n \ell_i(\theta)$, where $\ell_i = \log f(y_i|\theta)$. The MLE for θ is solution of the equation

$$\frac{\partial \ell}{\partial \theta} = 0.$$

The following two properties (see Exercise 3.9) hold:

$$\text{(i)}\ \mathbb{E}\left(\frac{\partial \ell_i}{\partial \theta}\right) = 0 \text{ and } \quad \text{(ii)}\ \mathbb{E}\left(\frac{\partial^2 \ell_i}{\partial \theta^2}\right) + \mathbb{V}\text{ar}\left(\frac{\partial \ell}{\partial \theta}\right) = 0. \qquad (3.4)$$

For the exponential family of distributions,

$$\ell_i = \ell(y_i, \theta, \phi) = \frac{y\theta - b(\theta)}{\phi} + c(y, \phi),$$

and $\frac{\partial \ell}{\partial \theta} = \frac{y - b'(\theta)}{\phi}$ and $\frac{\partial^2 \ell}{\partial \theta^2} = -\frac{b''(\theta)}{\phi}$. By properties (i) and (ii) from (3.4), if Y has pdf $f(y|\theta)$, then $\mathbb{E}(Y) = \mu = b'(\theta)$ and $\mathbb{V}\text{ar}(Y) = b''(\theta)\phi$. The function $b''(\theta)$ is called *variance function* and denoted by $V(\mu)$ (because θ depends on μ).

The *unit deviance* is defined as

$$d_i(y_i, \mu) = 2\int_{\mu_i}^{y_i} \frac{y_i - u}{V(u)} du,$$

and the total deviance, a measure of the distance between y and μ, is defined as

$$D(y, \mu) = \sum_{i=1}^{n} w_i d(y_i, \mu),$$

where the summation is over the data and w_i are the prior weights. The quantity $D(y, \mu)/\phi$ is called the scaled deviance. For the normal distribution, the deviance is equivalent to the residual sum-of-squares, $\sum_{i=1}^{n}(y_i - \mu)^2$.

3.4 EXERCISES

3.1. With $n = 10$ observations and $x = 2$ observed successes in i.i.d. trials, construct 99% two-sided confidence intervals for the unknown binomial parameter p using the three methods discussed in this section (exact method, Wald method, Wilson method). Compare your results.

3.2. From a manufacturing process, $n = 25$ items are manufactured. Let X be the number of defectives found in the lot. Construct a $\alpha = 0.01$ level test to see if the proportion of defectives is greater than 10%. What are your assumptions?

3.3. Derive the MLE for μ with an i.i.d. sample of exponential random variables, and compare the confidence interval based on the Fisher information to an exact confidence interval based on the chi-square distribution.

3.4. A single parameter ("shape" parameter) Pareto distribution ($\mathcal{P}a(1, \alpha)$ on p. 23) has density function given by $f(x|\alpha) = \alpha/x^{\alpha+1}$, $x \geq 1$.

For a given experiment, researchers believe that in Pareto model the shape parameter α exceeds 1, and that the first moment $\mathbb{E}X = \alpha/(\alpha-1)$ is finite.

(i) What is the moment-matching estimator of parameter α? Moment matching estimators are solutions of equations in which theoretical moments are replaced empirical counterparts. In this case, the moment-matching equation is $\bar{X} = \alpha/(\alpha - 1)$.

(ii) What is the maximum likelihood estimator (MLE) of α?

(iii) Calculate the two estimators when $X_1 = 2, X_2 = 4$ and $X_3 = 3$ are observed.

3.5. Write a MATLAB simulation program to estimate the true coverage probability of a two-sided 90% Wald confidence interval for the case in which $n = 10$ and $p = 0.5$. Repeat the simulation at $p = 0.9$. Repeat the $p = 0.9$ case but instead use the Wilson interval. To estimate, generate 1000 random binomial outcomes and count the proportion of time the confidence interval contains the true value of p. Comment on your results.

3.6. Show that the confidence interval (for μ) derived from the likelihood ratio in the last example of the chapter is equivalent to the Fisher interval.

3.7. Let $X_1, \ldots, X_n$ be i.i.d. $\mathcal{P}(\lambda)$, and Y_k be the number of $X_1, \ldots, X_n$ equal to k. Derive the conditional distribution of Y_k given $T = \sum X_i = t$.

3.8. Consider the following i.i.d. sample generated from $F(x)$:

$$\{2.5, 5.0, 8.0, 8.5, 10.5, 11.5, 20\}.$$

Graph the empirical distribution and estimate the probability $P(8 \leq X \leq 10)$, where X has distribution function $F(x)$.

3.9. Prove the equations in (3.4): (i) $\mathbb{E}(\frac{\partial \ell_i}{\partial \theta}) = 0$, (ii) $\mathbb{E}(\frac{\partial^2 \ell_i}{\partial \theta^2}) + \mathbb{V}\text{ar}(\frac{\partial \ell}{\partial \theta}) = 0$.

REFERENCES

Agresti, A. (1998), "Approximate is Better than "Exact" for Interval Estimation of Binomial Proportions," *American Statistician*, 52, 119-126.

Clopper, C. J., and Pearson, E. S. (1934), "The Use of Confidence or Fiducial Limits Illustrated in the Case of the Binomial," *Biometrika*, 26, 404-413.

Lehmann, E. L., and Casella, G. (1998), *Theory of Point Estimation*. New York: Springer Verlag.

Wilks, S. S. (1938), "The Large-Sample Distribution of the Likelihood Ratio for Testing Composite Hypotheses," *Annals of Mathematical Statistics*, 9, 60-62.

Wilson, E. B. (1927), "Probability Inference, the Law of Succession, and Statistical Inference," *Journal of the American Statistical Association*, 22, 209-212.

4
Bayesian Statistics

> To anyone sympathetic with the current neo-Bernoullian neo-Bayesian Ramseyesque Finettist Savageous movement in statistics, the subject of testing goodness of fit is something of an embarrassment.
>
> F. J. Anscombe (1962)

4.1 THE BAYESIAN PARADIGM

There are several paradigms for approaching statistical inference, but the two dominant ones are *frequentist* (sometimes called classical or traditional) and *Bayesian*. The overview in the previous chapter covered mainly classical approaches. According to the Bayesian paradigm, the unobservable parameters in a statistical model are treated as random. When no data are available, a *prior distribution* is used to quantify our knowledge about the parameter. When data are available, we can update our prior knowledge using the conditional distribution of parameters, given the data. The transition from the prior to the posterior is possible via the Bayes theorem. Figure 4.1(a) shows a portrait of the Reverend Thomas Bayes whose posthumously published essay gave impetus to alternative statistical approaches (Bayes, 1763). His signature is shown in Figure 4.1(b).

Suppose that before the experiment our prior distribution describing θ is $\pi(\theta)$. The data are coming from the assumed model (likelihood) which depends on the parameter and is denoted by $f(x|\theta)$. Bayes theorem updates the prior

Fig. 4.1 The Reverend Thomas Bayes (1702–1761); (b) Bayes' signature.

$\pi(\theta)$ to the posterior by accounting for the data x,

$$\pi(\theta|x) = \frac{f(x|\theta)\pi(\theta)}{m(x)}, \tag{4.1}$$

where $m(x)$ is a normalizing constant, $m(x) = \int_\Theta f(x|\theta)\pi(\theta)d\theta$.

Once the data x are available, θ is the only unknown quantity and the posterior distribution $\pi(\theta|x)$ completely describes the uncertainty. There are two key advantages of Bayesian paradigm: (i) once the uncertainty is expressed via the probability distribution and the statistical inference can be automated, it follows a conceptually simple recipe, and (ii) available prior information is coherently incorporated into the statistical model.

4.2 INGREDIENTS FOR BAYESIAN INFERENCE

The *model* for a typical observation X conditional on unknown parameter θ is the density function $f(x|\theta)$. As a function of θ, $f(x|\theta) = L(\theta)$ is called a *likelihood.* The functional form of f is fully specified up to a parameter θ. According to the *likelihood principle,* all experimental information about the data must be contained in this likelihood function.

The parameter θ, with values in the parameter space Θ, is considered a random variable. The random variable θ has a distribution $\pi(\theta)$ called the prior distribution. This prior describes uncertainty about the parameter before data are observed. If the prior for θ is specified up to a parameter τ, $\pi(\theta|\tau)$, τ is called a *hyperparameter.*

Our goal is to start with this prior information and update it using the data to make the best possible estimator of θ. We achieve this through the likelihood function to get $\pi(\theta|x)$, called the *posterior* distribution for θ, given

$X = x$. Accompanying its role as the basis to Bayesian inference, the posterior distribution has been a source for an innumerable accumulation of tacky "butt" jokes by unimaginative statisticians with low-brow sense of humor, such as the authors of this book, for example.

To find $\pi(\theta|x)$, we use Bayes rule to divide *joint* distribution for X and θ ($h(x,\theta) = f(x|\theta)\pi(\theta)$) by the *marginal* distribution $m(x)$, which can be obtained by integrating out parameter θ from the joint distribution $h(x,\theta)$,

$$m(x) = \int_{\Theta} h(x,\theta)d\theta = \int_{\Theta} f(x|\theta)\pi(\theta)d\theta.$$

The marginal distribution is also called the *prior predictive* distribution. Finally we arrive at an expression for the posterior distribution $\pi(\theta|x)$:

$$\pi(\theta|x) = \frac{h(x,\theta)}{m(x)} = \frac{f(x|\theta)\pi(\theta)}{m(x)} = \frac{f(x|\theta)\pi(\theta)}{\int_{\Theta} f(x|\theta)\pi(\theta)d\theta}.$$

The following table summarizes the notation:

Likelihood	$f(x\|\theta)$
Prior Distribution	$\pi(\theta)$
Joint Distribution	$h(x,\theta) = f(x\|\theta)\pi(\theta)$
Marginal Distribution	$m(x) = \int_{\Theta} f(x\|\theta)\pi(\theta)d\theta$
Posterior Distribution	$\pi(\theta\|x) = f(x\|\theta)\pi(\theta)/m(x)$

Example 4.1 Normal Likelihood with Normal Prior. The normal likelihood and normal prior combination is important as it is often used in practice. Assume that an observation X is normally distributed with mean θ and known variance σ^2. The parameter of interest, θ, has a normal distribution as well with hyperparameters μ and τ^2. Starting with our Bayesian model of $X|\theta \sim \mathcal{N}(\theta,\sigma^2)$ and $\theta \sim \mathcal{N}(\mu,\tau^2)$, we will find the marginal and posterior distributions.

The exponent ζ in the joint distribution $h(x,\theta)$ is

$$\zeta = -\frac{1}{2\sigma^2}(x-\theta)^2 - \frac{1}{2\tau}(\theta-\mu)^2.$$

After straightforward but somewhat tedious algebra, ζ can be expressed as

$$\zeta = -\frac{1}{2\rho}\left(\theta - \rho\left(\frac{x}{\sigma^2} + \frac{\mu}{\tau^2}\right)\right)^2 - \frac{1}{2(\sigma^2+\tau^2)}(x-\mu)^2,$$

where

$$\rho = \frac{\sigma^2 \tau^2}{\sigma^2 + \tau^2}.$$

Recall that $h(x, \theta) = f(x|\theta)\pi(\theta) = \pi(\theta|x)m(x)$, so the marginal distribution simply resolves to $X \sim \mathcal{N}(\mu, \sigma^2 + \tau^2)$ and the posterior distribution comes out to be

$$\theta | X \sim \mathcal{N}\left(\frac{\tau^2}{\sigma^2 + \tau^2}X + \frac{\sigma^2}{\sigma^2 + \tau^2}\mu, \frac{\sigma^2 \tau^2}{\sigma^2 + \tau^2}\right).$$

If $X_1, X_2, \ldots, X_n$ are observed instead of a single observation X, then the sufficiency of $\bar{X}$ implies that the Bayesian model for θ is the same as for X with σ^2/n in place of σ^2. In other words, the Bayesian model is

$$\bar{X}|\theta \sim \mathcal{N}\left(\theta, \frac{\sigma^2}{n}\right) \text{ and } \theta \sim \mathcal{N}(\mu, \tau^2),$$

producing

$$\theta | \bar{X} \sim \mathcal{N}\left(\frac{\tau^2}{\frac{\sigma^2}{n} + \tau^2}\bar{X} + \frac{\frac{\sigma^2}{n}}{\frac{\sigma^2}{n} + \tau^2}\mu, \frac{\frac{\sigma^2}{n}\tau^2}{\frac{\sigma^2}{n} + \tau^2}\right).$$

Notice that the posterior mean

$$\frac{\tau^2}{\frac{\sigma^2}{n} + \tau^2}\bar{X} + \frac{\frac{\sigma^2}{n}}{\frac{\sigma^2}{n} + \tau^2}\mu \tag{4.2}$$

is a weighted linear combination of the MLE $\bar{X}$ and the prior mean μ with weights

$$\lambda = \frac{n\tau^2}{\sigma^2 + n\tau^2}, \quad 1 - \lambda = \frac{\sigma^2}{\sigma^2 + n\tau^2}.$$

When the sample size n increases, $\lambda \to 1$, and the influence of the prior mean diminishes. On the other hand when n is small and our prior opinion about μ is strong (i.e., τ^2 is small) the posterior mean is close to the prior mean μ. We will see later several more cases in which the posterior mean is a linear combination of a frequentist estimate and the prior mean.

For instance, suppose 10 observations are coming from $\mathcal{N}(\theta, 100)$. Assume that the prior on θ is $\mathcal{N}(20, 20)$. Using the numerical example in the MATLAB code below, the posterior is $\mathcal{N}(6.8352, 6.6667)$. These three densities are shown in Figure 4.2.

```
>> dat=[2.9441,-13.3618,7.1432,16.2356,-6.9178,8.5800,...
        12.5400,-15.9373,-14.4096,5.7115];
>> [m,v] = BA_nornor2(dat,100,20,20)
```

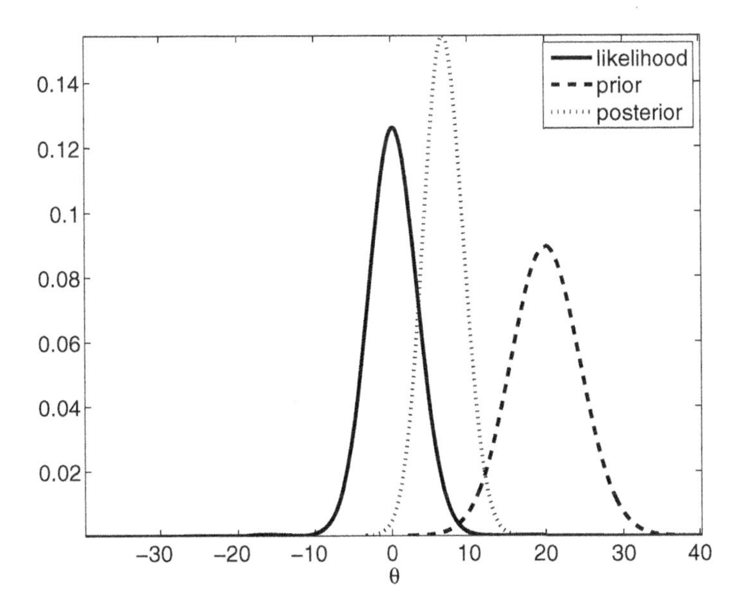

Fig. 4.2 The normal $\mathcal{N}(\theta, 100)$ likelihood, $\mathcal{N}(20, 20)$ prior, and posterior for data $\{2.9441, -13.3618, \ldots, 5.7115\}$.

4.2.1 Quantifying Expert Opinion

Bayesian statistics has become increasingly popular in engineering, and one reason for its increased application is that it allows researchers to input expert opinion as a catalyst in the analysis (through the prior distribution). Expert opinion might consist of subjective inputs from experienced engineers, or perhaps a summary judgment of past research that yielded similar results.

Example 4.2 Prior Elicitation for Reliability Tests. Suppose each of n independent reliability tests a machine reveals either a successful or unsuccessful outcome. If θ represents the reliability of the machine, let X be the number of successful missions the machine experienced in n independent trials. X is distributed binomial with parameters n (known) and θ (unknown). We probably won't expect an expert to quantify their uncertainty about θ directly into a prior distribution $\pi(\theta)$. Perhaps the researcher can elicit information such as the expected value and standard deviation of θ. If we suppose the prior distribution for θ is $\mathcal{B}e(\alpha, \beta)$, where the hyper-parameters α and β are known, then

$$\pi(\theta) = \frac{1}{B(\alpha, \beta)} \theta^{\alpha-1}(1-\theta)^{\beta-1}, \; 0 \leq \theta \leq 1.$$

With $X|\theta \sim \mathcal{B}in(n,\theta)$, the joint, marginal, and posterior distributions are

$$\begin{aligned} h(x,\theta) &= \frac{\binom{n}{x}}{B(\alpha,\beta)}\theta^{\alpha+x-1}(1-\theta)^{n-x+\beta-1},\ 0 \le \theta \le 1, x = 0,1,\ldots,n. \\ m(x) &= \frac{\binom{n}{x}B(x+\alpha, n-x+\beta)}{B(\alpha,\beta)},\ x = 0,1,\ldots,n. \\ \pi(\theta|x) &= \frac{1}{B(x+\alpha,n-x+\beta)}\theta^{\alpha+x-1}(1-\theta)^{n-x+\beta-1},\ 0 \le \theta \le 1. \end{aligned}$$

It is easy to see that the posterior distribution is $\mathcal{B}e(\alpha+x, n-x+\beta)$. Suppose the experts suggest that the previous version of this machine was "reliable 93% of the time, plus or minus 2%". We might take $\mathbb{E}(\theta) = 0.93$ and insinuate that $\sigma_\theta = 0.04$ (or $\mathbb{V}\text{ar}(\theta) = 0.0016$), using two-sigma rule as an argument. From the beta distribution,

$$\mathbb{E}\theta = \frac{\alpha}{\alpha+\beta} \quad \text{and} \quad \mathbb{V}\text{ar}\theta = \frac{\alpha\beta}{(\alpha+\beta)^2(\alpha+\beta+1)}.$$

We can actually solve for α and β as a function of the expected value μ and variance σ^2, as in (2.2),

$$\alpha = \mu(\mu - \mu^2 - \sigma^2)/\sigma^2, \quad \text{and} \quad \beta = (1-\mu)(\mu - \mu^2 - \sigma^2)/\sigma^2.$$

In this example, $(\mu, \sigma^2) = (0.93,\ 0.0016)$ leads to $\alpha = 36.91$ and β=2.78. To update the data X, we will use a $\mathcal{B}e(36.91, 2.78)$ distribution for a prior on θ. Consider the weight given to the expert in this example. If we observe one test only and the machine happened to fail, our posterior distribution is then $\mathcal{B}e(36.91, 3.78)$, which has a mean equal to 0.9071. The MLE for the average reliability is obviously zero, with with such precise information elicited from the expert, the posterior is close to the prior. In some cases when you do not trust your expert, this might be unsettling and less informative priors may be a better choice.

4.2.2 Point Estimation

The posterior is the ultimate experimental summary for a Bayesian. The location measures (especially the mean) of the posterior are of great importance. The posterior mean represents the most frequently used Bayes estimator for the parameter. The posterior mode and median are less commonly used alternative Bayes estimators.

An objective way to choose an estimator from the posterior is through a penalty or loss function $L(\hat{\theta}, \theta)$ that describes how we penalize the discrepancy of the estimator $\hat{\theta}$ from the parameter θ. Because the parameter is viewed as a random variable, we seek to minimize *expected* loss, or *posterior risk*:

$$R(\hat{\theta}, x) = \int L(\hat{\theta}, \theta)\pi(\theta|x)d\theta.$$

For example, the estimator based on the common squared-error loss $L(\hat{\theta}, \theta) = (\hat{\theta} - \theta)^2$ minimizes $\mathbb{E}((\hat{\theta} - \theta)^2)$, where expectation is taken over the posterior distribution $\pi(\theta|X)$. It's easy to show that the estimator turns out to be the posterior expectation. Similar to squared-error loss, if we use absolute-error loss $L(\hat{\theta}, \theta) = |\hat{\theta} - \theta|$, the Bayes estimator is the posterior median.

The posterior mode maximizes the posterior density the same way MLE is maximizing the likelihood. The *generalized MLE* maximizes $\pi(\theta|X)$. Bayesians prefer the name MAP (maximum aposteriori) estimator or simply posterior mode. The MAP estimator is popular in Bayesian analysis in part because it is often computationally less demanding than the posterior mean or median. The reason is simple; to find the maximum, the posterior need not to be fully specified because $\text{argmax}_\theta \pi(\theta|x) = \text{argmax}_\theta f(x|\theta)\pi(\theta)$, that is, one simply maximizes the product of likelihood and the prior.

In general, the posterior mean will fall between the MLE and the the prior mean. This was demonstrated in Example 4.1. As another example, suppose we flipped a coin four times and tails showed up on all 4 occasions. We are interested in estimating probability of heads, θ, in a Bayesian fashion. If the prior is $\mathcal{U}(0, 1)$, the posterior is proportional to $\theta^0(1 - \theta)^4$ which is beta $\mathcal{B}e(1, 5)$. The posterior mean *shrinks* the MLE toward the expected value of the prior (1/2) to get $\hat{\theta}_B = 1/(1 + 5) = 1/6$, which is a more reasonable estimator of θ then the MLE.

Example 4.3 Binomial-Beta Conjugate Pair. Suppose $X|\theta \sim \mathcal{B}in(n, \theta)$. If the prior distribution for θ is $\mathcal{B}e(\alpha, \beta)$, the posterior distribution is $\mathcal{B}e(\alpha + x, n - x + \beta)$. Under squared error loss $L(\hat{\theta}, \theta) = (\hat{\theta} - \theta)^2$, the Bayes estimator of θ is the expected value of the posterior

$$\hat{\theta}_B = \frac{\alpha + x}{(\alpha + x)(\beta + n - x)} = \frac{\alpha + x}{\alpha + \beta + n}.$$

This is actually a weighted average of MLE, X/n, and the prior mean $\alpha/(\alpha + \beta)$. Notice that, as n becomes large, the posterior mean is getting close to MLE, because the weight $n/(n + \alpha + \beta)$ tends to 1. On the other hand, when α is large, the posterior mean is close to the prior mean. Large α indicates small prior variance (for fixed β, the variance of $\mathcal{B}e(\alpha, \beta)$ behaves as $O(1/\alpha^2)$) and the prior is concentrated about its mean. Recall the Example 4.2; after one machine trial failure the posterior distribution mean changed from 0.93 (the prior mean) to 0.9071, shrinking only slightly toward the MLE (which is zero).

Example 4.4 Jeremy's IQ. Jeremy, an enthusiastic Georgia Tech student, spoke in class and posed a statistical model for his scores on standard IQ tests. He thinks that, in general, his scores are normally distributed with unknown mean θ and the variance of 80. Prior (and expert) opinion is that the IQ of Georgia Tech students, θ, is a normal random variable, with mean 110 and the variance 120. Jeremy took the test and scored 98. The traditional estimator of θ would be $\hat{\theta} = X = 98$. The posterior is $\mathcal{N}(102.8, 48)$, so the Bayes estimator of Jeremy's IQ score is $\hat{\theta}_B = 102.8$.

Example 4.5 Poisson-Gamma Conjugate Pair. Let $X_1, \dots, X_n$, given θ are Poisson $\mathcal{P}(\theta)$ with probability mass function

$$f(x_i|\theta) = \frac{\theta^{x_i}}{x_i!} e^{-\theta},$$

and $\theta \sim \mathcal{G}(\alpha, \beta)$ is given by $\pi(\theta) \propto \theta^{\alpha-1} e^{-\beta\theta}$. Then,

$$\pi(\theta|X_1, \dots, X_n) = \pi(\theta|\sum X_i) \propto \theta^{\sum X_i + \alpha - 1} e^{-(n+\beta)\theta},$$

which is $\mathcal{G}(\sum_i X_i + \alpha, n + \beta)$. The mean is $\mathbb{E}(\theta|X) = (\sum X_i + \alpha)/(n + \beta)$, and it can be represented as a weighted average of the MLE and the prior mean:

$$\mathbb{E}\theta|X = \frac{n}{n+\beta} \frac{\sum X_i}{n} + \frac{\beta}{n+\beta} \frac{\alpha}{\beta}.$$

4.2.3 Conjugate Priors

We have seen two convenient examples for which the posterior distribution remained in the same family as the prior distribution. In such a case, the effect of likelihood is only to "update" the prior parameters and not to change prior's functional form. We say that such priors are *conjugate* with the likelihood. Conjugacy is popular because of its mathematical convenience; once the conjugate pair likelihood/prior is found, the posterior is calculated with relative ease. In the years BC[1] and pre-MCMC era (see Chapter 18), conjugate priors have been extensively used (and overused and misused) precisely because of this computational convenience. Nowadays, the general agreement is that simple conjugate analysis is of limited practical value because, given the likelihood, the conjugate prior has limited modeling capability.

There are many univariate and multivariate instances of conjugacy. The following table provides several cases. For practice you may want to workout the posteriors in the table.

[1] For some, the BC era signifies *Before Christ,* rather than *Before Computers.*

Table 4.2 Some conjugate pairs. Here $\mathbf{X}$ stands for a sample of size n, $X_1, \ldots, X_n$.

Likelihood	Prior	Posterior
$X\|\theta \sim \mathcal{N}(\theta, \sigma^2)$	$\theta \sim \mathcal{N}(\mu, \tau^2)$	$\theta\|X \sim \mathcal{N}\left(\frac{\tau^2}{\sigma^2+\tau^2}X + \frac{\sigma^2}{\sigma^2+\tau^2}\mu, \frac{\sigma^2\tau^2}{\sigma^2+\tau^2}\right)$
$X\|\theta \sim \mathcal{B}(n, \theta)$	$\theta \sim \mathcal{B}e(\alpha, \beta)$	$\theta\|X \sim \mathcal{B}e(\alpha + x, n - x + \beta)$
$\mathbf{X}\|\theta \sim \mathcal{P}(\theta)$	$\theta \sim \mathcal{G}amma(\alpha, \beta)$	$\theta\|\mathbf{X} \sim \mathcal{G}amma(\sum_i X_i + \alpha, n + \beta)$.
$\mathbf{X}\|\theta \sim \mathcal{NB}(m, \theta)$	$\theta \sim \mathcal{B}e(\alpha, \beta)$	$\theta\|\mathbf{X} \sim \mathcal{B}e(\alpha + mn, \beta + \sum_{i=1}^n x_i)$
$X \sim \mathcal{G}amma(n/2, 1/(2\theta))$	$\theta \sim \mathcal{IG}(\alpha, \beta)$	$\theta\|X \sim \mathcal{IG}(n/2 + \alpha, x/2 + \beta)$
$\mathbf{X}\|\theta \sim \mathcal{U}(0, \theta)$	$\theta \sim \mathcal{P}a(\theta_0, \alpha)$	$\theta\|\mathbf{X} \sim \mathcal{P}a(\max\{\theta_0, X_1, \ldots, X_n\}, \alpha + n)$
$X\|\theta \sim \mathcal{N}(\mu, \theta)$	$\theta \sim \mathcal{IG}(\alpha, \beta)$	$\theta\|X \sim \mathcal{IG}(\alpha + 1/2, \beta + (\mu - X)^2/2)$
$X\|\theta \sim \mathcal{G}amma(\nu, \theta)$	$\theta \sim \mathcal{G}a(\alpha, \beta)$	$\theta\|X \sim \mathcal{G}amma(\alpha + \nu, \beta + x)$

4.2.4 Interval Estimation: Credible Sets

Bayesians call interval estimators of model parameters *credible sets*. Naturally, the measure used to assess the credibility of an interval estimator is the posterior distribution. Students learning concepts of classical confidence intervals (CIs) often err by stating that "the probability that the CI interval $[L, U]$ contains parameter θ is $1 - \alpha$". The correct statement seems more convoluted; one needs to generate data from the underlying model many times and for each generated data set to calculate the CI. The proportion of CIs covering the unknown parameter "tends to" $1 - \alpha$. The Bayesian interpretation of a credible set C is arguably more natural: The probability of a parameter belonging to the set C is $1 - \alpha$. A formal definition follows.

Assume the set C is a subset of Θ. Then, C is *credible set* with credibility $(1 - \alpha)100\%$ if

$$P(\theta \in C|X) = \mathbb{E}(I(\theta \in C)|X) = \int_C \pi(\theta|x)d\theta \geq 1 - \alpha.$$

If the posterior is discrete, then the integral is a sum (using the counting measure) and

$$P(\theta \in C|X) = \sum_{\theta_i \in C} \pi(\theta_i|x) \geq 1 - \alpha.$$

This is the definition of a $(1 - \alpha)100\%$ credible set, and for any given posterior distribution such a set is not unique.

For a given credibility level $(1-\alpha)100\%$, the shortest credible set has obvious appeal. To minimize size, the sets should correspond to highest posterior probability density areas (HPDs).

Definition 4.1 *The $(1-\alpha)100\%$ HPD credible set for parameter θ is a set C, subset of Θ of the form*

$$C = \{\theta \in \Theta | \pi(\theta|x) \geq k(\alpha)\},$$

where $k(\alpha)$ is the largest constant for which

$$P(\theta \in C|X) \geq 1 - \alpha.$$

Geometrically, if the posterior density is cut by a horizontal line at the hight $k(\alpha)$, the set C is projection on the θ axis of the part of line that lies below the density.

Example 4.6 Jeremy's IQ, Continued. Recall Jeremy, the enthusiastic Georgia Tech student from Example 4.4, who used Bayesian inference in modeling his IQ test scores. For a score $X|\theta$ he was using a $\mathcal{N}(\theta, 80)$ likelihood, while the prior on θ was $\mathcal{N}(110, 120)$. After the score of $X = 98$ was recorded, the resulting posterior was normal $\mathcal{N}(102.8, 48)$.

Here, the MLE is $\hat{\theta} = 98$, and a 95% confidence interval is $[98-1.96\sqrt{80},\ 98+1.96\sqrt{80}] = [80.4692, 115.5308]$. The length of this interval is approximately 35. The Bayesian counterparts are $\hat{\theta} = 102.8$, and $[102.8 - 1.96\sqrt{48},\ 102.8 + 1.96\sqrt{48}] = [89.2207, 116.3793]$. The length of 95% credible set is approximately 27. The Bayesian interval is shorter because the posterior variance is smaller than the likelihood variance; this is a consequence of the incorporation of information. The construction of the credible set is illustrated in Figure 4.3.

4.2.5 Bayesian Testing

Bayesian tests amount to comparison of posterior probabilities of the parameter regions defined by the two hypotheses.

Assume that Θ_0 and Θ_1 are two non-overlapping subsets of the parameter space Θ. We assume that Θ_0 and Θ_1 partition Θ, that is, $\Theta_1 = \Theta_0^c$, although cases in which $\Theta_1 \neq \Theta_0^c$ are easily formulated. Let $\theta \in \Theta_0$ signify the null hypothesis H_0 and let $\theta \in \Theta_1 = \Theta_0^c$ signify the alternative hypothesis H_1:

$$H_0 : \theta \in \Theta_0 \qquad\qquad H_1 : \theta \in \Theta_1.$$

Given the information from the posterior, the hypothesis with higher posterior probability is selected.

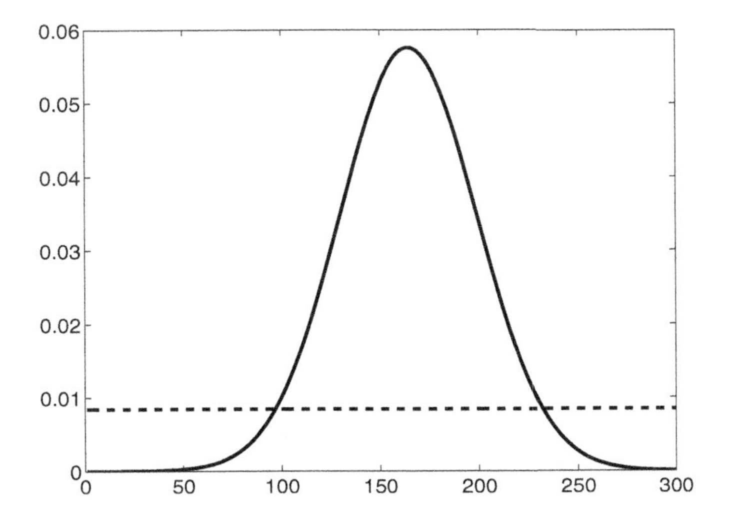

Fig. 4.3 Bayesian credible set based on $\mathcal{N}(102.8, 48)$ density.

Example 4.7 We return again to Jeremy (Examples 4.4 and 4.6) and consider the posterior for the parameter θ, $\mathcal{N}(102.8, 48)$. Jeremy claims he had a bad day and his genuine IQ is at least 105. After all, he is at Georgia Tech! The posterior probability of $\theta \geq 105$ is

$$p_0 = P^{\theta|X}(\theta \geq 105) = P\left(Z \geq \frac{105 - 102.8}{\sqrt{48}}\right) = 1 - \Phi(0.3175) = 0.3754,$$

less that 38%, so his claim is rejected. Posterior odds in favor of H_0 are $0.3754/(1-0.3754)=0.4652$, less than 50%.

We can represent the prior and posterior odds in favor of the hypothesis H_0, respectively, as

$$\frac{\pi_0}{\pi_1} = \frac{P^\theta(\theta \in \Theta_0)}{P^\theta(\theta \in \Theta_1)} \quad \text{and} \quad \frac{p_0}{p_1} = \frac{P^{\theta|X}(\theta \in \Theta_0)}{P^{\theta|X}(\theta \in \Theta_1)}.$$

The *Bayes factor* in favor of H_0 is the ratio of corresponding posterior to prior odds,

$$B_{01}^{\pi}(x) = \frac{\frac{P(\theta \in \Theta_0|X)}{P(\theta \in \Theta_1|X)}}{\frac{P(\theta \in \Theta_0)}{P(\theta \in \Theta_1)}} = \frac{p_0/p_1}{\pi_0/\pi_1}. \tag{4.3}$$

When the hypotheses are simple (i.e., $H_0 : \theta = \theta_0$ vs. $H_1 : \theta = \theta_1$), and the prior is just the two point distribution $\pi(\theta_0) = \pi_0$ and $\pi(\theta_1) = \pi_1 = 1 - \pi_0$,

Table 4.3 Treatment of H_0 According to the Value of log-Bayes Factor.

$0 \leq \log B_{10}(x) \leq 0.5$	evidence against H_0 is **poor**
$0.5 \leq \log B_{10}(x) \leq 1$	evidence against H_0 is **substantial**
$1 \leq \log B_{10}(x) \leq 2$	evidence against H_0 is **strong**
$\log B_{10}(x) > 2$	evidence against H_0 is **decisive**

then the Bayes factor in favor of H_0 becomes the likelihood ratio:

$$B_{01}^{\pi}(x) = \frac{\frac{P^{\theta|X}(\theta \in \Theta_0)}{P^{\theta|X}(\theta \in \Theta_1)}}{\frac{P^{\theta}(\theta \in \Theta_0)}{P^{\theta}(\theta \in \Theta_1)}} = \frac{f(x|\theta_0)\pi_0}{f(x|\theta_1)\pi_1} / \frac{\pi_0}{\pi_1} = \frac{f(x|\theta_0)}{f(x|\theta_1)}.$$

If the prior is a mixture of two priors, ξ_0 under H_0 and ξ_1 under H_1, then the Bayes factor is the ratio of two marginal (prior-predictive) distributions generated by ξ_0 and ξ_1. Thus, if $\pi(\theta) = \pi_0 \xi_0(\theta) + \pi_1 \xi_2(\theta)$ then,

$$B_{01}^{\pi}(x) = \frac{\frac{\int_{\Theta_0} f(x|\theta)\pi_0 \xi_0(\theta) d\theta}{\int_{\Theta_1} f(x|\theta)\pi_1 \xi_1(\theta) d\theta}}{\frac{\pi_0}{\pi_1}} = \frac{m_0(x)}{m_1(x)}.$$

The Bayes factor measures relative change in prior odds once the evidence is collected. Table 4.3 offers practical guidelines for Bayesian testing of hypotheses depending on the value of log-Bayes factor. One could use $B_{01}^{\pi}(x)$ of course, but then $a \leq \log B_{10}(x) \leq b$ becomes $-b \leq \log B_{01}(x) \leq -a$. Negative values of the log-Bayes factor are handled by using symmetry and changed wording, in an obvious way.

4.2.5.1 Bayesian Testing of Precise Hypotheses Testing precise hypotheses in Bayesian fashion has a considerable body of research. Berger (1985), pp. 148–157, has a comprehensive overview of the problem and provides a wealth of references. See also Berger and Sellke (1984) and Berger and Delampady (1987).

If the priors are continuous, testing precise hypotheses in Bayesian fashion is impossible because with continuous priors and posteriors, the probability of a singleton is 0. Suppose $X|\theta \sim f(x|\theta)$ is observed and we are interested in testing

$$H_0 : \theta = \theta_0 \quad v.s. \quad H_1 : \theta \neq \theta_0.$$

The answer is to have a prior that has a point mass at the value θ_0 with prior weight π_0 and a spread distribution $\xi(\theta)$ which is the prior under H_1 that has prior weight $\pi_1 = 1 - \pi_0$. Thus, the prior is the 2-point mixture

$$\pi(\theta) = \pi_0 \delta_{\theta_0} + \pi_1 \xi(\theta),$$

where δ_{θ_0} is Dirac mass at θ_0.

The marginal density for X is

$$m(x) = \pi_0 f(x|\theta_0) + \pi_1 \int f(x|\theta)\xi(\theta)d\theta = \pi_0 f(x|\theta_0) + \pi_1 m_1(x).$$

The posterior probability of $\theta = \theta_0$ is

$$\pi(\theta_0|x) = f(x|\theta_0)\pi_0/m(x) = \frac{f(x|\theta_0)\pi_0}{\pi_0 f(x|\theta_0) + \pi_1 m_1(x)} = \left(1 + \frac{\pi_1}{\pi_0} \cdot \frac{m_1(x)}{f(x|\theta_0)}\right)^{-1}.$$

4.2.6 Bayesian Prediction

Statistical prediction fits naturally into the Bayesian framework. Suppose $Y \sim f(y|\theta)$ is to be observed. The posterior predictive distribution of Y, given observed $X = x$ is

$$f(y|x) = \int_\Theta f(y|\theta)\pi(\theta|x)d\theta.$$

For example, in the normal distribution example, the predictive distribution of Y, given $X_1, \ldots, X_n$ is

$$Y|\bar{X} \sim \mathcal{N}\left(\frac{\tau^2}{\frac{\sigma^2}{n} + \tau^2}\bar{X} + \frac{\frac{\sigma^2}{n}}{\frac{\sigma^2}{n} + \tau^2}\mu,\ \sigma^2 + \frac{\frac{\sigma^2}{n}\tau^2}{\frac{\sigma^2}{n} + \tau^2}\right). \qquad (4.4)$$

Example 4.8 Martz and Waller (1985) suggest that Bayesian reliability inference is most helpful in applications where little system failure data exist, but past data from like systems are considered relevant to the present system. They use an example of heat exchanger reliability, where the lifetime X is the failure time for heat exchangers used in refining gasoline. From past research and modeling in this area, it is determined that X has a Weibull distribution with $\kappa = 3.5$. Furthermore, the scale parameter λ is considered to be in the interval $0.5 \le \lambda \le 1.5$ with no particular value of λ considered more likely than others.

From this argument, we have

$$\pi(\lambda) = \begin{cases} 1 & 0.5 \le \lambda \le 1.5 \\ 0 & \text{otherwise} \end{cases}$$

$$f(x|\lambda) = \kappa\lambda^\kappa x^{\kappa-1} e^{-(x\lambda)^\kappa}$$

where $\kappa = 3.5$. With n=9 observed failure times (measured in years of service) at (0.41, 0.58, 0.75, 0.83, 1.00, 1.08, 1.17, 1.25, 1.35), the likelihood is

$$f(x_1, \ldots, x_9|\lambda) \propto \lambda^p \left(\prod_{i=1}^{9} {x_i}^{2.5} \right) e^{-\lambda^{3.5}(\sum {x_i}^{3.5})},$$

so the sufficient statistic is

$$\sum_{i=1}^{n} {x_i}^{3.5} = 10.16.$$

The resulting posterior distribution is not distributed Weibull (like the likelihood) or uniform (like the prior). It can be expressed as

$$\pi(\lambda|x_1, \ldots, x_9) = \begin{cases} (1621.39)\lambda^9 e^{-10.16\lambda^{3.5}} & 0.5 \le \lambda \le 1.5 \\ 0 & \text{otherwise,} \end{cases}$$

and has expected value of $\lambda_B = 0.6896$. Figure 4.4(a) shows the posterior density From the prior distribution, $\mathbb{E}(\lambda) = 1$, so our estimate of λ has decreased in the process of updating the prior with the data.

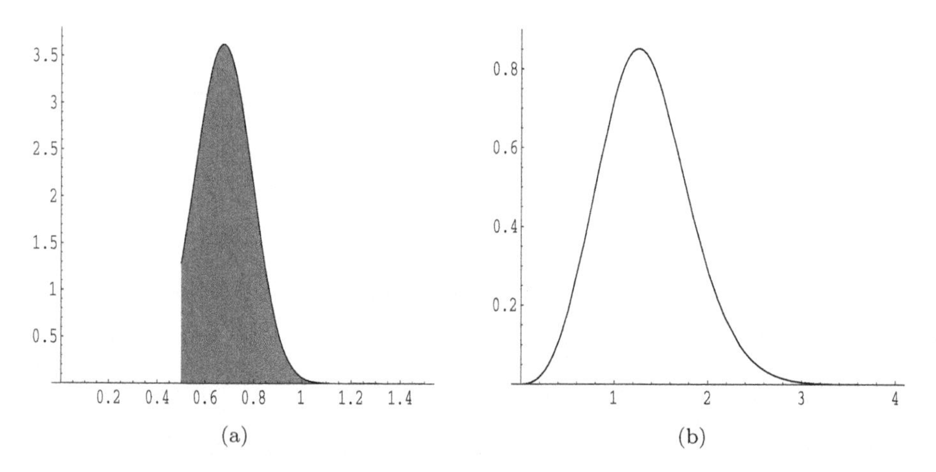

Fig. 4.4 (a) Posterior density for λ; (b) Posterior predictive density for heat-exchanger lifetime.

Estimation of λ was not the focus of this study; the analysts were interested in predicting future lifetime of a generic (randomly picked) heat exchanger. Using the predictive density from (4.4),

$$f(y|x) = \int_{0.5}^{1.5} \left(3.5\lambda^{3.5} y^{2.5} e^{-(\lambda y)^{3.5}} \right) \left(1621.39\lambda^9 e^{-10.16\lambda^{3.5}} \right) d\lambda.$$

The predictive density is a bit messy, but straightforward to work with. The plot of the density in Figure 4.4(b) shows how uncertainty is gauged for the lifetime of a new heat-exchanger. From $f(y|x)$, we might be interested in predicting early failure by the new item; for example, a 95% lower bound for heat-exchanger lifetime is found by computing the lower 0.05-quantile of $f(y|x)$, which is approximately 0.49.

4.3 BAYESIAN COMPUTATION AND USE OF WINBUGS

If the selection of an adequate prior was the major conceptual and modeling challenge of Bayesian analysis, the major implementational challenge is computation. When the model deviates from the conjugate structure, finding the posterior distribution and the Bayes rule is all but simple. A closed form solution is more an exception than the rule, and even for such exceptions, lucky mathematical coincidences, convenient mixtures, and other tricks are needed to uncover the explicit expression.

If the classical statistics relies on optimization, Bayesian statistics relies on integration. The marginal needed for the posterior is an integral

$$m(x) = \int_\Theta f(x|\theta)\pi(\theta)d\theta,$$

and the Bayes estimator of $h(\theta)$, with respect to the squared error loss is a ratio of integrals,

$$\delta_\pi(x) = \int_\Theta h(\theta)\pi(\theta|x)d\theta = \frac{\int_\Theta h(\theta)f(x|\theta)\pi(\theta)d\theta}{\int_\Theta f(x|\theta)\pi(\theta)d\theta}.$$

The difficulties in calculating the above Bayes rule come from the facts that (i) the posterior may not be representable in a finite form, and (ii) the integral of $h(\theta)$ does not have a closed form even when the posterior distribution is explicit.

The last two decades of research in Bayesian statistics contributed to broadening the scope of Bayesian models. Models that could not be handled before by a computer are now routinely solved. This is done by *Markov chain Monte Carlo* (MCMC) Methods, and their introduction to the field of statistics revolutionized Bayesian statistics.

The Markov chain Monte Carlo (MCMC) methodology was first applied in statistical physics, (Metropolis et al., 1953). Work by Gelfand and Smith (1990) focused on applications of MCMC to Bayesian models. The principle of MCMC is simple: to sample randomly from a target probability distribution one designs a Markov chain whose stationary distribution is the target distribution. By simulating long runs of such a Markov chain, the target distribution can be well approximated. Various strategies for constructing appropriate Markov chains that simulate form the desired distribution are possible: Metropolis-Hastings, Gibbs sapmler, slice sampling, perfect sampling, and many specialized techniques. They are beyond the scope of this text and the interested reader is directed to Robert (2001), Robert and Casella (2004), and Chen, Shao, and Ibrahim (2000), for an overview and a comprehensive treatment.

We will use WinBUGS for doing Bayesian inference on non conjugate models. Appendix B offers a brief introduction to the front-end of WinBUGS. Three volumes of examples are standard addition to the software, in the Examples menu of WinBUGS, see Spiegelhalter, Thomas, Best, and Gilks (1996). It is recommended that you go over some of those in detail because they illustrate the functionality and real modeling power of WinBUGS. A wealth of examples on Bayesian modeling strategies using WinBUGS can be found in the monographs of Congdon (2001, 2003, 2005). The following example demonstrates the simulation power of WinBUGS, although it involves approximating probabilities of complex events and has nothing to do with Bayesian inference.

Example 4.9 Paradox DeMere in WinBUGS. In 1654 the Chevalier de Mere asked Blaise Pascal (1623–1662) the following question: *In playing a game with three dice why the sum 11 is advantageous to sum 12 when both are results of six possible outcomes?* Indeed, there are six favorable triplets for each of the sums **11** and **12**,

11:	(1, 4, 6), (1, 5, 5), (2, 3, 6), (2, 4, 5), (3, 3, 5), (3, 4, 4)
12:	(1, 5, 6), (2, 4, 6), (2, 5, 5), (3, 3, 6), (3, 4, 5), (4, 4, 4)

The solution to this "paradox" deMere is simple. By taking into account all possible permutations of the triples, the sum 11 has 27 favorable permutations while the sum 12 has 25 favorable permutation. But what if 300 fair dice are rolled and we are interested if the sum 1111 is advantageous to the sum 1112? Exact solution is unappealing, but the probabilities can be well approximated by WinBUGS model `demere1`.

```
model demere1;
{
for (i in 1:300) {
dice[i] ~ dcat(p.dice[]);
}
is1111 <- equals(sum(dice[]),1111)
is1112 <- equals(sum(dice[]),1112)
}
```

The data are

```
list(p.dice=c(0.1666666, 0.1666666,
0.1666667, 0.1666667, 0.1666667, 0.1666667) )
```

and the initial values are generated. After five million rolls, WinBUGS outputs `is1111 = 0.0016` and `is1112 = 0.0015`, so the sum of 1111 is advantageous to the sum of 1112.

Example 4.10 Jeremy in WinBUGS. We will calculate a Bayes estimator for Jeremy's true IQ using BUGS. Recall, the model in Example 4.4 was $X \sim \mathcal{N}(\theta, 80)$ and $\theta \sim \mathcal{N}(100, 120)$. In WinBUGS we will use the precision parameters $1/120 = 0.00833$ and $1/80 = 0.0125$.

```
#Jeremy in WinBUGS
model{
x ~ dnorm( theta, tau)
theta ~ dnorm( 110, 0.008333333)
}
#data
list( tau=0.0125, x=98)
#inits
list(theta=100)
```

Below is the summary of MCMC output.

node	mean	sd	MC error	2.5%	median	97.5%
θ	102.8	6.917	0.0214	89.17	102.8	116.3

Because this is a conjugate normal/normal model, the exact posterior distribution, $\mathcal{N}(102.8, 48)$, was easy to find, (see Example 4.4). Note that in simulations, the MCMC approximation, when rounded, coincides with the exact posterior mean. The MCMC variance of θ is $6.917^2 = 47.84489$, close to the exact posterior variance of 48.

4.4 EXERCISES

4.1. A lifetime X (in years) of a particular machine is modeled by an exponential distribution with unknown failure rate parameter θ. The lifetimes of $X_1 = 5, X_2 = 6$, and $X_3 = 4$ are observed, and assume that an expert believes that θ should have exponential distribution as well and that, on average θ should be $1/3$.

(i) Write down the MLE of θ for those observations.

(ii) Elicit a prior according to the expert's beliefs.

(iii) For the prior in (ii), find the posterior. Is the problem conjugate?

(iv) Find the Bayes estimator $\hat{\theta}_{Bayes}$, and compare it with the MLE estimator from (i). Discuss.

4.2. Suppose $X = (X_1, \ldots, X_n)$ is a sample from $\mathcal{U}(0, \theta)$. Let θ have Pareto $\mathcal{P}a(\theta_0, \alpha)$ distribution. Show that the posterior distribution is $\mathcal{P}a(\max\{\theta_0, x_1, \ldots, x_n\}\ \alpha + n)$.

4.3. Let $X \sim \mathcal{G}(n/2, 2\theta)$, so that X/θ is χ^2_n. Let $\theta \sim \mathcal{IG}(\alpha, \beta)$. Show that the posterior is $\mathcal{IG}(n/2 + \alpha, (x/2 + \beta^{-1})^{-1})$.

4.4. If $X = (X_1, \ldots, X_n)$ is a sample from $\mathcal{NB}(m, \theta)$ and $\theta \sim \mathcal{B}e(\alpha, \beta)$, show that the posterior for θ is beta $\mathcal{B}e(\alpha + mn, \beta + \sum_{i=1}^n x_i)$.

4.5. In Example 4.5 on p. 54, show that the marginal distribution is negative binomial.

4.6. What is the Bayes factor B_{01}^{π} in Jeremy's case (Example 4.7)? Test H_0 is using the Bayes factor and wording from the Table 4.3. Argue that the evidence against H_0 is poor.

4.7. Assume $X|\theta \sim \mathcal{N}(\theta, \sigma^2)$ and $\theta \sim \pi(\theta) = 1$. Consider testing $H_0 : \theta \leq \theta_0$ v.s. $H_1 : \theta > \theta_0$. Show that $p_0 = P^{\theta|X}(\theta \leq \theta_0)$ is equal to the classical p-value.

4.8. Show that the Bayes factor is $B_{01}^{\pi}(x) = f(x|\theta_0)/m_1(x)$.

4.9. Show that

$$p_0 = \pi(\theta_0|x) \geq \left[1 + \frac{\pi_1}{\pi_0} \cdot \frac{r(x)}{f(x|\theta_0)}\right]^{-1},$$

where $r(x) = \sup_{\theta \neq \theta_0} f(x|\theta)$. Usually, $r(x) = f(x|\hat{\theta}_{\text{MLE}})$, where $\hat{\theta}_{mle}$ is MLE estimator of θ. The Bayes factor $B_{01}^{\pi}(x)$ is bounded from below:

$$B_{01}^{\pi}(x) \geq \frac{f(x|\theta_0)}{r(x)} = \frac{f(x|\theta_0)}{f(x|\hat{\theta}_{\text{MLE}})}.$$

4.10. Suppose $X = -2$ was observed from the population distributed as $N(0, 1/\theta)$ and one wishes to estimate the parameter θ. (Here θ is the reciprocal of variance σ^2 and is called the *precision parameter.* The precision parameter is used in WinBUGS to parameterize the normal distribution). A classical estimator of θ (e.g., the MLE) does exist, but one may be disturbed to estimate $1/\sigma^2$ based on a single observation. Suppose the analyst believes that the prior on θ is $\mathcal{G}amma(1/2, 3)$.

(i) What is the MLE of θ?

(ii) Find the posterior distribution and the Bayes estimator of θ. If the prior on θ is $\mathcal{G}amma(\alpha, \beta)$, represent the Bayes estimator as weighted average (sum of weights = 1) of the prior mean and the MLE.

(iii) Find a 95% HPD Credible set for θ.

(iv) Test the hypothesis $H_0 : \theta \leq 1/4$ versus $H_1 : \theta > 1/4$.

4.11. *The Lindley (1957) Paradox.* Suppose $\bar{y}|\theta \sim N(\theta, 1/n)$. We wish to test $H_0 : \theta = 0$ versus the two sided alternative. Suppose a Bayesian puts the prior $P(\theta = 0) = P(\theta \neq 0) = 1/2$, and in the case of the alternative, the 1/2 is uniformly spread over the interval $[-M/2, M/2]$. Suppose $n = 40,000$ and $\bar{y} = 0.01$ are observed, so $\sqrt{n}\ \bar{y} = 2$. The classical statistician rejects H_0 at level $\alpha = 0.05$. Show that posterior odds in favor of H_0 are 11 if $M = 1$, indicating that a Bayesian statistician strongly favors H_0, according to Table 4.3.

4.12. This exercise concerning Bayesian binary regression with a probit model using WinBUGS is borrowed from David Madigan's Bayesian Course Site. Finney (1947) describes a binary regression problem with data of size $n = 39$, two continuous predictors x_1 and x_2, and a binary response y. Here are the data in BUGS-ready format:

```
list(n=39,x1=c(3.7,3.5,1.25,0.75,0.8,0.7,0.6,1.1,0.9,0.9,0.8,0.55,0.6,1.4,
 0.75,2.3,3.2,0.85,1.7,1.8,0.4,0.95,1.35,1.5,1.6,0.6,1.8,0.95,1.9,1.6,2.7,
 2.35,1.1,1.1,1.2,0.8,0.95,0.75,1.3),
x2=c(0.825,1.09,2.5,1.5,3.2,3.5,0.75,1.7,0.75,0.45,0.57,2.75,3.0,2.33,3.75,
 1.64,1.6,1.415,1.06,1.8,2.0,1.36,1.35,1.36,1.78,1.5,1.5,1.9,0.95,0.4,0.75,
 0.03,1.83,2.2,2.0,3.33,1.9,1.9,1.625),
y=c(1,1,1,1,1,1,0,0,0,0,0,0,0,1,1,1,1,1,0,1,0,0,0,0,1,0,1,0,1,0,1,0,0,1,1,
 1,0,0,1))
```

The objective is to build a predictive model that predicts y from x_1 and x_2. Proposed approach is the probit model: $P(y = 1|x_1, x_2) = \Phi(\beta_0 + \beta_1\ x_1 + \beta_2\ x_2)$ where Φ is the standard normal CDF.

(i) Use WinBUGS to compute posterior distributions for β_0, β_1 and β_2 using diffuse normal priors for each.

(ii) Suppose instead of the diffuse normal prior for β_i, $i = 0, 1, 2$, you use a normal prior with mean zero and variance v_i, and assume the v_is are independently exponentially distributed with some hyperparameter γ. Fit this model using BUGS. How different are the two posterior distributions from this exercise?

4.13. The following WinBUGS code flips a coin, the outcome H is coded by 1 and tails by 0. Mimic the following code to simulate a rolling of a fair die.

```
#coin.bug:
model coin;
{
flip12 ~ dcat(p.coin[])
coin <- flip12 - 1
}
#coin.dat:
list(p.coin=c(0.5, 0.5))
# just generate initials
```

4.14. The highly publicized (recent TV reports) *in vitro fertilization* success cases for women in their late fifties all involve donor's egg. If the egg is the woman's own, the story is quite different.

In vitro fertilization (IVF), one of the assisted reproductive technology (ART) procedures, involves extracting a woman's eggs, fertilizing the eggs in the laboratory, and then transferring the resulting embryos into the womans uterus through the cervix. Fertilization involves a specialized technique known as intracytoplasmic sperm injection (ICSI).

The table shows the live-birth success rate per transfer rate from the recipients' eggs, stratified by age of recipient. The data are for year 1999, published by US - Centers for Disease Control and Prevention (CDC): (`http://www.cdc.gov/reproductivehealth/ART99/index99.htm`)

Age (x)	24	25	26	27	28	29	30	31
Percentage (y)	38.7	38.6	38.9	41.4	39.7	41.1	38.7	37.6

Age (x)	32	33	34	35	36	37	38	39
Percentage(y)	36.3	36.9	35.7	33.8	33.2	30.1	27.8	22.7

Age (x)	40	41	42	43	44	45	46
Percentage(y)	21.3	15.4	11.2	9.2	5.4	3.0	1.6

Assume the change-point regression model

$$\begin{aligned} y_i &= \beta_0 + \beta_1 x_i + \epsilon_i, \; i = 1, \ldots, \tau \\ y_i &= \gamma_0 + \gamma_1 x_i + \epsilon_i, \; i = \tau + 1, \ldots, n \\ \epsilon_i &\sim \mathcal{N}(0, \sigma^2). \end{aligned}$$

(i) Propose priors (with possibly hyperpriors) on σ^2, β_0, β_1, γ_0, and γ_1.

(ii) Take discrete uniform prior on τ and write a WinBUGS program.

4.15. Is the cloning of humans moral? Recent Gallup Poll estimates that about 88% Americans opposed cloning humans. Results are based on telephone interviews with a randomly selected national sample of $n = 1000$ adults, aged 18 and older, conducted May 2-4, 2004. In these 1000 interviews, 882 adults opposed cloning humans.

(i) Write WinBUGS program to estimate the proportion p of people opposed to cloning humans. Use a non-informative prior for p.

(ii) Test the hypothesis that $p \leq 0.87$.

(iii) Pretend that the original poll had $n = 1062$ adults, i.e., results for 62 adults are missing. Estimate the number of people opposed to cloning among the 62 missing in the poll. *Hint:*

```
model {     anticlons ~  dbin(prob,npolled) ;
```

```
        lessthan87 <- step(prob-0.87)
        anticlons.missing ~ dbin(prob,nmissing)
        prob ~ dbeta(1,1)}
Data
list(anticlons=882,npolled= 1000, nmissing=62)
```

REFERENCES

Anscombe, F. J. (1962), "Tests of Goodness of Fit," *Journal of the Royal Statistical Society (B)*, 25, 81-94.

Bayes, T. (1763), "An Essay Towards Solving a Problem in the Doctrine of Chances," *Philosophical Transactions of the Royal Society, London*, 53, 370-418.

Berger, J. O. (1985), *Statistical Decision Theory and Bayesian Analysis*, Second Edition, New York: Springer-Verlag.

Berger, J. O., and Delampady, M. (1987), "Testing Precise Hypothesis," *Statistical Science*, 2, 317-352.

Berger, J. O., and Selke, T. (1987), "Testing a Point Null Hypothesis: The Irreconcilability of p-values and Evidence (with Discussion)", *Journal of American Statistical Association*, 82, 112-122.

Chen, M.-H., Shao, Q.-M., and Ibrahim, J. (2000), *Monte Carlo Methods in Bayesian Computation,* New York: Springer Verlag.

Congdon, P. (2001), *Bayesian Statistical Modelling*, Hoboken, NJ: Wiley.

Congdon, P. (2003), *Applied Bayesian Models*, Hoboken, NJ: Wiley.

Congdon, P. (2005), *Bayesian Models for Categorical Data*, Hoboken, NJ: Wiley.

Finney, D. J. (1947), "The Estimation from Individual Records of the Relationship Between Dose and Quantal Response," *Biometrika*, 34, 320-334.

Gelfand, A. E., and Smith, A. F. M. (1990), "Sampling-based Approaches to Calculating Marginal Densities," *Journal of American Statistical Association*, 85, 398-409.

Lindley, D. V. (1957), "A Statistical Paradox," *Biometrika*, 44, 187-192.

Madigan, D. `http://stat.rutgers.edu/ madigan/bayes02/`. A Web Site for Course on Bayesian Statistics.

Martz, H., and Waller, R. (1985), *Bayesian Reliability Analysis*, New York: Wiley.

Metropolis, N., Rosenbluth, A., Rosenbluth, M., Teller, A., and Teller, E. (1953), "Equation of State Calculations by Fast Computing Machines," *The Journal of Chemical Physics*, 21, 1087-1092.

Robert, C. (2001), *The Bayesian Choice: From Decision-Theoretic Motivations to Computational Implementation*, Second Edition, New York: Springer Verlag.

Robert, C. and Casella, G. (2004), *Monte Carlo Statistical Methods*, Second Edition, New York: Springer Verlag.

Spiegelhalter, D. J., Thomas, A., Best, N. G., and Gilks, W. R. (1996), "BUGS Examples Volume 1," Version 0.5. Cambridge: Medical Research Council Biostatistics Unit (PDF).

5

Order Statistics

The early bird gets the worm, but the second mouse gets the cheese.

Steven Wright

Let $X_1, X_2, \ldots, X_n$ be an independent sample from a population with absolutely continuous cumulative distribution function F and density f. The continuity of F implies that $P(X_i = X_j) = 0$, when $i \neq j$ and the sample could be ordered with strict inequalities,

$$X_{1:n} < X_{2:n} < \cdots < X_{n-1:n} < X_{n:n}, \tag{5.1}$$

where $X_{i:n}$ is called the ith *order statistic* (out of n). The *range* of the data is $X_{n:n} - X_{1:n}$, where $X_{n:n}$ and $X_{1:n}$ are, respectively, the sample maximum and minimum. The study of order statistics permeates through all areas of statistics, including nonparametric. There are several books dedicated just to probability and statistics related to order statistics; the textbook by David and Nagaraja (2003) is a deservedly popular choice.

The marginal distribution of $X_{i:n}$ is not the same as X_i. Its distribution function $F_{i:n}(t) = P(X_{i:n} \leq t)$ is the probability that *at least* i out of n observations from the original sample are no greater than t, or

$$F_{i:n}(t) = P(X_{i:n} \leq t) = \sum_{k=i}^{n} \binom{n}{k} F(t)^k \left(1 - F(t)\right)^{n-k}.$$

If F is differentiable, it is possible to show that the corresponding density function is

$$f_{i:n}(t) = i\binom{n}{i} F(t)^{i-1}\left(1 - F(t)\right)^{n-i} f(t). \tag{5.2}$$

Example 5.1 Recall that for any continuous distribution F, the transformed sample $F(X_1), \ldots, F(X_n)$ is distributed $\mathcal{U}(0,1)$. Similarly, from (5.2) the distribution of $F(X_{i:n})$ is $\mathcal{B}e(i, n-i+1)$. Using the MATLAB code below, the densities are graphed in Figure 5.1.

```
>> x=0:0.025:1;
>> for i=1,5
>>  plot(betapdf(x,i,6-i))
>> hold all
>> end
```

Example 5.2 Reliability Systems. In reliability, series and parallel systems are building blocks for system analysis and design. A *series system* is one that works only if all of its components are working. A *parallel system* is one that fails only if all of its components fail. If the lifetimes of a n-component system $(X_1, \ldots, X_n)$ are i.i.d. distributed, then if the system is in series, the system lifetime is $X_{1:n}$. On the other hand, for a parallel system, the lifetime is $X_{n:n}$.

5.1 JOINT DISTRIBUTIONS OF ORDER STATISTICS

Unlike the original sample $(X_1, X_2, \ldots, X_n)$, the set of order statistics is inevitably dependent. If the vector $(X_1, X_2, \ldots, X_n)$ has a joint density

$$f_{1,2,\ldots,n}(x_1, x_2, \ldots, x_n) = \prod_{i=1}^{n} f(x_i),$$

then the joint density for the order statistics, $f_{1,2,\ldots,n:n}(x_1, ..., x_n)$ is

$$f_{1,2,\ldots,n:n}(x_1, ..., x_n) = \begin{cases} n! \prod_{i=1}^{n} f(x_i) & x_1 < x_2 < \cdots < x_n \\ 0 & \text{otherwise.} \end{cases} \tag{5.3}$$

To understand why this is true, consider the conditional distribution of the order statistics $\mathbf{y} = (x_{1:n}, x_{2:n}, \ldots, x_{n:n})$ given $\mathbf{x} = (x_1, x_2, \ldots, x_n)$. Each one of the $n!$ permutations of $(X_1, X_2, \ldots, X_n)$ are equal in probability, so computing $f_{\mathbf{y}} = \int f_{\mathbf{y}|\mathbf{x}} dF_{\mathbf{x}}$ is incidental. The joint density can also be derived using a Jacobian transformation (see Exercise 5.3).

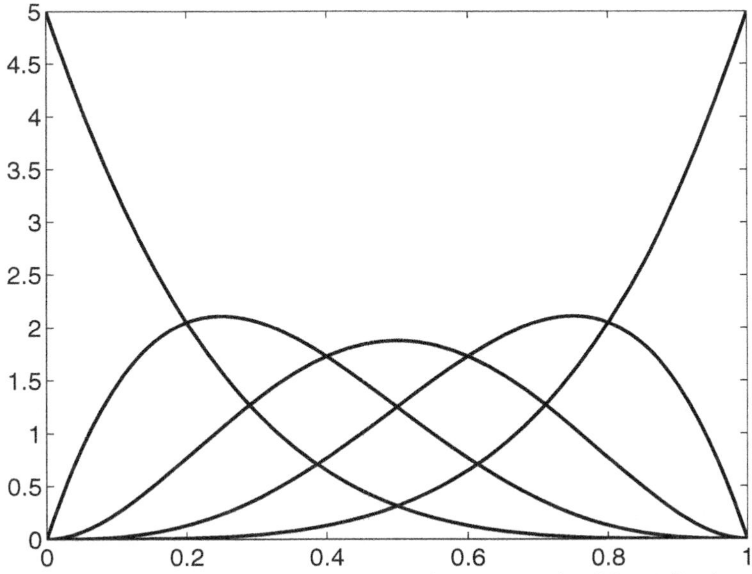

Fig. 5.1 Distribution of order statistics from a sample of five $\mathcal{U}(0,1)$.

From (5.3) we can obtain the distribution of any subset of order statistics. The joint distribution of $X_{r:n}, X_{s:n}, 1 \leq r < s \leq n$ is defined as

$$F_{r,s:n}(x_r, x_s) = P(X_{r:n} \leq x_r, X_{s:n} \leq x_s),$$

which is the probability that at least r out of n observations are at most x_r, *and* at least s of n observations are at most x_s. The probability that *exactly* i observations are at most x_r and j are at most x_s is

$$\frac{n!}{(i-1)!(j-i)!(n-j)!} F(x_r)^i \left(F(x_s) - F(x_r)\right)^{j-i} \left(1 - F(x_s)\right)^{n-j},$$

where $-\infty < x_r < x_s < \infty$; hence

$$\begin{aligned} F_{r,s:n}(x_r, x_s) &= \sum_{j=s}^{n} \sum_{i=r}^{s} \frac{n!}{(i-1)!(j-i)!(n-j)!} \times \\ & \quad F(x_r)^i \left(F(x_s) - F(x_r)\right)^{j-i} \left(1 - F(x_s)\right)^{n-j}. \end{aligned} \quad (5.4)$$

If F is differentiable, it is possible to formulate the joint density of two order statistics as

$$
\begin{aligned}
f_{r,s:n}(x_r, x_s) &= \frac{n!}{(r-1)!(s-r-1)!(n-s)!} \times \qquad (5.5)\\
& F(x_r)^{r-1}\left(F(x_s)-F(x_r)\right)^{s-r-1}\left(1-F(x_s)\right)^{n-s} f(x_r)f(x_s).
\end{aligned}
$$

Example 5.3 Sample Range. The range of the sample, R, defined before as $X_{n:n} - X_{1:n}$, has density

$$
f_R(u) = \int_{-\infty}^{\infty} n(n-1)[F(v) - F(v-u)]^{n-2} f(v-u) f(v) dv. \qquad (5.6)
$$

To find $f_R(u)$, start with the joint distribution of $(X_{1:n}, X_{n:n})$ in (5.5),

$$
f_{1,n:n}(y_1, y_n) = n(n-1)[F(y_n) - F(y_1)]^{n-2} f(y_1) f(y_n).
$$

and make the transformation

$$
\begin{aligned}
u &= y_n - y_1\\
v &= y_n.
\end{aligned}
$$

The Jacobian of this transformation is 1, and $y_1 = v - u, y_n = v$. Plug y_1, y_n into the joint distribution $f_{1,n:n}(y_1, y_n)$ and integrate out v to arrive at (5.6). For the special case in which $F(t) = t$, the probability density function for the sample range simplifies to

$$
f_R(u) = n(n-1)u^{n-2}(1-u), \quad 0 < u < 1.
$$

5.2 SAMPLE QUANTILES

Recall that for a distribution F, the pth quantile (x_p) is the value x such that $F(x) = p$, if the distribution is continuous, and more generally, such that $F(x) \geq p$ and $P(X \geq x) \geq 1 - p$, if the distribution is arbitrary. For example, if the distribution F is discrete, there may not be any value x for which $F(x) = p$.

Analogously, if $X_1, \ldots, X_n$ represents a sample from F, the pth *sample quantile* $(\hat{x}_p)$ is a value of x such that $100p\%$ of the sample is smaller than x. This is also called the $100p\%$ *sample percentile.* With large samples, there is a number $1 \leq r \leq n$ such that $X_{r:n} \approx x_p$. Specifically, if n is large enough so that $p(n+1) = r$ for some $r \in \mathbb{Z}$, then $\hat{x}_p = X_{r:n}$ because there would be $r-1$ values smaller than $\hat{x}_p$ in the sample, and $n - r$ larger than it.

If $p(n+1)$ is not an integer, we can consider estimating the population quantile by an inner point between two order statistics, say $X_{r:n}$ and $X_{(r+1):n}$, where $F(X_{r:n}) < p - \epsilon$ and $F(X_{(r+1):n}) > p + \epsilon$ for some small $\epsilon > 0$. In this case, we can use a number that interpolates the value of $\hat{x}_p$ using the line between $(X_{r:n}, r/(n+1))$ and $(X_{(r+1):n}, (r+1)/(n+1))$:

$$\hat{x}_p = (-p(n+1) + r + 1)\, X_{r:n} + (p(n+1) - r)\, X_{(r+1):n}. \tag{5.7}$$

Note that if $p = 1/2$ and n is an even number, then $r = n/2$ and $r+1 = n/2+1$, and $\hat{x}_p = (X_{\frac{n}{2}:n} + X_{(\frac{n}{2}+1):n})/2$. That is, the sample median is the average of the two middle sample order statistics.

We note that there are alternative definitions of sample quantile in the literature, but they all have the same large sample properties.

5.3 TOLERANCE INTERVALS

Unlike the confidence interval, which is constructed to contain an unknown parameter with some specified degree of uncertainty (say, $1 - \gamma$), a *tolerance interval* contains at least a proportion p of the population with probability γ. That is, a tolerance interval is a confidence interval for a distribution. Both p, the proportion of coverage, and $1 - \gamma$, the uncertainty associated with the confidence statement, are predefined probabilities. For instance, we may be 95% confident that 90% of the population will fall within the range specified by a tolerance interval.

Order statistics play an important role in the construction of tolerance intervals. From a sample $X_1, \dots, X_n$ from (continuous) distribution F, two statistics $T_1 < T_2$ represent a 100γ percent tolerance interval for $100p$ percent of the distribution F if

$$P\left(F(T_2) - F(T_1) \geq p\right) \geq \gamma.$$

Obviously, the distribution $F(T_2) - F(T_1)$ should not depend on F. Recall that for an order statistic $X_{i:n}$, $U_{i:n} \equiv F(X_{i:n})$ is distributed $\mathcal{B}e(i, n-i+1)$. Choosing T_1 and T_2 from the set of order statistics satisfies the requirements of the tolerance interval and the computations are not difficult.

One-sided tolerance intervals are related to confidence intervals for quantiles. For instance, a 90% upper tolerance bound for 95% of the population is identical to a 90% one-sided confidence interval for $x_{0.95}$, the 0.95 quantile of the distribution. With a sample of $x_1, \dots, x_n$ from F, a γ interval for $100p\%$ of the population would be constructed as $(-\infty, x_{r:n})$ for some $r \in \{1, \dots, n\}$.

Here are four simple steps to help determine r:

1. We seek r so that $P(-\infty < x_p < X_{r:n}) = \gamma = P(X_{r:n} > x_p)$
2. At most $r-1$ out of n observations are less than x_p
3. Let Y = number of observations less than x_p. so that $Y \sim \mathcal{B}in(n,p)$ if x_p is the pth quantile
4. Find r large enough so that $P(Y \le r-1) = \gamma$.

Example 5.4 A 90% upper confidence bound for the 75th percentile (or *upper quartile*) is found by assigning Y= number of observations less than $x_{0.75}$, where $Y \sim \mathcal{B}in(n, 0.75)$. Let $n = 20$. Note $P(Y \le 16) = 0.7748$ and $P(Y \le 17) = 0.9087$, so $r - 1 = 17$. The 90% upper bound for $x_{0.75}$, which is equivalent to a 90% upper tolerance bound for 75% of the population, is $x_{18:20}$ (the third largest observation out of 20).

For large samples, the normal approximation allows us to generate an upper bound more simply. For the upper bound $x_{r:n}$, r is approximated with

$$\tilde{r} = np + z_\gamma \sqrt{np(1-p)}.$$

In the example above, with $n = 20$ (of course, this is not exactly what we think of as "large"), $\tilde{r} = 20(0.75) + 1.28\sqrt{0.75(0.25)20} = 17.48$. According to this rule, $x_{17:20}$ is insufficient for the approximate interval, so $x_{18:20}$ is again the upper bound.

Example 5.5 Sample Range. From a sample of n, what is the probability that $100p\%$ of the population lies within the sample range $(X_{1:n}, X_{n:n})$?

$$P\left(F(X_{n:n}) - F(X_{1:n}) \ge p\right) = 1 - P\left(U_n < p\right)$$

where $U_n = U_{n:n} - U_{1:n}$. From (5.6) it was shown that $U_n \sim \mathcal{B}e(n-1, 2)$. If we let $\gamma = P(U_n \ge p)$, then γ, the tolerance coefficient can be solved

$$1 - \gamma = np^{n-1} - (n-1)p^n.$$

Example 5.6 The tolerance interval is especially useful in compliance monitoring at industrial sites. Suppose one is interested in maximum contaminant levels (MCLs). The tolerance interval already takes into account the fact that some values will be high. So if a few values exceed the MCL standard, a site may still not be in violation (because the calculated tolerance interval may still be lower than the MCL). But if too many values are above the MCL, the calculated tolerance interval will extend beyond the acceptable standard.

As few as three data points can be used to generate a tolerance interval, but the EPA recommends having at least eight points for the interval to have any usefulness (EPA/530-R-93-003).

Example 5.7 How large must a sample size n be so that at least 75% of the contamination levels are between $X_{2:n}$ and $X_{(n-1):n}$ with probability of at least 0.95? If we follow the approach above, the distribution of $V_n = U_{(n-1):n} - U_{2:n}$ is $\mathcal{B}e\left((n-1)-2, n-(n-1)+2+1\right) = \mathcal{B}e(n-3, 4)$. We need n so that $P(V_n \geq 0.75) =$ `betainc(0.25,4,n-3)` ≥ 0.95 which occurs as long as $n \geq 29$.

5.4 ASYMPTOTIC DISTRIBUTIONS OF ORDER STATISTICS

Let $X_{r:n}$ be r^{th} order statistic in a sample of size n from a population with an absolutely continuous distribution function F having a density f. Let $r/n \to p$, when $n \to \infty$. Then

$$\sqrt{\frac{n}{p(1-p)}} f(x_p)(X_{r:n} - x_p) \Longrightarrow \mathcal{N}(0,1),$$

where x_p is p^{th} quantile of F, i.e., $F(x_p) = p$.

Let $X_{r:n}$ and $X_{s:n}$ be r^{th} and s^{th} order statistics $(r < s)$ in the sample of size n. Let $r/n \to p_1$ and $s/n \to p_2$, when $n \to \infty$. Then, for large n,

$$\begin{pmatrix} X_{r:n} \\ X_{s:n} \end{pmatrix} \overset{\text{appr}}{\sim} \mathcal{N}\left(\begin{bmatrix} x_{p_1} \\ x_{p_2} \end{bmatrix}, \Sigma \right),$$

where

$$\Sigma = \begin{bmatrix} p_1(1-p_1)[f(x_{p_1})]^{-2}/n & p_1(1-p_2)/[nf(x_{p_1})f(x_{p_2})]^{-1} \\ p_1(1-p_2)/[nf(x_{p_1})f(x_{p_2})]^{-1} & p_2(1-p_2)[f(x_{p_2})]^{-2}/n \end{bmatrix}$$

and x_{p_i} is ${p_i}^{th}$ quantile of F.

Example 5.8 Let $r = n/2$ so we are estimating the population median with $\hat{x}_{.50} = x_{(n/2):n}$. If $f(x) = \theta \exp(-\theta x)$, for $x > 0$, then $x_{0.50} = \ln(2)/\theta$ and

$$\sqrt{n}\left(\hat{x}_{0.50} - x_{0.50}\right) \Longrightarrow \mathcal{N}\left(0, \theta^{-2}\right).$$

5.5 EXTREME VALUE THEORY

Earlier we equated a series system lifetime (of n i.i.d. components) with the sample minimum $X_{1:n}$. The limiting distribution of the minima or maxima are not so interesting, e.g., if X has distribution function F, $X_{1:n} \to x_0$, where $x_0 = \inf_x\{x : F(x) > 0\}$. However, the *standardized limit* is more interesting. For an example involving sample maxima, with $X_1, ..., X_n$ from an exponential distribution with mean 1, consider the asymptotic distribution of $X_{n:n} - \log(n)$:

$$\begin{aligned} P(X_{n:n} - \log(n) \le t)) &= P(X_{n:n} \le t + \log(n)) = [1 - \exp\{-t - \log(n)\}]^n \\ &= [1 - e^{-t}n^{-1}]^n \to \exp\{-e^{-t}\}. \end{aligned}$$

This is because $(1 + \alpha/n)^n \to e^{\alpha}$ as $n \to \infty$. This distribution, a special form of the Gumbel distribution, is also called the *extreme-value distribution.*

Extreme value theory states that the standardized series system lifetime converges to one of the three following distribution types F^* (not including scale and location transformation) as the number of components increases to infinity:

$$\begin{array}{lrcll} \text{Gumbel} & F^*(x) & = & \exp(-\exp(-x)), & -\infty < x < \infty \\ \text{Fréchet} & F^*(x) & = & \begin{cases} \exp(-x^{-a}), & x > 0,\ a > 0 \\ 0, & x \le 0 \end{cases} & \\ \text{Negative Weibull} & F^*(x) & = & \begin{cases} \exp(-(-x)^{a}), & x < 0,\ a > 0 \\ 0, & x \ge 0 \end{cases} & \end{array}$$

5.6 RANKED SET SAMPLING

Suppose a researcher is sent out to Leech Lake, Minnesota, to ascertain the average weight of Walleye fish caught from that lake. She obtains her data by stopping the fishermen as they are returning to the dock after a day of fishing. In the time the researcher waited at the dock, three fishermen arrived, each with their daily limit of three Walleye. Because of limited time, she only has time to make one measurement with each fisherman, so at the end of her field study, she will get three measurements.

McIntyre (1952) discovered that with this forced limitation on measurements, there is an efficient way of getting information about the population mean. We might assume the researcher selected the fish to be measured randomly for each of the three fishermen that were returning to shore. McIntyre found that if she instead inspected the fish visually and selected them non-randomly, the data could beget a better estimator for the mean. Specifically, suppose the researcher examines the three Walleye from the first fisherman and selects the smallest one for measurement. She measures the second smallest from the next batch, and the largest from the third batch.

Opposed to a simple random sample (SRS), this *ranked set sample* (RSS) consists of independent order statistics which we will denote by $X_{[1:3]}$, $X_{[2:3]}$, $X_{[3:3]}$. If $\bar{X}$ is the sample mean from a SRS of size n, and $\bar{X}_{RSS}$ is the mean of a ranked set sample $X_{[1:n]}, \ldots, X_{[n:n]}$, it is easy to show that like $\bar{X}$, $\bar{X}_{RSS}$ is an unbiased estimator of the population mean. Moreover, it has smaller variance. That is, $\mathbb{V}\mathrm{ar}(\bar{X}_{RSS}) \leq \mathbb{V}\mathrm{ar}(\bar{X})$.

This property is investigated further in the exercises. The key is that variances for order statistics are generally smaller than the variance of the i.i.d. measurements. If you think about the SRS estimator as a linear combination of order statistics, it differs from the linear combination of order statistics from a RSS by its covariance structure. It seems apparent, then, that the expected value of $\bar{X}_{RSS}$ must be the same as the expected value of a $\bar{X}_{RSS}$.

The sampling aspect of RSS has received the most attention. Estimators of other parameters can be constructed to be more efficient than SRS estimators, including nonparametric estimators of the CDF (Stokes and Sager, 1988). The book by Chen, Bai, and Sinha (2003) is a comprehensive guide about basic results and recent findings in RSS theory.

5.7 EXERCISES

5.1. In MATLAB: Generate a sequence of 50 uniform random numbers and find their range. Repeat this procedure $M = 1000$ times; you will obtain 1000 ranges for 1000 sequences of 50 uniforms. Next, simulate 1000 percentiles from a beta $\mathcal{B}e(49, 2)$ distribution for $p = (1 : 1000)/1001$. Use M-file `betainv(p, 49, 2)`. Produce a histogram for both sets of data, comparing the ordered ranges and percentiles of their theoretical distribution, $\mathcal{B}e(49, 2)$.

5.2. For a set of i.i.d. data from a continuous distribution $F(x)$, derive the probability density function of the order statistic $X_{i:n}$ in (5.2).

5.3. For a sample of $n = 3$ observations, use a Jacobian transformation to derive the joint density of the order statistics, $X_{1:3}, X_{2:3}, X_{3:3}$.

5.4. Consider a system that is composed of n identical components that have independent life distributions. In reliability, a *k-out-of-n system* is one for which at least k out of n components must work in order for the system to work. If the components have lifetime distribution F, find the distribution of the system lifetime and relate it to the order statistics of the component lifetimes.

5.5. In 2003, the lab of Human Computer Interaction and Health Care Informatics at the Georgia Institute of Technology conducted empirical research on the performance of patients with Diabetic Retinopathy. The experiment included 29 participants placed either in the control group (without Diabetic Retinopathy) or the treatment group (with Diabetic Retinopathy). The visual acuity data of all participants are listed below. Normal visual acuity is 20/20, and 20/60 means a person sees at 20 feet what a normal person sees at 60 feet.

20/20	20/20	20/20	20/25	20/15	20/30	20/25	20/20
20/25	20/80	20/30	20/25	20/30	20/50	20/30	20/20
20/15	20/20	20/25	20/16	20/30	20/15	20/15	20/25

The data of five participants were excluded from the table due to their failure to meet the requirement of the experiment, so 24 participants are counted in all. In order to verify if the data can represent the visual acuity of the general population, a 90% upper tolerance bound for 80% of the population is calculated.

5.6. In MATLAB, repeat the following $M = 10000$ times.

- Generate a normal sample of size $n = 100$, $X_1, \ldots, X_{100}$.
- For a two-sided tolerance interval, fix the coverage probability as $p = 0.8$, and use the random interval $(X_{5:100}, X_{95:100})$. This interval will cover the proportion $F_X(X_{95:100}) - F_X(X_{5:100}) = U_{95:100} - U_{5:100}$ of the normal population.
- Count how many times in M runs $U_{95:100} - U_{5:100}$ exceeds the preassigned coverage p? Use this count to estimate γ.
- Compare the simulation estimator of γ with the theory, $\gamma = 1$ - `betainc(p, s-r, (n+1)-(s-r))`.

  What if instead of normal sample you used an exponentially distributed sample?

5.7. Suppose that components of a system are distributed i.i.d. $\mathcal{U}(0,1)$ lifetime. By standardizing with $1/n$ where n are the number of components in the system, find the limiting lifetime distribution of a parallel system as the number of components increases to infinity.

5.8. How large of a sample is needed in order for the sample range to serve as a 99% tolerance interval that contains 90% of the population?

5.9. How large must the sample be in order to have 95% confidence that at least 90% of the population is less than $X_{(n-1):n}$?

5.10. For a large sample of i.i.d. randomly generated $\mathcal{U}(0,1)$ variables, compare the asymptotic distribution of the sample mean with that of the sample median.

5.11. Prove that a ranked set sample mean is unbiased for estimating the population mean by showing that $\sum_{i=1}^{n} \mathbb{E}(X_{[i:n]}) = n\mu$. In the case the underlying data are generated from $\mathcal{U}(0,1)$, prove that the sample variance for the RSS mean is strictly less than that of the sample mean from a SRS.

5.12. Find a 90% upper tolerance interval for the 99th percentile of a sample of size n=1000.

5.13. Suppose that N items, labeled by sequential integers as $\{1, 2, \ldots, N\}$, constitute the population. Let $X_1, X_2, \ldots, X_n$ be a sample of size n (without repeating) from this population and let $X_{1:n}, \ldots, X_{n:n}$ be the order statistics. It is of interest to estimate the size of population, N.

This theoretical scenario is a basis for several interesting popular problems: tramcars in San Francisco, captured German tanks, maximal lottery number, etc. The most popular is the German tanks story, featured in *The Guardian* (2006). The full story is quite interesting, but the bottom line is to estimate total size of production if five German tanks with "serial numbers" 12, 33, 37, 78, and 103 have been captured by Allied forces.

(i) Show that the distribution of $X_{i:n}$ is

$$P(X_{i:n} = k) = \frac{\binom{k-1}{i-1}\binom{N-k}{n-i}}{\binom{N}{n}}, \quad k = i, i+1, \ldots, N-n+1.$$

(ii) Using the identity $\sum_{k=i}^{N-n+i} \binom{k-1}{i-1}\binom{N-k}{n-i} = \binom{N}{n}$ and distribution from (i), show that $\mathbb{E}X_{i:n} = i(N+1)/(n+1)$.

(iii) Show that the estimator $Y_i = (n+1)/iX_{i:n} - 1$ is unbiased for estimating N for any $i = 1, 2, \ldots, n$. Estimate number of tanks N on basis of Y_5 from the observed sample $\{12, 33, 37, 78, 103\}$.

REFERENCES

Chen, Z., Bai, Z., and Sinha, B. K. (2003), *Ranked Set Sampling: Theory and Applications,* New York: Springer Verlag.

David, H. A. and Nagaraj, H. N. (2003), *Order Statistics*, Third Edition, New York: Wiley.

McIntyre, G. A. (1952), "A method for unbiased selective sampling using ranked sets," *Australian Journal of Agricultural Research*, 3, 385-390.

Stokes, S. L., and Sager, T. W. (1988), Characterization of a Ranked-Set Sample with Application to Estimating Distribution Functions, *Journal of the American Statistical Association*, 83, 374-381.

The Guardian (2006), "Gavyn Davies Does the Maths: How a Statistical Formula Won the War," Thursday, July 20, 2006.

6

Goodness of Fit

> Believe nothing just because a so-called wise person said it.
> Believe nothing just because a belief is generally held.
> Believe nothing just because it is said in ancient books.
> Believe nothing just because it is said to be of divine origin.
> Believe nothing just because someone else believes it.
> Believe only what you yourself test and judge to be true.
>
> paraphrased from the Buddha

Modern experiments are plagued by well-meaning assumptions that the data are distributed according to some "textbook" CDF. This chapter introduces methods to test the merits of a hypothesized distribution in fitting the data. The term *goodness of fit* was coined by Pearson in 1902, and refers to statistical tests that check the quality of a model or a distribution's fit to a set of data. The first measure of goodness of fit for general distributions was derived by Kolmogorov (1933). Andrei Nikolaevich Kolmogorov (Figure 6.1 (a)), perhaps the most accomplished and celebrated Soviet mathematician of all time, made fundamental contributions to probability theory, including test statistics for distribution functions – some of which bear his name. Nikolai Vasil'yevich Smirnov (Figure 6.1 (b)), another Soviet mathematician, extended Kolmogorov's results to two samples.

In this section we emphasize objective tests (with p-values, etc.) and later we analyze *graphical* methods for testing goodness of fit. Recall the empirical distribution functions from p. 34. The *Kolmogorov statistic* (sometimes called

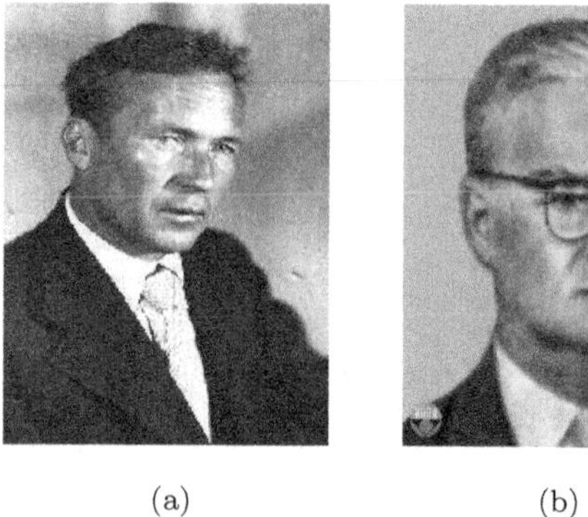

(a) (b)

Fig. 6.1 (a) Andrei Nikolaevich Kolmogorov (1905–1987); (b) Nikolai Vasil'yevich Smirnov (1900–1966)

the Kolmogorov-Smirnov test statistic)

$$D_n = \sup_t |F_n(t) - F(t)|$$

is a basis to many nonparametric goodness-of-fit tests for distributions, and this is where we will start.

6.1 KOLMOGOROV-SMIRNOV TEST STATISTIC

Let $X_1, X_2, \ldots, X_n$ be a sample from a population with continuous, but unknown CDF F. As in (3.1), let $F_n(x)$ be the empirical CDF based on the sample. To test the hypothesis

$$H_0 : F(x) = F_0(x), \ (\forall x)$$

versus the alternative

$$H_1 : F(x) \neq F_0(x),$$

we use the modified statistics $\sqrt{n}D_n = \sup_x \sqrt{n}|F_n(x) - F_0(x)|$ calculated from the sample as

$$\sqrt{n}D_n = \sqrt{n}\ \max\{\max_i |F_n(X_i) - F_0(X_i)|, \max_i |F_n(X_i{}^-) - F_0(X_i)|\}.$$

This is a simple discrete optimization problem because F_n is a step function and F_0 is nondecreasing so the maximum discrepancy between F_n and F_0 occurs at the observation points or at their left limits. When the hypothesis H_0 is true, the statistic $\sqrt{n}D_n$ is distributed free of F_0. In fact, Kolmogorov

(1933) showed that under H_0,

$$P(\sqrt{n}D_n \leq d) \Longrightarrow H(d) = 1 - 2\sum_{j=1}^{\infty}(-1)^{j-1}e^{-2j^2d^2}.$$

In practice, most Kolmogorov-Smirnov (KS) tests are two sided, testing whether the F is equal to F_0, the distribution postulated by H_0, or not. Alternatively, we might test to see if the distribution is larger or smaller than a hypothesized F_0. For example, to find out if X is stochastically smaller than Y ($F_X(x) \geq F_Y(x)$), the two one-sided alternatives that can be tested are

$$H_{1,-} : F_X(x) \leq F_0(x) \quad \text{or} \quad H_{1,+} : F_X(x) \geq F_0(x).$$

Appropriate statistics for testing $H_{1,-}$ and $H_{1,+}$ are

$$\sqrt{n}D_n^- \equiv -\inf_x \sqrt{n}(F_n(x) - F_0(x)),$$

$$\sqrt{n}D_n^+ \equiv \sup_x \sqrt{n}(F_n(x) - F_0(x)),$$

which are calculated at the sample values as

$$\begin{aligned} \sqrt{n}D_n^- &= \sqrt{n}\max\{\max_i(F_0(X_i) - F_n(X_i{}^-)), 0\} \text{ and} \\ \sqrt{n}D_n^+ &= \sqrt{n}\max\{\max_i(F_n(X_i) - F_0(X_i)), 0\}. \end{aligned}$$

Obviously, $D_n = \max\{D_n^-, D_n^+\}$. In terms of order statistics,

$$\begin{aligned} D_n^+ &= \max\{\max_i(F_n(X_i) - F_0(X_i)), 0\} = \max\{\max_i(i/n - F_0(X_{i:n}), 0\} \quad \text{and} \\ D_n^- &= \max\{\max_i(F_0(X_{i:n} - (i-1)/n), 0\}. \end{aligned}$$

Under H_0, the distributions of D_n^+ and D_n^- coincide. Although conceptually straightforward, the derivation of the distribution for D_n^+ is quite involved. Under H_0, for $c \in (0,1)$, we have

$$\begin{aligned} P(D_n^+ < c) &= P(i/n - U_{i:n} < c, \quad \text{for all } i = 1, 2, \ldots, n) \\ &= P(U_{i:n} > i/n - c, \quad \text{for all } i = 1, 2, \ldots, n) \\ &= \int_{1-c}^{1}\int_{\frac{n-1}{n}-c}^{1} \cdots \int_{\frac{2}{n}-c}^{1}\int_{\frac{1}{n}-c}^{1} f(u_1, \ldots, u_n)du_1 \ldots du_n, \end{aligned}$$

where $f(u_1, \ldots, u_n) = n!\mathbf{1}(0 < u_1 < \cdots < u_n < 1)$ is the joint density of n order statistics from $\mathcal{U}(0,1)$.

Birnbaum and Tingey (1951) derived a more computationally friendly representation; if c is the observed value of D_n^+ (or D_n^-), then the p-value for testing H_0 against the corresponding one sided alternative is

$$P(\sqrt{n}D_n^+ > c) = (1-c)^n + c \sum_{j=1}^{\lfloor n(1-c) \rfloor} \binom{n}{j} (1-c-j/n)^{n-j}(c+j/n)^{j-1}.$$

This is an exact p-value. When the sample size n is large (enough so that the error of order $O(n^{-3/2})$ can be tolerated), an approximation can be used:

$$P\left[\frac{(6nD_n^+ + 1)^2}{18n} > x\right] = e^{-x}\left(1 - \frac{2x^2 - 4x - 1}{18n}\right) + O\left(n^{-3/2}\right).$$

To obtain the p-value approximation, take $x = (6nc+1)^2/(18n)$, where c is the observed D_n^+ (or D_n^-) and plug in the right-hand-side of the above equation.

Table 6.4, taken from Miller (1956), lists quantiles of D_n^+ for values of $n \leq 40$. The D_n^+ values refer to the one-sided test, so for the two sided test, we would reject H_0 at level α if $D_n^+ > k_n(1-\alpha/2)$, where $k_n(1-\alpha)$ is the tabled quantile under α. If $n > 40$, we can approximate these quantiles $k_n(\alpha)$ as

k_n	$1.07/\sqrt{n}$	$1.22/\sqrt{n}$	$1.36/\sqrt{n}$	$1.52/\sqrt{n}$	$1.63/\sqrt{n}$
α	0.10	0.05	0.025	0.01	0.005

Later, we will discuss alternative tests for distribution goodness of fit. The KS test has advantages over exact tests based on the χ^2 goodness-of-fit statistic (see Chapter 9), which depend on an adequate sample size and proper interval assignments for the approximations to be valid. The KS test has important limitations, too. Technically, it only applies to continuous distributions. The KS statistic tends to be more sensitive near the center of the distribution than at the tails. Perhaps the most serious limitation is that the distribution must be fully specified. That is, if location, scale, and shape parameters are estimated from the data, the critical region of the KS test is no longer valid. It typically must be determined by simulation.

Example 6.1 With 5 observations $\{0.1, 0.14, 0.2, 0.48, 0.58\}$, we wish to test H_0: Data are distributed $\mathcal{U}(0,1)$ versus H_1: Data are not distributed $\mathcal{U}(0,1)$. We check F_n and $F_0(x) = x$ at the five points of data along with their left-hand limits. $|F_n(x_i) - F_0(x_i)|$ equals (0.1, 0.26, 0.4, 0.32, 0.42) at $i = 1, \dots, 5$, and $|F_n(x_i{}^-) - F_0(x_i)|$ equals (0.1, 0.06, 0.2, 0.12, 0.22), so that $D_n = 0.42$. According to the table, $k_5(.10) = 0.44698$. This is a two-sided test, so the test statistic is not rejectable at $\alpha = 0.20$. This is due more to the lack of sample size than the evidence presented by the five observations.

Example 6.2 Galaxy velocity data, available on the book's website, was analyzed by Roeder (1990), and consists of the velocities of 82 distant galaxies, diverging from our own galaxy. A mixture model was applied to describe the underlying distribution. The first hypothesized fit is the normal distribution,

Table 6.4 Upper Quantiles for Kolmogorov–Smirnov Test Statistic.

n	$\alpha = .10$	$\alpha = .05$	$\alpha = .025$	$\alpha = .01$	$\alpha = .005$
1	.90000	.95000	.97500	.99000	.99500
2	.68377	.77639	.84189	.90000	.92929
3	.56481	.63604	.70760	.78456	.82900
4	.49265	.56522	.62394	.68887	.73424
5	.44698	.50945	.56328	.62718	.66853
6	.41037	.46799	.51926	.57741	.61661
7	.38148	.43607	.48342	.53844	.57581
8	.35831	.40962	.45427	.50654	.54179
9	.33910	.38746	.43001	.47960	.51332
10	.32260	.36866	.40925	.45662	.48893
11	.30829	.35242	.39122	.43670	.46770
12	.29577	.33815	.37543	.41918	.44905
13	.28470	.32549	.36143	.40362	.43247
14	.27481	.31417	.34890	.38970	.41762
15	.26588	.30397	.33760	.37713	.40420
16	.25778	.29472	.32733	.36571	.39201
17	.25039	.28627	.31796	.35528	.38086
18	.24360	.27851	.30936	.34569	.37062
19	.23735	.27136	.30143	.33685	.36117
20	.23156	.26473	.29408	.32866	.35241
21	.22617	.25858	.28724	.32104	.34427
22	.22115	.25283	.28087	.31394	.33666
23	.21645	.24746	.27490	.30728	.32954
24	.21205	.24242	.26931	.30104	.32286
25	.20790	.23768	.26404	.29516	.31657
26	.20399	.23320	.25907	.28962	.31064
27	.20030	.22898	.25438	.28438	.30502
28	.19680	.22497	.24993	.27942	.29971
29	.19348	.22117	.24571	.27471	.29466
30	.19032	.21756	.24170	.27023	.28987
31	.18732	.21412	.23788	.26596	.28530
32	.18445	.21085	.23424	.26189	.28094
33	.18171	.20771	.23076	.25801	.27677
34	.17909	.20472	.22743	.25429	.27279
35	.17659	.20185	.22425	.25073	.26897
36	.17418	.19910	.22119	.24732	.26532
37	.17188	.19646	.21826	.24404	.26180
38	.16966	.19392	.21544	.24089	.25843
39	.16753	.19148	.21273	.23786	.25518
40	.16547	.18913	.21012	.23494	.25205

specifically $\mathcal{N}(21, (\sqrt{21})^2)$, and the KS distance ($\sqrt{n}D_n = 1.6224$ with p-value of 0.0103. The following mixture of normal distributions with five components was also fit to the data:

$$\hat{F} = 0.1\Phi(9, 0.5^2) + 0.02\Phi(17, (\sqrt{0.8})^2) + 0.4\Phi(20, (\sqrt{5})^2) \\ +0.4\Phi(23, (\sqrt{8})^2) + 0.05\Phi(33, (\sqrt{2})^2),$$

where $\Phi(\mu, \sigma)$ is the CDF for the normal distribution. The KS statistics is $\sqrt{n}D_n = 1.1734$ and corresponding p-value is 0.1273. Figure 6.2 plots the the CDF of the transformed variables $\hat{F}(X)$, so a good fit is indicated by a straight line. Recall, if $X \sim F$, than $F(X) \sim U\ \mathcal{U}(0,1)$ and the straight line is, in fact, the CDF of $\mathcal{U}(0,1)$, $F(x) = x, 0 \le x \le 1$. Panel (a) shows the fit for the $\mathcal{N}(21, (\sqrt{21})^2)$ model while panel (b) shows the fit for the mixture model. Although not perfect itself, the mixture model shows significant improvement over the single normal model.

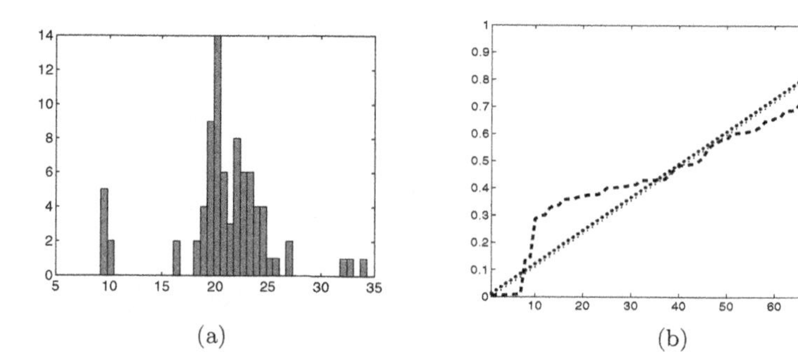

Fig. 6.2 Fitted distributions: (a) $\mathcal{N}(21, (\sqrt{21})^2)$ and (b) Mixture of Normals.

6.2 SMIRNOV TEST TO COMPARE TWO DISTRIBUTIONS

Smirnov (1939a, 1939b) extended the KS test to compare two distributions based on independent samples from each population. Let $X_1, X_2, \ldots, X_m$ and $Y_1, Y_2, \ldots, Y_n$ be two independent samples from populations with unknown CDFs F_X and G_Y. Let $F_m(x)$ and $G_n(x)$ be the corresponding empirical distribution functions.

We would like to test

$$H_0 : F_X(x) = G_Y(x) \ \ \forall x \quad \text{versus} \quad H_1 : F_X(x) \neq G_Y(x) \text{ for some } x.$$

We will use the analog of the KS statistic D_n:

$$D_{m,n} = \sup_x |F_m(x) - G_n(x)|, \tag{6.1}$$

where $D_{m,n}$ can be simplified (in terms of programming convenience) to

$$D_{m,n} = \max_i\{|F_m(Z_i) - G_n(Z_i)|\}$$

and $Z = Z_1, \ldots, Z_{m+n}$ is the *combined* sample $X_1, \ldots, X_m, Y_1, \ldots, Y_n$. $D_{m,n}$ will be large if there is a cluster of values from one sample after the samples are combined. The imbalance can be equivalently measured in how the *ranks* of one sample compare to those of the other after they are joined together. That is, values from the samples are not directly relevant except for how they are ordered when combined. This is the essential nature of rank tests that we will investigate later in the next chapter.

The two-distribution test extends simply from two-sided to one-sided. The one-sided test statistics are $D^+_{m,n} = \sup_x(F_m(x) - G_n(x))$ or $D^-_{m,n} = \sup_x(G_n(x) - F_m(x))$. Note that the ranks of the two groups of data determine the supremum difference in (6.1), and the values of the data determine only the position of the jumps for $G_n(x) - F_m(x)$.

Example 6.3 For the test of $H_1 : F_X(x) > G_Y(x)$ with $n = m = 2$, there are $\binom{4}{2} = 6$ different sample representations (with equal probability):

sample order	$D^+{}_{m,n}$
$X < X < Y < Y$	1
$X < Y < X < Y$	1/2
$X < Y < Y < X$	1/2
$Y < X < X < Y$	1/2
$Y < X < Y < X$	0
$Y < Y < X < X$	0

The distribution of the test statistic is

$$P(D_{2,2} = d) = \begin{cases} 1/3 & \text{if } d = 0 \\ 1/2 & \text{if } d = 1/2 \\ 1/6 & \text{if } d = 1. \end{cases}$$

If we reject H_0 in the case $D_{2,2} = 1$ (for $H_1 : F_X(x) > G_Y(x)$) then our type-I error rate is $\alpha = 1/6$.

If $m = n$ in general, the null distribution of the test statistic simplifies to

$$P(D^+_{n,n} > d) = P(D^-_{n,n} > d) = \frac{\binom{2n}{\lfloor n(d+1) \rfloor}}{\binom{2n}{n}},$$

where $\lfloor a \rfloor$ denotes the greatest integer $\leq a$. For two sided tests, this is doubled to obtain the p-value. If m and n are large ($m, n > 30$) and of comparable

Table 6.5 Tail Probabilities for Smirnov Two-Sample Test.

One-sided test	$\alpha = 0.05$	$\alpha = 0.025$	$\alpha = 0.01$	$\alpha = 0.005$
Two-sided test	$\alpha = 0.10$	$\alpha = 0.05$	$\alpha = 0.02$	$\alpha = 0.01$
	$1.22\sqrt{\frac{m+n}{mn}}$	$1.36\sqrt{\frac{m+n}{mn}}$	$1.52\sqrt{\frac{m+n}{mn}}$	$1.63\sqrt{\frac{m+n}{mn}}$

size, then an approximate distribution can be used:

$$P\left(\sqrt{\frac{mn}{m+n}}D_{m,n} \le d\right) \approx 1 - 2\sum_{k=1}^{\infty} e^{-2k^2d^2}.$$

A simpler large sample approximation, given in Table 6.5 works effectively if m and n are both larger than, say, 50.

Example 6.4 Suppose we have $n = m = 4$ with data $(x_1, x_2, x_3, x_4) = (16, 4, 7, 21)$ and $(y_1, y_2, y_3, y_4) = (56, 31, 15, 19)$. For the Smirnov test of $H_1 : F \neq G$, the only thing important about the data is how they are ranked within the group of eight combined observations:

$$x_{1:4} < x_{2:4} < y_{1:4} < x_{3:4} < y_{2:4} < x_{4:4} < y_{3:4} < y_{4:4}.$$

$|F_n - G_m|$ is never larger than 1/2, achieved in intervals (7,15), (16,19), (21, 31). The p-value for the two-sided test is

$$p\text{-value} = \frac{2\binom{2\times 4}{[4\times 1.5]}}{\binom{8}{4}} = \frac{2\binom{8}{6}}{\binom{8}{4}} = \frac{56}{70} = 0.80.$$

Example 6.5 Figure 6.3 shows the EDFs for two samples of size 100. One is generated from normal data, and the other from exponential data. They have identical mean ($\mu = 10$) and variance ($\sigma^2 = 100$). The MATLAB m-file

`kstest` and `kstest2`

both can be used for the two-sample test. The MATLAB code shows the p-value is 0.0018. If we compared the samples using a two-sample t-test, the significance value is 0.313 because the t-test is testing only the means, and not the distribution (which is assumed to be normal). Note that $\sup_x |F_m(x) - G_n(x)| = 0.26$, and according to Table 6.5, the 0.99 quantile for the two-sided test is 0.2305.

```
>> xn=randgauss(10,100,100);
>> ne=randexpo(.1,100)
>> cdfplot(xn)
>> hold on
```

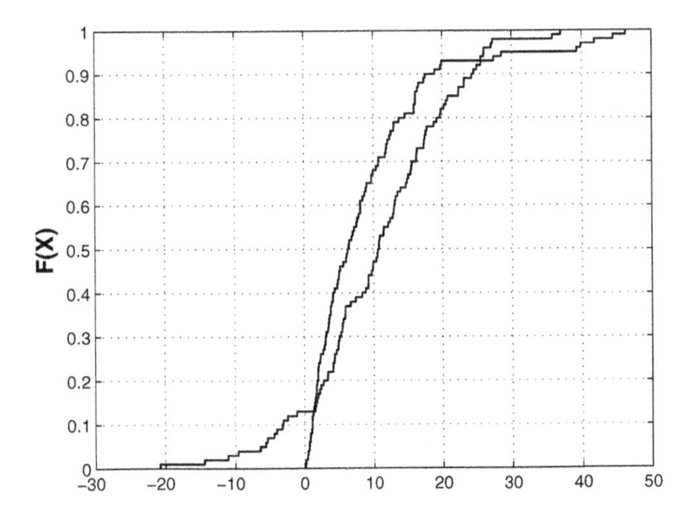

Fig. 6.3 EDF for samples of $n = m = 100$ generated from normal and exponential with $\mu = 10$ and $\sigma^2 = 100$.

```
Current plot held
>> cdfplot(ne)
>> [h,p,ks2]=kstest2(xn,ne)
    h =      1
    p =    0.0018
    ks2 = 0.2600
>> [h,p,ci]=ttest2(ne,xn)
    h =      0
    p =      0.3130
    ci =     -3.8992      1.2551
```

6.3 SPECIALIZED TESTS FOR GOODNESS OF FIT

In this section, we will go over some of the most important goodness-of-fit tests that were made specifically for certain distributions such as the normal or exponential. In general, there is not a clear ranking on which tests below are best and which are worst, but they all have clear advantages over the less-specific KS test.

Table 6.6 Null Distribution of Anderson-Darling Test Statistic: Modifications of A^2 and Upper Tail Percentage Points

Modification A^*, A^{**}	Upper Tail Probability α			
	0.10	0.05	0.025	0.01
(a) Case 0: Fully specified $\mathcal{N}(\mu, \sigma^{\in})$	1.933	2.492	3.070	3.857
(b) Case 1: $\mathcal{N}(\mu, \sigma^{\in})$, only σ^2 known	0.894	1.087	1.285	1.551
Case 2: σ^2 estimated by s^2, μ known	1.743	2.308	2.898	3.702
Case 3: μ and σ^2 estimated, A^*	0.631	0.752	0.873	1.035
(c) Case 4: $\mathcal{E}\text{xp}(\theta)$, A^{**}	1.062	1.321	1.591	1.959

6.3.1 Anderson-Darling Test

Anderson and Darling (1954) looked to improve upon the Kolmogorov-Smirnov statistic by modifying it for distributions of interest. The Anderson-Darling test is used to verify if a sample of data came from a population with a specific distribution. It is a modification of the KS test that accounts for the distribution and test and gives more attention to the tails. As mentioned before, the KS test is distribution free, in the sense that the critical values do not depend on the specific distribution being tested. The Anderson-Darling test makes use of the specific distribution in calculating the critical values. The advantage is that this sharpens the test, but the disadvantage is that critical values must be calculated for each hypothesized distribution.

The statistics for testing $H_0 : F(x) = F_0(x)$ versus the two sided alternative is $A^2 = -n - S$, where

$$S = \sum_{i=1}^{n} \frac{2i-1}{n} \left[\log F_0(X_{i:n}) + \log(1 - F_0(X_{n+1-i:n}))\right].$$

Tabulated values and formulas have been published (Stephens, 1974, 1976) for the normal, lognormal, and exponential distributions. The hypothesis that the distribution is of a specific form is rejected if the test statistic, A^2 (or modified A^*, A^*) is greater than the critical value given in Table 6.6. Cases 0, 1, and 2 do not need modification, i.e., observed A^2 is directly compared to those in Table. Case 3 and (c) compare a modified A^2 (A^* or A^{**}) to the critical values in Table 6.6. In (b), $A^* = A^2(1 + \frac{0.75}{n} + \frac{2.25}{n^2})$, and in (c), $A^{**} = A^2(1 + \frac{0.3}{n})$.

Example 6.6 The following example has been used extensively in testing for normality. The weights of 11 men (in pounds) are given: 148, 154, 158, 160, 161, 162, 166, 170, 182, 195, and 236. The sample mean is 172 and sample standard deviation is 24.952. Because mean and variance are estimate, this refers to Case 3 in Table 6.6. The standardized observations are $w_1 = (148 - 172)/24.952 = -0.9618, \ldots, w_{11} = 2.5649$. and

$z_1 = \Phi(w_1) = 0.1681, \ldots, z_{11} = 0.9948$. Next we calculate $A^2 = 0.9468$ and modify it as $A^* = A^2(1 + 0.75/11 + 0.25/121) = 1.029$. From the table we see that this is significant at all levels except for $\alpha = 0.01$, e.g., the null hypothesis of normality is rejected at level $\alpha = 0.05$. Here is the corresponding MATLAB code:

```
>> weights = [148, 154, 158, 160, 161, 162, 166, 170, 182, 195, 236];
>> n = length(weights); ws = (weights - mean(weights))/std(weights);
>> zs = 1/2 + 1/2*erf(ws/sqrt(2));
   % transformation to uniform o.s.
   % calculation of anderson-darling s=0; for i = 1:n
>> s = s + (2*i-1)/n * (log(zs(i)) + log(1-zs(n+1-i)));
>> a2 = -n - s;
>> astar = a2 * (1 + 0.75/n + 2.25/n^2 );
```

Example 6.7 Weight is one of the most important quality characteristics of the positive plate in storage batteries. Each positive plate consists of a metal frame inserted in an acid-resistant bag (called 'oxide holder') and the empty space in the bag is filled with active material, such as powdered lead oxide. About 75% of the weight of a positive plate consists of the filled oxide. It is also known from past experience that variations in frame and bag weights are negligible. The distribution of the weight of filled plate weights is, therefore, an indication of how good the filling process has been. If the process is perfectly controlled, the distribution should be normal, centered around the target; whereas departure from normality would indicate lack of control over the filling operation.

Weights of 97 filled plates (chosen at random from the lot produced in a shift) are measured in grams. The data are tested for normality using the Anderson-Darling test. The data and the MATLAB program written for this part are listed in Appendix A. The results in the MATLAB program list $A^2 = 0.8344$ and $A^* = 0.8410$.

6.3.2 Cramér-Von Mises Test

The Cramér-Von Mises test measures the weighted distance between the empirical CDF F_n and postulated CDF F_0. Based on a squared-error function, the test statistic is

$$\omega_n^2(\psi(F_0)) = \int_{-\infty}^{\infty} (F_n(x) - F_0(x))^2 \, \psi(F_0(x)) dF_0(x). \qquad (6.2)$$

There are several popular choices for the (weight) functional ψ. When $\psi(x) = 1$, this is the "standard" Cramér-Von Mises statistic $\omega_n^2(1) = \omega_n^2$, in which case

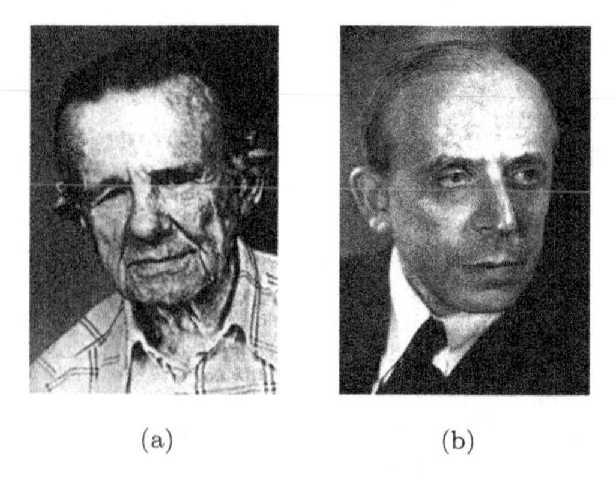

(a) (b)

Fig. 6.4 Harald Cramér (1893–1985); Richard von Mises (1883–1953).

the test statistic becomes

$$n\omega_n^2 = \frac{1}{12n} + \sum_{i=1}^{n}\left(F_0(X_{i:n}) - \frac{2i-1}{2n}\right)^2.$$

When $\psi(x) = x^{-1}(1-x)^{-1}$, $\omega_n^2(1/(F_0(1-F_0))) = A^2/n$, and A^2 is the Anderson-Darling statistic. Under the hypothesis $H_0 : F = F_0$, the asymptotic distribution of $\omega_n^2(\psi(F))$ is

$$\lim_{n\to\infty} P(n\omega_n^2 < x) = \frac{1}{\sqrt{2x}}\sum_{j=0}^{\infty}\frac{\Gamma(j+1/2)}{\Gamma(1/2)\Gamma(j+1)}\sqrt{4j+1}$$
$$\times \exp\left\{-\frac{(4j+1)^2}{16x}\right\}\cdot\left[J_{-1/4}\left(\frac{(4j+1)^2}{16x}\right) - J_{1/4}\left(\frac{(4j+1)^2}{16x}\right)\right],$$

where $J_k(z)$ is the modified Bessel function (in MATLAB: `bessel(k,z)`).

In MATLAB, the particular Cramér-Von Mises test for *normality* can be applied to a sample x with the function

$$\texttt{mtest}(\texttt{x}, \alpha),$$

where the weight function is one and α must be less than 0.10. The MATLAB code below shows how it works. Along with the simple "reject or not" output, the m-file also produces a graph (Figure 6.5) of the sample EDF along with the $\mathcal{N}(0,1)$ CDF. *Note: the data are assumed to be standardized.* The output of 1 implies we do not reject the null hypothesis ($H_0 : \mathcal{N}(0,1)$) at the entered α level.

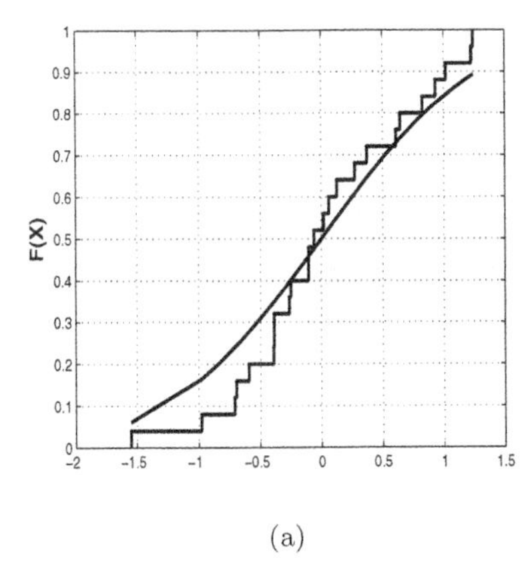

(a)

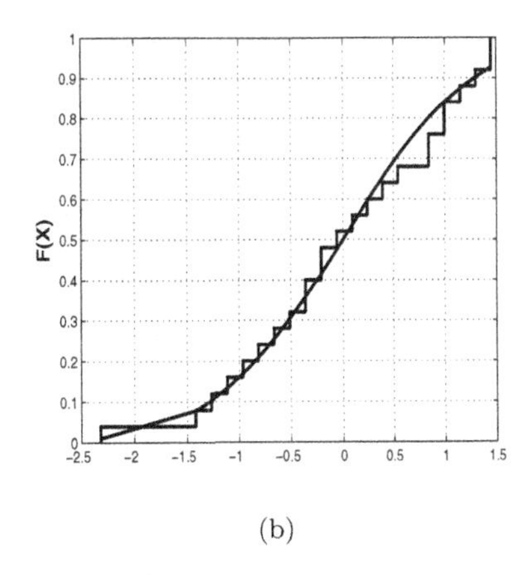

(b)

Fig. 6.5 Plots of EDF versus $\mathcal{N}(0,1)$ CDF for $n = 25$ observations of $\mathcal{N}(0,1)$ data and standardized $\mathcal{B}in(100, 0.5)$ data.

```
  >> x = rand_nor(0,1,25,1)
  >> mtest(x',0.05)
ans =
     1
  >> y = rand_bin(100,0.5,25)
  >> y2 = (y-mean(y))/std(y)
  >> mtest(y2,0.05)
ans =
     1
```

6.3.3 Shapiro-Wilk Test for Normality

The Shapiro-Wilk (Shapiro and Wilk, 1965) test calculates a statistic that tests whether a random sample, $X_1, X_2, \ldots, X_n$ comes from a normal distribution. Because it is custom made for the normal, this test has done well in comparison studies with other goodness of fit tests (and far outperforms the Kolmogorov-Smirnov test) if normally distributed data are involved.

The test statistic (W) is calculated as

$$W = \frac{\left(\sum_{i=1}^{n} a_i X_{i:n}\right)^2}{\sum_{i=1}^{n}(X_i - \bar{X})^2},$$

where the $X_{1:n} < X_{2:n} < \cdots < X_{n:n}$ are the ordered sample values and the a_i are constants generated from the means, variances and covariances of the order statistics of a sample of size n from a normal distribution (see Table 6.8). If H_0 is true, W is close to one; otherwise, $W < 1$ and we reject H_0

(a) (b)

Fig. 6.6 (a) Samuel S. Shapiro; (b) Martin Bradbury Wilk, born 1922.

for small values of W. Table 6.7 lists Shapiro-Wilk test statistic quantiles for sample sizes up to $n = 39$.

The weights a_i are defined as the components of the vector

$$a = M'V^{-1}\left((M'V^{-1})(V^{-1}M)\right)^{-1/2}$$

where M denotes the expected values of standard normal order statistic for a sample of size n, and V is the corresponding covariance matrix. While some of these values are tabled here, most likely you will see the test statistic (and critical value) listed in computer output.

Example 6.8 For $n = 5$, the coefficients a_i given in Table 6.8 lead to

$$W = \frac{(0.6646(x_{5:5} - x_{1:5}) + 0.2413(x_{4:5} - x_{2:5}))^2}{\sum(x_i - \bar{x})^2}.$$

If the data resemble a normally distributed set, then the numerator will be approximately to $\sum(x_i - \bar{x})^2$, and $W \approx 1$. Suppose $(x_1, \ldots, x_5) = (-2, -1, 0, 1, 2)$, so that $\sum(x_i - \bar{x})^2 = 10$ and $W = 0.1(0.6646[2 - (-2)] + 0.2413[1 - (-1)])^2 = 0.987$. From Table 6.7, $w_{0.10} = 0.806$, so our test statistic is clearly not significant, In fact, $W \approx w_{0.95} = 0.986$, so the critical value (p-value) for this goodness-of-fit test is nearly 0.95. Undoubtedly the perfect symmetry of the invented sample is a cause for this.

6.3.4 Choosing a Goodness of Fit Test

At this point, several potential goodness of fit tests have been introduced with nary a word that recommends one over another. There are several other specialized tests we have not mentioned, such as the Lilliefors tests (for exponentiality and normality), the D'Agostino-Pearson test, and the Bowman-Shenton test. These last two tests are extensions of the Shapiro-Wilk test.

Table 6.7 Quantiles for Shapiro-Wilk Test Statistic

	α								
n	0.01	0.02	0.05	0.10	0.50	0.90	0.95	0.98	0.99
3	0.753	0.756	0.767	0.789	0.959	0.998	0.999	1.000	1.000
4	0.687	0.707	0.748	0.792	0.935	0.987	0.992	0.996	0.997
5	0.686	0.715	0.762	0.806	0.927	0.979	0.986	0.991	0.993
6	0.713	0.743	0.788	0.826	0.927	0.974	0.981	0.986	0.989
7	0.730	0.760	0.803	0.838	0.928	0.972	0.979	0.985	0.988
8	0.749	0.778	0.818	0.851	0.932	0.972	0.978	0.984	0.987
9	0.764	0.791	0.829	0.859	0.935	0.972	0.978	0.984	0.986
10	0.781	0.806	0.842	0.869	0.938	0.972	0.978	0.983	0.986
11	0.792	0.817	0.850	0.876	0.940	0.973	0.979	0.984	0.986
12	0.805	0.828	0.859	0.883	0.943	0.973	0.979	0.984	0.986
13	0.814	0.837	0.866	0.889	0.945	0.974	0.979	0.984	0.986
14	0.825	0.846	0.874	0.895	0.947	0.975	0.980	0.984	0.986
15	0.835	0.855	0.881	0.901	0.950	0.975	0.980	0.984	0.987
16	0.844	0.863	0.887	0.906	0.952	0.976	0.981	0.985	0.987
17	0.851	0.869	0.892	0.910	0.954	0.977	0.981	0.985	0.987
18	0.858	0.874	0.897	0.914	0.956	0.978	0.982	0.986	0.988
19	0.863	0.879	0.901	0.917	0.957	0.978	0.982	0.986	0.988
20	0.868	0.884	0.905	0.920	0.959	0.979	0.983	0.986	0.988
21	0.873	0.888	0.908	0.923	0.960	0.980	0.983	0.987	0.989
22	0.878	0.892	0.911	0.926	0.961	0.980	0.984	0.987	0.989
23	0.881	0.895	0.914	0.928	0.962	0.981	0.984	0.987	0.989
24	0.884	0.898	0.916	0.930	0.963	0.981	0.984	0.987	0.989
25	0.888	0.901	0.918	0.931	0.964	0.981	0.985	0.988	0.989
26	0.891	0.904	0.920	0.933	0.965	0.982	0.985	0.988	0.989
27	0.894	0.906	0.923	0.935	0.965	0.982	0.985	0.988	0.990
28	0.896	0.908	0.924	0.936	0.966	0.982	0.985	0.988	0.990
29	0.898	0.910	0.926	0.937	0.966	0.982	0.985	0.988	0.990
30	0.900	0.912	0.927	0.939	0.967	0.983	0.985	0.988	0.900
31	0.902	0.914	0.929	0.940	0.967	0.983	0.986	0.988	0.990
32	0.904	0.915	0.930	0.941	0.968	0.983	0.986	0.988	0.990
33	0.906	0.917	0.931	0.942	0.968	0.983	0.986	0.989	0.990
34	0.908	0.919	0.933	0.943	0.969	0.983	0.986	0.989	0.990
35	0.910	0.920	0.934	0.944	0.969	0.984	0.986	0.989	0.990
36	0.912	0.922	0.935	0.945	0.970	0.984	0.986	0.989	0.990
37	0.914	0.924	0.936	0.946	0.970	0.984	0.987	0.989	0.990
38	0.916	0.925	0.938	0.947	0.971	0.984	0.987	0.989	0.990
39	0.917	0.927	0.939	0.948	0.971	0.984	0.987	0.989	0.991

Table 6.8 Coefficients for the Shapiro-Wilk Test

n	i=1	i=2	i=3	i=4	i=5	i=6	i=7	i=8
2	0.7071							
3	0.7071	0.0000						
4	0.6872	0.1677						
5	0.6646	0.2413	0.0000					
6	0.6431	0.2806	0.0875					
7	0.6233	0.3031	0.1401	0.0000				
8	0.6052	0.3164	0.1743	0.0561				
9	0.5888	0.3244	0.1976	0.0947	0.0000			
10	0.5739	0.3291	0.2141	0.2141	0.1224	0.0399		
11	0.5601	0.3315	0.2260	0.1429	0.0695	0.0000		
12	0.5475	0.3325	0.2347	0.1586	0.0922	0.0303		
13	0.5359	0.3325	0.2412	0.1707	0.1099	0.0539	0.0000	
14	0.5251	0.3318	0.2460	0.1802	0.1240	0.0727	0.0240	
15	0.5150	0.3306	0.2495	0.1878	0.1353	0.0880	0.0433	0.0000
16	0.5056	0.3290	0.2521	0.1939	0.1447	0.1005	0.0593	0.0196

Obviously, the specialized tests will be more powerful than an omnibus test such as the Kolmogorov-Smirnov test. D'Agostino and Stephens (1986) warn

> ...for testing for normality, the Kolmogorov-Smirnov test is only a historical curiosity. It should never be used. It has poor power in comparison to [specialized tests such as Shapiro-Wilk, D'Agostino-Pearson, Bowman-Shenton, and Anderson-Darling tests].

These top-performing tests fail to distinguish themselves across a broad range of distributions and parameter values. Statistical software programs often list two or more test results, allowing the analyst to choose the one that will best support their research grants.

There is another way, altogether different, for testing the fit of a distribution to the data. This is detailed in the upcoming section on probability plotting. One problem with all of the analytical tests discussed thus far involves the large sample behavior. As the sample size gets large, the test can afford to be pickier about what is considered a departure from the hypothesized null distribution F_0. In short, your data might look normally distributed to you, for all practical purposes, but if it is not *exactly* normal, the goodness of fit test will eventually find this out. Probability plotting is one way to avoid this problem.

6.4 PROBABILITY PLOTTING

A probability plot is a graphical way to show goodness of fit. Although it is more subjective than the analytical tests (e.g., Kolmogorov-Smirnov, Anderson-Darling, Shapiro-Wilk), it has important advantages over them. First, it allows the practitioner to see what observations of the data are in agreement (or disagreement) with the hypothesized distribution. Second, while no significance level is attached to the plotted points, the analytical tests can be misleading with large samples (this will be illustrated below). There is no such problem with large samples in probability plotting – the bigger the sample the better.

The plot is based on transforming the data with the hypothesized distribution. After all, if $X_1, \dots, X_n$ have distribution F, we know $F(X_i), \dots, F(X_n)$ are $\mathcal{U}(0,1)$. Specifically, if we find a transformation with F that linearizes the data, we can find a linear relationship to plot.

Example 6.9 Normal Distribution. If Φ represents the CDF of the standard normal distribution function, then the quantile for a normal distribution with parameters (μ, σ^2) can be written as

$$x_p = \mu + \Phi^{-1}(p)\sigma.$$

The plot of x_p versus $\Phi^{-1}(p)$ is a straight line. If the line shows curvature, we know Φ^{-1} was not the right inverse-distribution that transformed the percentile to the normal quantile.

A vector consisting of 1000 generated variables from $\mathcal{N}(0,1)$ and 100 from $\mathcal{N}(0.1,1)$ is tested for normality. For this case, we used the Cramér-Von Mises Test using the MATLAB procedure `mtest(`z`,` α`)`. We input a vector z of data to test, and α represents the test level. The plot in Figure 6.4(a) shows the EDF of the 1100 observations versus the best fitting normal distribution. In this case, the Cramér-Von Mises Test rejects the hypothesis that the data are normally distributed at level $\alpha = 0.001$. But the data are not discernably non-normal for all practical purposes. The probability plot in Figure 6.4(b) is constructed with the MATLAB function

`probplot`

and confirms this conjecture.

As the sample size increases, the goodness of fit tests grow increasingly sensitive to slight perturbations in the normality assumption. In fact, the Cramér-Von Mises test has correctly found the non-normality in the data that was generated by a normal mixture.

```
>> [x]=randgauss(0,1,1000);
>> [y]=randgauss(0.1,1,100);
>> [z]=[x,y];
```

```
>> [gg]=mtest(z,.001)

   gg =
          0

>> probplot(z)
```

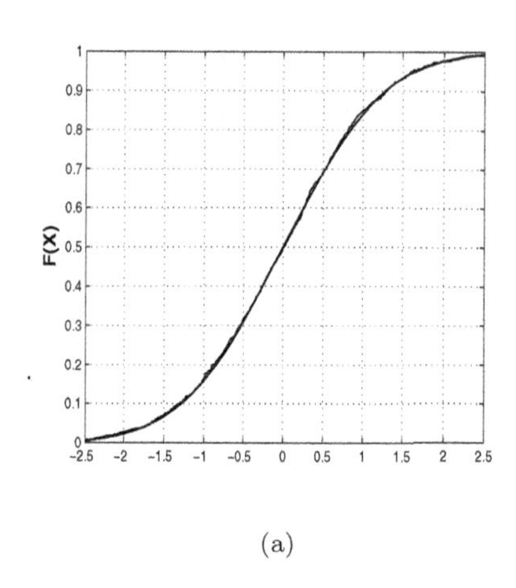

(a)

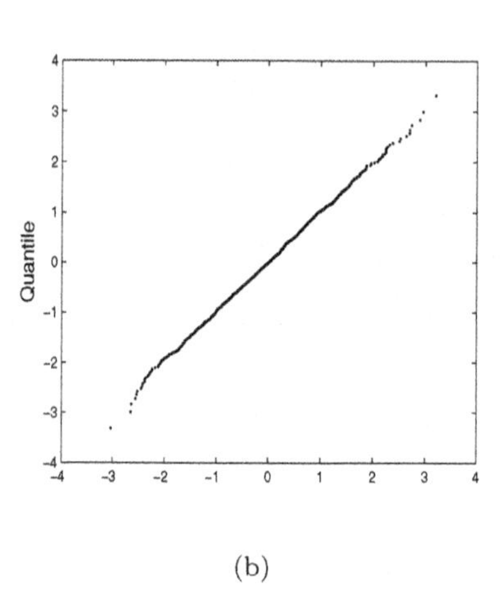

(b)

Fig. 6.7 (a) Plot of EDF vs. normal CDF, (b) normal probability plot.

Example 6.10 Thirty observations were generated from a normal distribution. The MATLAB function `qqweib` constructs a probability plot for Weibull data. The Weibull probability plot in Figure 6.8 shows a slight curvature which suggests the model is misfit. To linearize the Weibull CDF, if the CDF is expressed as $F(x) = 1 - \exp(-(x/\gamma)^\beta)$, then

$$\ln(x_p) = \frac{1}{\beta}\ln(-\ln(1-p)) + \ln(\gamma).$$

The plot of $\ln(x_p)$ versus $\ln(-\ln(1-p))$ is a straight line determined by the two parameters β^{-1} and $\ln(\gamma)$. The MATLAB procedure `qqweib` also reports the the scale parameter *scale* and the shape parameter *shape*, estimated by the method of least-squares.

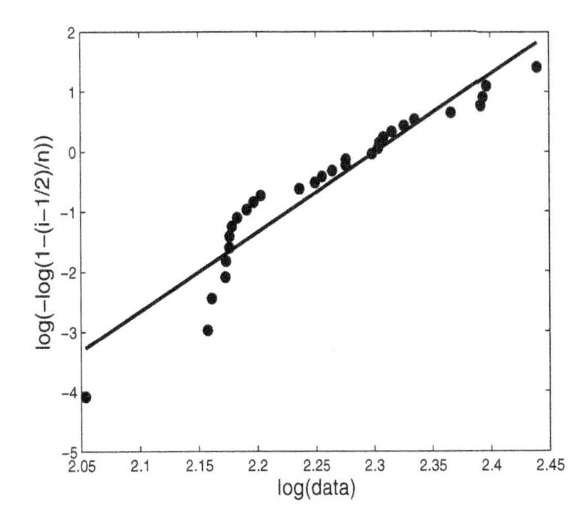

Fig. 6.8 Weibull probability plot of 30 observations generated from a normal distribution.

```
>> [x]=randgauss(10,1,30);
>> [shape,scale]=qqweib(x)

shape =
   13.2094

scale =

    9.9904
>>
```

Example 6.11 Quantile-Quantile Plots. For testing the equality of two distributions, the graphical analog to the Smirnov test is the Quantile-Quantile Plot, or q-q plot. The MATLAB function qqplot($x, y, *$) plots the empirical quantiles of the vector x versus that of y. The third argument is optional and represents the plotting symbol to use in the q-q plot. If the plotted points veer away from the 45° reference line, evidence suggests the data are generated by populations with different distributions. Although the q-q plot leads to subjective judgment, several aspects of the distributions can be compared graphically. For example, if the two distributions differ only by a location shift ($F(x) = G(x + \delta)$), the plot of points will be parallel to the reference line.

Many practitioners use the q-q plot as a probability plot by replacing the second sample with the quantiles of the hypothesized distribution. Three other MATLAB functions for probability plotting are listed below, but they use the q-q plot moniker. The argument `symbol` is optional in all three.

`qqnorm(x,symbol)`	Normal probability plot
`qqweib(x,symbol)`	Weibull probability plot
`qqgamma(x,symbol)`	Gamma probability plot

In Figure 6.9, the q-q plots are displayed for the random generated data in the MATLAB code below. The standard `qqplot` MATLAB outputs (scatterplot and dotted line fit) are enhanced by dashed line $y = x$ representing identity of two distributions. In each case, a distribution is plotted against $\mathcal{N}(100, 10^2)$ data. The first case (a) represents $\mathcal{N}(120, 10^2)$ and the points appear parallel to the reference line because the only difference between the two distributions is a shift in the mean. In (b) the second distribution is distributed $\mathcal{N}(100, 40^2)$. The only difference is in variance, and this is reflected in the slope change in the plot. In the cases (c) and (d), the discrepancy is due to the lack of distribution fit; the data in (c) are generated from the t-distribution with 1 degree of freedom, so the tail behavior is much different than that of the normal distribution. This is evident in the left and right end of the q-q plot. In (d), the data are distributed gamma, and the illustrated difference between the two samples is more clear.

```
>> x=rand_nor(100,10,30,1);
>> y1=rand_nor(120,10,30,1);   qqplot(x,y1)
>> y2=rand_nor(100,40,30,1);   qqplot(x,y2)
>> y3=100+10*rand_t(1,30,1);   qqplot(x,y3)
>> y4=rand_gamma(200,2,30,1); qqplot(x,y4)
```

6.5 RUNS TEST

A chief concern in the application of statistics is to find and understand patterns in data apart from the randomness (noise) that obscures them. While humans are good at deciphering and interpreting patterns, we are much less able to detect randomness. For example, if you ask any large group of people to randomly choose an integer from one to ten, the numbers seven and four are chosen nearly half the time, while the endpoints (one, ten) are rarely chosen. Someone trying to think of a random number in that range imagines something toward the middle, but not exactly in the middle. Anything else just doesn't look "random" to us.

In this section we use statistics to look for randomness in a simple string of dichotomous data. In many examples, the runs test will not be the most efficient statistical tool available, but the runs test is intuitive and easier

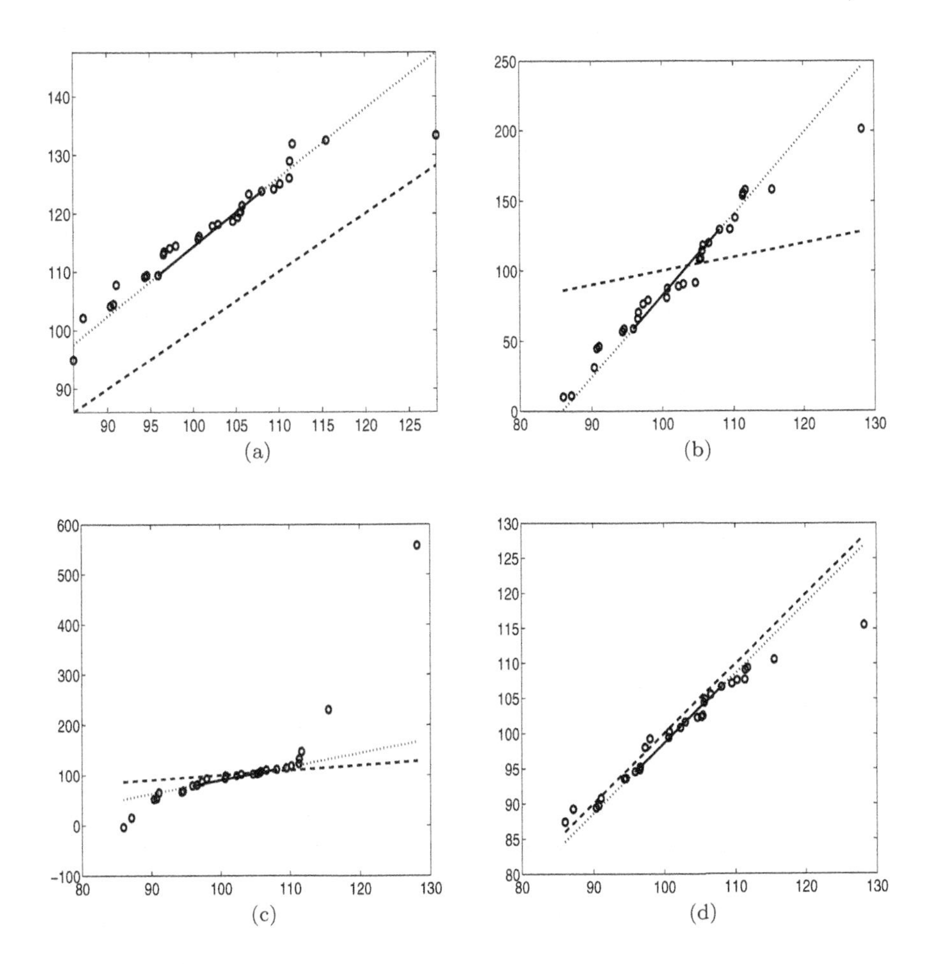

Fig. 6.9 Data from $\mathcal{N}(100, 10^2)$ are plotted against data from (a) $\mathcal{N}(120, 10^2)$, (b) $\mathcal{N}(100, 40^2)$, (c) t_1 and (d) $\mathcal{G}amma(200, 2)$. The standard **qqplot** MATLAB outputs (scatterplot and dotted line fit) are enhanced by dashed line $y = x$ representing identity of two distributions.

to interpret than more computational tests. Suppose items from the sample $X_1, X_2, \ldots, X_n$ could be classified as type 1 or type 2. If the sample is random, the 1's and 2's are well mixed, and any clustering or pattern in 1's and 2's is violating the hypothesis of randomness. To decide whether or not the pattern is random, we consider the statistic R, defined as the number of homogenous runs in a sequence of ones and twos. In other words R represents the number of times the symbols change in the sequence (including the first one). For example, $R = 5$ in this sequence of $n = 11$:

$$1\ 2\ 2\ 2\ 1\ 1\ 2\ 2\ 1\ 1\ 1.$$

Obviously if there were only two runs in that sequence, we could see the pattern where the symbols are separated right and left. On the other hand if $R = 11$, the symbols are intermingling in a non-random way. If R is too large, the sequence is showing anti-correlation, a repulsion of same symbols, and zig-zag behavior. If R is too small, the sample is suggesting trends, clustering and groupings in the order of the dichotomous symbols. If the null hypothesis claims that the pattern of randomness exists, then if R is either too big or too small, the alternative hypothesis of an existing trend is supported.

Assume that a dichotomous sequence has n_1 ones and n_2 twos, $n_1 + n_2 = n$. If R is the number of subsequent runs, then if the hypothesis of randomness is true (*sequence is made by random selection of 1's and 2's from the set containing n_1 1's and n_2 2's*), then

$$f_R(r) = \begin{cases} \dfrac{2\binom{n_1-1}{r/2-1}\cdot\binom{n_2-1}{r/2-1}}{\binom{n}{n_1}} & \text{if } r \text{ is even,} \\[2ex] \dfrac{\binom{n_1-1}{(r-1)/2}\binom{n_2-1}{(r-3)/2}+\binom{n_1-1}{(r-3)/2}\binom{n_2-1}{(r-1)/2}}{\binom{n}{n_1}} & \text{if } r \text{ is odd,} \end{cases}$$

for $r = 2, 3, \ldots, n$. Here is a hint for solving this: first note that the number of ways to put n objects into r groups *with no cell being empty* is $\binom{n-1}{r-1}$.

The null hypothesis is that the sequence is random, and alternatives could be one-sided and two sided. Also, under the hypotheses of randomness the symbols 1 and 2 are interchangeable and without loss of generality we assume that $n_1 \leq n_2$. The first three central moments for R (under the hypothesis of randomness) are,

$$\begin{aligned} \mu_R &= 1 + \frac{2n_1n_2}{n}, \\ {\sigma_R}^2 &= \frac{2n_1n_2(2n_1n_2 - n)}{n^2(n-1)}, \quad \text{and} \\ E(R - \mu_R)^3 &= -\frac{2n_1n_2(n_2 - n_1)^2(4n_1n_2 - 3n)}{n^3(n-1)(n-2)}, \end{aligned}$$

and whenever $n_1 > 15$ and $n_2 > 15$ the normal distribution can be used to to approximate lower and upper quantiles. Asymptotically, when $n_1 \to \infty$ and $\epsilon \leq n_1/(n_1+n_2) \leq 1-\epsilon$ (for some $0 < \epsilon < 1$),

$$P(R \leq r) = \Phi\left(\frac{r + 0.5 - \mu_R}{\sigma_R}\right) + O(n_1^{-1/2}).$$

The hypothesis of randomness is rejected at level α if the number of runs is either too small (smaller than some $g(\alpha, n_1, n_2)$) or too large (larger than some $G(\alpha, n_1, n_2)$). Thus there is no statistical evidence to reject H_0 if

$$g(\alpha, n_1, n_2) < R < G(\alpha, n_1, n_2).$$

Based on the normal approximation, critical values are

$$\begin{aligned} g(\alpha, n_1, n_2) &\approx \lfloor \mu_R - z_\alpha \sigma_R - 0.5 \rfloor \\ G(\alpha, n_1, n_2) &\approx \lfloor \mu_R + z_\alpha \sigma_R + 0.5 \rfloor. \end{aligned}$$

For the two-sided rejection region, one should calculate critical values with $z_{\alpha/2}$ instead of z_α. One-sided critical regions, again based on the normal approximation, are values of R for which

$$\frac{R - \mu_R + 0.5}{\sigma_R} \leq -z_\alpha$$
$$\frac{R - \mu_R - 0.5}{\sigma_R} \geq z_\alpha$$

while the two-sided critical region can be expressed as

$$\frac{(R - \mathbb{E}R)^2}{\sigma_R{}^2} \geq \left(z_{\alpha/2} + \frac{1}{2\sigma_R}\right)^2.$$

When the ratio n_1/n_2 is small, the normal approximation becomes unreliable. If the exact test is still too cumbersome for calculation, a better approximation is given by

$$P(R \leq r) \approx I_{1-x}(N - r + 2, r - 1) = I_x(r - 1, N - r + 2),$$

where $I_x(a, b)$ is the incomplete beta function (see Chapter 2) and

$$x = 1 - \frac{n_1 n_2}{n(n-1)} \quad \text{and} \quad N = \frac{(n-1)(2n_1 n_2 - n)}{n_1(n_1 - 1) + n_2(n_2 - 1)}.$$

Critical values are then approximated by $g(\alpha, n_1, n_2) \approx \lfloor g^* \rfloor$ and $G(\alpha, n_1, n_2) \approx$

$1 + \lfloor G^* \rfloor$, where g^* and G^* are solutions to

$$I_{1-x}(N - g^* + 2, g^* - 1) = \alpha,$$
$$I_x(G^* - 1, N - G^* + 3) = \alpha.$$

Example 6.12 The tourism officials in Santa Cruz worried about global worming and El Niño effect, compared daily temperatures (7/1/2003 - 7/21/2003) with averages of corresponding daily temperatures in 1993-2002. If the temperature in year 2003 is above the same day average in 1993-2002, then symbol A is recorded, if it is below, the symbol B is recorded. The following sequence of 21 letters was obtained:

$$AAABBAA|AABAABA|AAABBBB.$$

We wish to test the hypothesis of random direction of deviation from the average temperature against the alternative of non-randomness at level $\alpha =$ 5%. The MATLAB procedure for computing the test is `runs_test`.

```
>> cruz = [1 1 1 2 2 1 1 1 1 2 1 1 2 1 1 1 1 2 2 2 2];
>> [problow, probup, nruns, expectedruns] = runs_test(cruz)
    runones = 4
    runtwos = 4
    trun = 8
    n1 = 13
    n2 = 8
    n =  21
    problow =  0.1278
    probup = 0.0420
    nruns =  8
    expectedruns = 10.9048
```

If observed number of runs is LESS than expected, `problow` is

$$P(R = 2) + \cdots + P(R = nruns)$$

and `probup` is

$$P(R = n - nruns + 2) + \cdots + P(R = n).$$

Alternatively, if $nruns$ is LARGER than expected, then `problow` is

$$P(R = 2) + \cdots + P(R = n - nruns + 2)$$

and `probup` is

$$P(R = nruns) + \cdots + P(R = n).$$

In this case, the number of runs (8) was less than expected (10.9048), and the probability of seeing 8 or fewer runs in a random scattering is 0.1278. But this

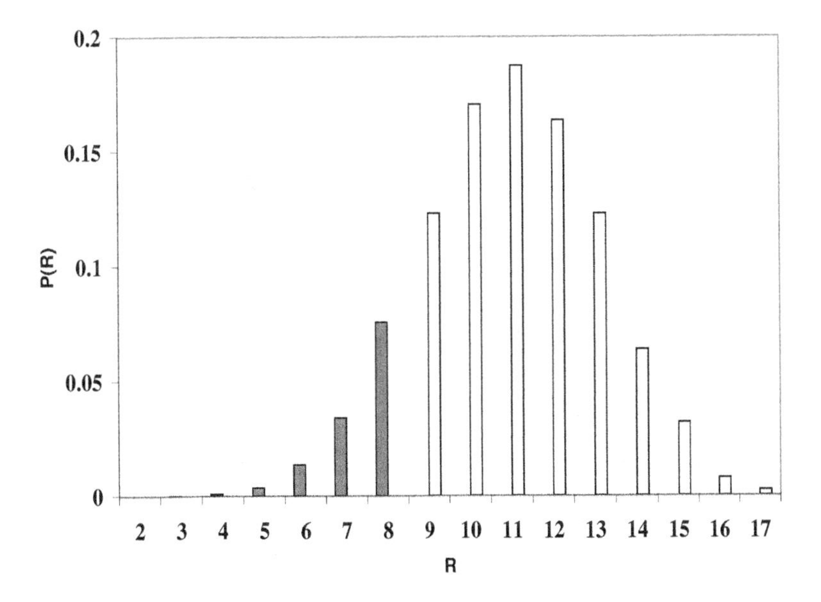

Fig. 6.10 Probability distribution of runs under H_0.

is a two-sided test. This MATLAB test implies we should use $P(R \geq n - n_2 + 2) = P(R \geq 15) = 0.0420$ as the "other tail" to include in the critical region (which would make the p-value equal to 0.1698). But using $P(R \geq 15)$ is slightly misleading, because there is no symmetry in the null distribution of R; instead, we suggest using `2*problow` $= 0.2556$ as the critical value for a two-sided test.

Example 6.13 The following are 30 time lapses, measured in minutes, between eruptions of Old Faithful geyser in Yellowstone National Park. In the MATLAB code below, `forruns` stores 2 if the temperature is below average, otherwise stores 1. The expected number of runs (15.9333) is larger than what was observed (13), and the p-value for the two-sided runs test is 2*0.1678=0.3356.

```
>> oldfaithful = [68 63 66 63 61 44 60 62 71 62 62 55 62 67 73 ...
                  72 55 67 68 65 60 61 71 60 68 67 72 69 65 66];
>> mean(oldfaithful)
   ans =  64.1667
>> forruns = (oldfaithful - 64.1667 > 0) + 1
   forruns =
    2    1    2    1    1    1    1    1    2    1
    1    1    1    2    2    2    1    2    2    2
    1    1    2    1    2    2    2    2    2    2
>> [problow, probup, nruns, expectedruns] = runs_test(forruns)
```

```
runones = 6
runtwos = 7
trun = 13
n1 = 14
n2 = 16
n = 30
problow = 0.1804
probup =  0.1678
nruns = 13
expectedruns = 15.9333
```

Before we finish with the runs test, we are compelled to make note of its limitations. After its inception by Mood (1940), the runs test was used as a cure-all nonparametric procedure for a variety of problems, including two-sample comparisons. However, it is inferior to more modern tests we will discuss in Chapter 7. More recently, Mogull (1994) showed an anomaly of the one-sample runs test; it is unable to reject the null hypothesis for series of data with run length of two.

6.6 META ANALYSIS

Meta analysis is concerned with combining the inference from several studies performed under similar conditions and experimental design. From each study an "effect size" is derived before the effects are combined and their variability assessed. However, for optimal meta analysis, the analyst needs substantial information about the experiment such as sample sizes, values of the test statistics, the sampling scheme and the test design. Such information is often not provided in the published work. In many cases, only the p-values of particular studies are available to be combined.

Meta analysis based on p-values only is often called nonparametric or omnibus meta analysis because the combined inference dose not depend on the form of data, test statistics, or distributions of the test statistics. There are many situations in which such combination of tests is needed. For example, one might be interested in

(i) multiple t tests in testing equality of two treatments versus one sided alternative. Such tests often arise in function testing and estimation, fMRI, DNA comparison, etc;

(ii) multiple F tests for equality of several treatment means. The test may not involve the same treatments and parametric meta analysis may not be appropriate; or

(iii) multiple χ^2 tests for testing the independence in contingency tables (see Chapter 9). The table counts may not be given or the tables could be of different size (the same factor of interest could be given at different levels).

Most of the methods for combining the tests on basis of their p-values use the facts that, (1) under H_0 and assuming the test statistics have a continuous distribution, the p-values are uniform and (2) if G is a monotone CDF and $U \sim \mathcal{U}(0,1)$, then $G^{-1}(U)$ has distribution G. A nice overview can be found in Folks (1984) and the monograph by Hedges and Olkin (1985).

Tippet-Wilkinson Method. If the p-values from n studies, $p_1, p_2, \ldots, p_n$ are ordered in increasing order, $p_{1:n}, p_{2:n}, \ldots, p_{n:n}$, then, for a given k, $1 \leq k \leq n$, the k-th smallest p-value, $p_{k:n}$, is distributed $\mathcal{B}e(k, n-k+1)$ and

$$p = P(X \leq p_{k:n}), \quad X \sim \mathcal{B}e(k, n-k+1).$$

Beta random variables are related to the F distribution via

$$P\left(V \leq \frac{\alpha}{\alpha + \beta w}\right) = P(W \geq w),$$

for $V \sim \mathcal{B}e(\alpha, \beta)$ and $W \sim \mathcal{F}(2\beta, 2\alpha)$. Thus, the combined significance level p is

$$p = P\left(X \geq \frac{k}{n-k+1} \frac{1 - p_{k:n}}{p_{k:n}}\right)$$

where $X \sim \mathcal{F}(2(n-k+1), 2k)$. This single p represents a measure of the uniformity of $p_1, \ldots, p_n$ and can be thought as a combined p-value of all n tests. The nonparametric nature of this procedure is unmistakable. This method was proposed by Tippet (1931) with $k = 1$ and $k = n$, and later generalized by Wilkinson (1951) for arbitrary k between 1 and n. For $k = 1$, the test of level α rejects H_0 if $p_{1:n} \leq 1 - (1-\alpha)^{1/n}$.

Fisher's Inverse χ^2 Method. Maybe the most popular method of combining the p-values is Fisher's inverse χ^2 method (Fisher, 1932). Under H_0, the random variable $-2 \log p_i$ has χ^2 distribution with 2 degrees of freedom, so that $\sum_i \chi^2_{k_i}$ is distributed as χ^2 with $\sum_i k_i$ degrees of freedom. The combined p-value is

$$p = P\left(\chi^2_{2k} \geq -2 \sum_{i=1}^{n} \log p_i\right).$$

This test is, in fact, based on the product of all p-values due to the fact that

$$-2 \sum_i \log p_i = -2 \log \prod_i p_i.$$

Averaging p-Values by Inverse Normals. The following method for combining p-values is based on the fact that if $Z_1, Z_2, \ldots, Z_n$ are i.i.d. $\mathcal{N}(0,1)$, then $(Z_1 + Z_2 + \cdots + Z_n)/\sqrt{n}$ is distributed $\mathcal{N}(0,1)$, as well. Let Φ^{-1} denote the inverse function to the standard normal CDF Φ, and let $p_1, p_2, \ldots, p_n$ be the p-values to be averaged. Then the averaged p-value is

$$p = P\left(Z > \frac{\Phi^{-1}(1-p_1) + \ldots \Phi^{-1}(1-p_n)}{\sqrt{n}}\right),$$

where $Z \sim \mathcal{N}(0,1)$. This procedure can be extended by using weighted sums:

$$p = P\left(Z > \frac{\lambda_1 \Phi^{-1}(1-p_1) + \cdots + \lambda_n \Phi^{-1}(1-p_n)}{\sqrt{\lambda_1^2 + \cdots + \lambda_n^2}}\right).$$

There are several more approaches in combining the p-values. Good (1955) suggested use of weighted product

$$-2 \sum_i \log p_i = -2 \log \prod_i p_i^{\lambda_i},$$

but the distributional theory behind this statistic is complex. Mudholkar and George (1979) suggest transforming the p-values into logits, that is, $\text{logit}(p) = \log(p/(1-p))$. The combined p-value is

$$p \approx P\left(t_{5n+4} > \frac{-\sum_{i=1}^{n} \text{logit}(p_i)}{\sqrt{\pi^2 n/3}}\right).$$

As an alternative, Lancaster (1961) proposes a method based on inverse gamma distributions.

Example 6.14 This example is adapted from a presentation by Jessica Utts from University of California, Davis. Two scientists, Professors A and B, each have a theory they would like to demonstrate. Each plans to run a fixed number of Bernoulli trials and then test $H_0 : p = 0.25$ verses $H_1 : p > 0.25$.

Professor A has access to large numbers of students each semester to use as subjects. He runs the first experiment with 100 subjects, and there are 33 successes ($p = 0.04$). Knowing the importance of replication, Professor A then runs an additional experiment with 100 subjects. He finds 36 successes ($p = 0.009$).

Professor B only teaches small classes. Each quarter, she runs an experiment on her students to test her theory. Results of her ten studies are given in the table below.

At first glance professor A's theory has much stronger support. After all, the p-values are 0.04 and 0.009. None of the ten experiments of professor

B was found significant. However, if the results of the experiment for each professor are aggregated, Professor B actually demonstrated a higher level of success than Professor A, with 71 out of 200 as opposed to 69 out of 200 successful trials. The p-values for the combined trials are 0.0017 for Professor A and 0.0006 for Professor B.

n	# of successes	p-value
10	4	0.22
15	6	0.15
17	6	0.23
25	8	0.17
30	10	0.20
40	13	0.18
18	7	0.14
10	5	0.08
15	5	0.31
20	7	0.21

Now suppose that reports of the studies have been incomplete and only p-values are supplied. Nonparametric meta analysis performed on 10 studies of Professor B reveals an overall omnibus test significant. The MATLAB code for Fisher's and inverse-normal methods are below; the combined p-values for Professor B are 0.0235 and 0.021.

```
>> pvals = [0.22, 0.15, 0.23, 0.17, 0.20, 0.18, 0.14, 0.08, 0.31, 0.21];
>> fisherstat = - 2 * sum( log(pvals))
 fisherstat =
     34.4016
>> 1-chi2cdf(fisherstat, 2*10)
 ans =
     0.0235
>> 1 - normcdf( sum(norminv(1-pvals))/sqrt(length(pvals))  )
 ans =
     0.0021
```

6.7 EXERCISES

6.1. Derive the exact distribution of the Kolmogorov test statistic D_n for the case $n = 1$.

6.2. Go the NIST link below to download 31 measurements of polished window strength data for a glass airplane window. In reliability tests such as this one, researchers rely on parametric distributions to characterize the observed lifetimes, but the normal distribution is not commonly used. Does this data follow any well-known distribution? Use probability plotting to make your point.

`http://www.itl.nist.gov/div898/handbook/eda/section4/eda4291.htm`

6.3. Go to the NIST link below to download 100 measurements of the speed of light in air. This classic experiment was carried out by a U.S. Naval Academy teacher Albert Michelson is 1879. Do the data appear to be normally distributed? Use three tests (Kolmogorov, Anderson-Darling, Shapiro-Wilk) and compare answers.

`http://www.itl.nist.gov/div898/strd/univ/data/Michelso.dat`

6.4. Do those little peanut bags handed out during airline flights actually contain as many peanuts as they claim? From a box of peanut bags that have 14g label weights, fifteen bags are sampled and weighed: 16.4, 14.4, 15.5, 14.7, 15.6, 15.2, 15.2, 15.2, 15.3, 15.4, 14.6, 15.6, 14.7, 15.9, 13.9. Are the data approximately normal so that a t-test has validity?

6.5. Generate a sample $\mathcal{S}_0$ of size $m = 47$ from the population with normal $\mathcal{N}(3, 1)$ distribution. Test the hypothesis that the sample is standard normal $H_0 : F = F_0 = \mathcal{N}(0, 1)$ (not at $\mu = 3$) versus the alternative $H_1 : F < F_0$. You will need to use D_n^- in the test. Repeat this testing procedure (with new samples, of course) 1000 times. What proportion of p-values exceeded 5%?

6.6. Generate two samples of sizes $m = 30$ and $m = 40$ from $\mathcal{U}(0, 1)$. Square the observations in the second sample. What is the theoretical distribution of the squared uniforms? Next, "forget" that you squared the second sample and test by Smirnov test equality of the distributions. Repeat this testing procedure (with new samples, of course) 1000 times. What proportion of p-values exceeded 5%?

6.7. In MATLAB, generate two data sets of size $n = 10,000$: the first from $\mathcal{N}(0, 1)$ and the second from the t distribution with 5 degrees of freedom. These are your two samples to be tested for normality. Recall the asymptotic properties of order statistics from Chapter 5 and find the approximate distribution of $X_{[3000]}$. Standardize it appropriately (here $p = 0.3$, and $\mu = \texttt{norminv(0.3)} = -0.5244$, and find the two-sided p-values for the goodness-of-fit test of the normal distribution. If the testing is repeated 10 times, how many times will you reject the hypothesis of normality for the second, t distributed sequence? What if the degrees of freedom in the t sequence increase from 5 to 10; to 40? Comment.

6.8. For two samples of size $m = 2$ and $n = 4$, find the exact distribution of the Smirnov test statistics for the test of $H_0 : F(x) \le G(x)$ versus $H_1 : F(x) > G(x)$.

6.9. Let $X_1, X_2, \ldots, X_{n_1}$ be a sample from a population with distribution F_X and $Y_1, Y_2, \ldots, Y_{n_2}$ be a sample from distribution F_Y. If we are interested in testing $H_0 : F_X = F_Y$ one possibility is to use the runs test in the following way. Combine the two samples and let $Z_1, Z_2, \ldots, Z_{n_1+n_2}$ denote the respective order statistics. Let dichotomous variables 1 and 2 signify if Z is from the first or the second sample. Generate 50 $\mathcal{U}(0,1)$ numbers and 50 $\mathcal{N}(0,1)$ numbers. Concatenate and sort them. Keep track of each number's source by assigning 1 if the number came from the uniform distribution and 2 otherwise. Test the hypothesis that the distributions are the same.

6.10. Combine the p-values for Professor B from the meta-analysis example using the Tippet-Wilkinson method with the smallest p-value and Lancaster's Method.

6.11. Derive the exact distribution of the number of runs for $n = 4$ when there are $n_1 = n_2 = 2$ observations of ones and twos. Base your derivation on the exhausting all $\binom{4}{2}$ possible outcomes.

6.12. The link below connects you to the Dow-Jones Industrial Average (DJIA) closing values from 1900 to 1993. First column contains the date (yymmdd), second column contains the value. Use the runs test to see if there is a non-random pattern in the increases and decreases in the sequence of closing values. Consult

`http://lib.stat.cmu.edu/datasets/djdc0093`

6.13. Recall Exercise 5.1. Repeat the simulation and make a comparison between the two populations using `qqplot`. Because the sample range has a beta $\mathcal{B}e(49, 2)$. distribution, this should be verified with a straight line in the plot.

6.14. The table below displays the accuracy of meteorological forecasts for the city of Marietta, Georgia. Results are supplied for the month of February, 2005. If the forecast differed for the real temperature for more than $3°F$, the symbol 1 was assigned. If the forecast was in error limits $< 3°F$, the symbol 2 was assigned. Is it possible to claim that correct and wrong forecasts group at random?

2	2	2	2	2	2	2	2	2	2	2	1	1	1
1	1	2	2	1	1	2	2	2	2	2	1	2	2

6.15. Previous records have indicated that the total points of Olympic dives are normally distributed. Here are the records for *Men 10-meter Platform Preliminary* in 2004. Test the normality of the point distribution. For a computational exercise, generate 1000 sets of 33 normal observations with the same mean and variance as the diving point data.

Use the Smirnov test to see how often the p-value corresponding to the test of equal distributions exceeds 0.05. Comment on your results.

Rank	Name	Country	Points	Lag
1	HELM, Mathew	AUS	513.06	
2	DESPATIE, Alexandre	CAN	500.55	12.51
3	TIAN, Liang	CHN	481.47	31.59
4	WATERFIELD, Peter	GBR	474.03	39.03
5	PACHECO, Rommel	MEX	463.47	49.59
6	HU, Jia	CHN	463.44	49.62
7	NEWBERY, Robert	AUS	461.91	51.15
8	DOBROSKOK, Dmitry	RUS	445.68	67.38
9	MEYER, Heiko	GER	440.85	72.21
10	URAN-SALAZAR, Juan G.	COL	439.77	73.29
11	TAYLOR, Leon	GBR	433.38	79.68
12	KALEC, Christopher	CAN	429.72	83.34
13	GALPERIN, Gleb	RUS	427.68	85.38
14	DELL'UOMO, Francesco	ITA	426.12	86.94
15	ZAKHAROV, Anton	UKR	420.3	92.76
16	CHOE, Hyong Gil	PRK	419.58	93.48
17	PAK, Yong Ryong	PRK	414.33	98.73
18	ADAM, Tony	GER	411.3	101.76
19	BRYAN, Nickson	MAS	407.13	105.93
20	MAZZUCCHI, Massimiliano	ITA	405.18	107.88
21	VOLODKOV, Roman	UKR	403.59	109.47
22	GAVRIILIDIS, Ioannis	GRE	395.34	117.72
23	GARCIA, Caesar	USA	388.77	124.29
24	DURAN, Cassius	BRA	387.75	125.31
25	GUERRA-OLIVA, Jose Antonio	CUB	375.87	137.19
26	TRAKAS, Sotirios	GRE	361.56	151.5
27	VARLAMOV, Aliaksandr	BLR	361.41	151.65
28	FORNARIS, ALVAREZ Erick	CUB	351.75	161.31
29	PRANDI, Kyle	USA	346.53	166.53
30	MAMONTOV, Andrei	BLR	338.55	174.51
31	DELALOYE, Jean Romain	SUI	326.82	186.24
32	PARISI, Hugo	BRA	325.08	187.98
33	HAJNAL, Andras	HUN	305.79	207.27

6.16. Consider the Cramér von Mises test statistic with $\psi(x) = 1$. With a sample of $n = 1$, derive the test statistic distribution and show that it is maximized at $X = 1/2$.

6.17. Generate two samples $\mathcal{S}_1$ and $\mathcal{S}_2$, of sizes $m = 30$ and $m = 40$ from the uniform distribution. Square the observations in the second sample. What is the theoretical distribution of the squared uniforms? Next, "forget" that you squared the second sample and test equality of the distributions. Repeat this testing procedure (with new samples, of course) 1000 times. What proportion of p-values exceeded 5%?

6.18. Recall the Gumbel distribution (or *extreme value distribution*) from Chapter 5. Linearize the CDF of the Gumbel distribution to show how a probability plot could be constructed.

REFERENCES

Anderson, T. W., and Darling, D. A. (1954), "A Test of Goodness of Fit," *Journal of the American Statistical Association*, 49, 765-769.

Birnbaum, Z. W., and Tingey, F. (1951), "One-sided Confidence Contours for Probability Distribution Functions," *Annals of Mathematical Statistics,* 22, 592-596.

D'Agostino, R. B., and Stephens, M. A. (1986), *Goodness-of-Fit Techniques*, New York: Marcel Dekker.

Feller, W. (1948), On the Kolmogorov-Smirnov Theorems, *Annals of Mathematical Statistics,* 19, 177-189.

Fisher, R. A. (1932), *Statistical Methods for Research Workers*, 4th ed, Edinburgh, UK: Oliver and Boyd.

Folks, J. L. (1984), "Combination of Independent Tests," in *Handbook of Statistics* 4, Nonparametric Methods, Eds. P. R. Krishnaiah and P. K. Sen, Amsterdam, North-Holland: Elsevier Science, pp. 113-121.

Good, I. J. (1955), "On the Weighted Combination of Significance Tests," *Journal of the Royal Statistical Society (B)*, 17, 264265.

Hedges, L. V., and Olkin, I. (1985), *Statistical Methods for Meta-Analysis*, New York: Academic Press.

Kolmogorov, A. N. (1933), "Sulla Determinazione Empirica di Una Legge di Distribuzione," *Giornio Instituto Italia Attuari*, 4, 83-91.

Lancaster, H. O. (1961), "The Combination of Probabilities: An Application of Orthonormal Functions," *Australian Journal of Statistics*,3, 20-33.

Miller, L. H. (1956), "Table of percentage points of Kolmogorov Statistics," *Journal of the American Statistical Association*, 51, 111-121.

Mogull, R. G. (1994). "The one-sample runs test: A category of exception," *Journal of Educational and Behavioral Statistics* 19, 296-303.

Mood, A. (1940), "The distribution theory of runs," *Annals of Mathematical Statistics*, 11, 367–392.

Mudholkar, G. S., and George, E. O. (1979), "The Logit Method for Combining Probabilities," in *Symposium on Optimizing Methods in Statistics*, ed. J. Rustagi, New York: Academic Press, pp. 345-366.

Pearson, K. (1902), "On the Systematic Fitting of Curves to Observations and Measurements," *Biometrika*, 1, 265-303.

Roeder, K. (1990), "Density Estimation with Confidence Sets Exemplified by Superclusters and Voids in the Galaxies," *Journal of the American Statistical Association*, 85, 617-624.

Shapiro, S. S., and Wilk, M. B. (1965), "An Analysis of Variance Test for Normality (Complete Samples)," *Biometrika*, 52, 591-611.

Smirnov, N. V. (1939a), "On the Derivations of the Empirical Distribution Curve," *Matematicheskii Sbornik*, 6, 2-26.

_______ (1939b), "On the Estimation of the Discrepancy Between Empirical Curves of Distribution for Two Independent Samples," *Bulletin Moscow University*, 2, 3-16.

Stephens, M. A. (1974), "EDF Statistics for Goodness of Fit and Some Comparisons," *Journal of the American Statistical Association*,69, 730-737.

_______ (1976), "Asymptotic Results for Goodness-of-Fit Statistics with Unknown Parameters," *Annals of Statistics,* 4, 357-369.

Tippett, L. H. C. (1931), *The Method of Statistics*, 1st ed., London: Williams and Norgate.

Wilkinson, B. (1951), "A Statistical Consideration in Psychological Research," *Psychological Bulletin*, 48, 156-158.

7

Rank Tests

> Each of us has been doing statistics all his life, in the sense that each of us has been busily reaching conclusions based on empirical observations ever since birth.
>
> William Kruskal

All those old basic statistical procedures – the t-test, the correlation coefficient, the analysis of variance (ANOVA) – depend strongly on the assumption that the sampled data (or the sufficient statistics) are distributed according to a well-known distribution. Hardly the fodder for a nonparametrics text book. But for every classical test, there is a nonparametric alternative that does the same job with fewer assumptions made of the data. Even if the assumptions from a parametric model are modest and relatively non-constraining, they will undoubtedly be false in the most pure sense. Life, along with your experimental data, are too complicated to fit perfectly into a framework of i.i.d. errors and exact normal distributions.

Mathematicians have been researching ranks and order statistics since ages ago, but it wasn't until the 1940s that the idea of rank tests gained prominence in the statistics literature. Hotelling and Pabst (1936) wrote one of the first papers on the subject, focusing on rank correlations.

There are nonparametric procedures for one sample, for comparing two or more samples, matched samples, bivariate correlation, and more. The key to evaluating data in a nonparametric framework is to compare observations based on their *ranks* within the sample rather than entrusting the

(a)

(b)

(c)

Fig. 7.1 Frank Wilcoxon (1892–1965), Henry Berthold Mann (1905–2000), and Professor Emeritus Donald Ransom Whitney

actual data measurements to your analytical verdicts. The following table shows non-parametric counterparts to the well known parametric procedures (WSiRT/WSuRT stands for Wilcoxon Signed/Sum Rank Test).

PARAMETRIC	NON-PARAMETRIC
Pearson coefficient of correlation	Spearman coefficient of correlation
One sample t-test for the location	sign test, WSiRT
paired test t test	sign test, WSiRT
two sample t test	WSurT, Mann-Whitney
ANOVA	Kruskal-Wallis Test
Block Design ANOVA	Friedman Test

To be fair, it should be said that many of these nonparametric procedures come with their own set of assumptions. We will see, in fact, that some of them are rather obtrusive on an experimental design. Others are much less so. Keep this in mind when a nonparametric test is touted as "assumption free". Nothing in life is free.

In addition to properties of ranks and basic sign test, in this chapter we will present the following nonparametric procedures:

- **Spearman Coefficient:** Two-sample correlation statistic.
- **Wilcoxon Test:** One-sample median test (also see *Sign Test*).
- **Wilcoxon Sum Rank Test:** Two-sample test of distributions.
- **Mann-Whitney Test:** Two-sample test of medians.

7.1 PROPERTIES OF RANKS

Let $X_1, X_2, \ldots, X_n$ be a sample from a population with continuous CDF F_X. The nonparametric procedures are based on how observations within the sample are *ranked*, whether in terms of a parameter μ or another sample. The ranks connected with the sample $X_1, X_2, \ldots, X_n$ denoted as

$$r(X_1), r(X_2), \ldots, r(X_n),$$

are defined as

$$r(X_i) = \#\{X_j | X_j \leq X_i, j = 1, \ldots, n\}.$$

Equivalently, ranks can be defined via the *order statistics* of the sample, $r(X_{i:n}) = i$, or

$$r(X_i) = \sum_{j=1}^{n} I(X_i \geq X_j).$$

Since $X_1, \ldots, X_n$ is a random sample, it is true that $X_1, \ldots, X_n \stackrel{d}{=} X_{\pi_1}, \ldots, X_{\pi_n}$ where $\pi_1, \ldots, \pi_n$ is a permutation of $1, 2, \ldots, n$ and $\stackrel{d}{=}$ denotes equality in distribution. Consequently, $P(r(X_i) = j) = 1/n, \ 1 \leq j \leq n$, i.e., *ranks in an i.i.d. sample are distributed as discrete uniform random variables.* Corresponding to the data r_i, let $R_i = r(X_i)$, the rank of the random variable X_i.

From Chapter 2, the properties of integer sums lead to the following properties for ranks:

(i) $\mathbb{E}(R_i) = \sum_{j=1}^{n} \frac{j}{n} = \frac{n+1}{2}$.

(ii) $\mathbb{E}(R_i^2) = \sum_{j=1}^{n} \frac{j^2}{n} = \frac{n(n+1)(2n+1)}{6n} = \frac{(n+1)(2n+1)}{6}$

(iii) $\mathbb{V}\mathrm{ar}(R_i) = \frac{n^2-1}{12}$.

(iv) $\mathbb{E}(X_i R_i) = \frac{1}{n} \sum_{i=1}^{n} i\mathbb{E}(X_{i:n})$

where

$$\mathbb{E}(X_{r:n}) = F_X^{-1}\left(\frac{r}{n+1}\right) \quad \text{and}$$

$$\mathbb{E}(X_i R_i) = \mathbb{E}(\mathbb{E}(R_i X_i) | R_i = k) = \mathbb{E}(\mathbb{E}(k X_{k:n})) = \frac{1}{n} \sum_{i=1}^{n} i\mathbb{E}(X_{i:n}).$$

In the case of ties, it is customary to average the tied rank values. The MATLAB procedure rank does just that:

```
>> ranks([3 1 4 1 5 9 2 6 5 3 5 8 9])
ans =
 Columns 1 through 7
 4.5000   1.5000   6.0000   1.5000   8.0000   12.5000   3.0000

 Columns 8 through 13
 10.0000   8.0000   4.5000   8.0000   11.0000   12.5000
```

Property (iv) can be used to find the correlation between observations and their ranks. Such correlation depends on the sample size and the underlying distribution. For example, for $X \sim \mathcal{U}(0,1)$, $\mathbb{E}(X_i R_i) = (2n+1)/6$, which gives $\mathbb{C}\text{ov}(X_i, R_i) = (n-1)/12$ and $\mathbb{C}\text{orr}(X_i, R_i) = \sqrt{(n-1)/(n+1)}$.

With two samples, comparisons between populations can be made in a nonparametric way by comparing ranks for the combined ordered samples. Rank statistics that are made up of sums of indicator variables comparing items from one sample with those of the other are called *linear rank statistics.*

7.2 SIGN TEST

Suppose we are interested in testing the hypothesis H_0 that a population with continuous CDF has a median m_0 against one of the alternatives $H_1 : m > m_0$, $H_1 : m < m_0$ or $H_1 : m \neq m_0$. Designate the sign $+$ when $X_i > m_0$ (i.e., when the difference $X_i - m_0$ is positive), and the sign $-$ otherwise. For continuous distributions, the case $X_i = m$ (a tie) is theoretically impossible, although in practice ties are often possible, and this feature can be accommodated. For now, we assume the ideal situation in which the ties are not present.

> **Assumptions:** Actually, no assumptions are necessary for the sign test other than the data are at least ordinal

If m_0 is the median, i.e., if H_0 is true, then by definition of the median, $P(X_i > m_0) = P(X_i < m_0) = 1/2$. If we let T be the total number of $+$ signs, that is,

$$T = \sum_{i=1}^{n} I(X_i > m_0),$$

then $T \sim \mathcal{B}in(n, 1/2)$.

Let the level of test, α, be specified. When the alternative is $H_1 : m > m_0$, the critical values of T are integers larger than or equal to t_α, which is defined as the smallest integer for which

$$P(T \geq t_\alpha | H_0) = \sum_{t=t_\alpha}^{n} \binom{n}{t} \left(\frac{1}{2}\right)^n < \alpha.$$

Likewise, if the alternative is $H_1 : m < m_0$, the critical values of T are integers smaller than or equal to t'_α, which is defined as the largest integer for which

$$P(T \leq t'_\alpha | H_0) = \sum_{t=0}^{t'_\alpha} \binom{n}{t} \left(\frac{1}{2}\right)^n < \alpha.$$

If the alternative hypothesis is two-sided ($H_1 : m \neq m_0$), the critical values of T are integers smaller than or equal to $t'_{\alpha/2}$ and integers larger than or equal to $t_{\alpha/2}$, which are defined via

$$\sum_{t=0}^{t'_{\alpha/2}} \binom{n}{t} \left(\frac{1}{2}\right)^n < \alpha/2, \quad \text{and} \quad \sum_{t=t_{\alpha/2}}^{n} \binom{n}{t} \left(\frac{1}{2}\right)^n < \alpha/2.$$

If the value T is observed, then in testing against alternative $H_1 : m > m_0$, large values of T serve as evidence against H_0 and the p-value is

$$p = \sum_{i=T}^{n} \binom{n}{i} 2^{-n} = \sum_{i=0}^{n-T} \binom{n}{i} 2^{-n}.$$

When testing against the alternative $H_1 : m < m_0$, small values of T are critical and the p-value is

$$p = \sum_{i=0}^{T} \binom{n}{i} 2^{-n}.$$

When the hypothesis is the two-sided, take $T' = \min\{T, n - T\}$ and calculate p-value as

$$p = 2 \sum_{i=0}^{T'} \binom{n}{i} 2^{-n}.$$

7.2.1 Paired Samples

Consider now the case in which two samples are paired:

$$\{(X_1, Y_1), \ldots, (X_n, Y_n)\}.$$

Suppose we are interested in finding out whether the median of the population differences is 0. In this case we let $T = \sum_{i=1}^{n} I(X_i > Y_i)$, which is the total number of strictly positive differences.

For two population means it is true that the hypothesis of equality of means is equivalent to the hypothesis that the mean of the population differences is equal to zero. This is not always true for the test of medians. That is, if $D = X - Y$, then it is quite possible that $m_D \neq m_X - m_Y$. With the sign test we are not testing the *equality* of two medians, but whether the *median of the difference* is 0.

Under H_0: *equal population medians*, $\mathbb{E}(T) = \sum P(X_i > Y_i) = n/2$ and $\mathbb{V}\text{ar}(T) = n \cdot \mathbb{V}\text{ar}(I(X > Y)) = n/4$. With large enough n, T is approximately normal, so for the statistical test of H_1: *the medians are not equal*, we would reject H_0 if T is far enough away from $n/2$; that is,

$$z_0 = \frac{T - n/2}{\sqrt{n}/2} : \qquad \text{reject } H_0 \text{ if } |z_0| > z_{\alpha/2}.$$

Example 7.1 According to The Rothstein Catalog on Disaster Recovery, the median number of violent crimes per state dropped from the year 1999 to 2000. Of 50 states, if X_i is number of violent crimes in state i in 1999 and Y_i is the number for 2000, the median of sample differences is $X_i - Y_i$. This number decreased in 38 out of 50 states in one year. With $T = 38$ and $n = 50$, we find $z_0 = 3.67$, which has a p-value of 0.00012 for the one-sided test (medians decreased over the year) or .00024 for the two-sided test.

Example 7.2 Let X_1 and X_2 be independent random variables distributed as Poisson with parameters λ_1 and λ_2. We would like to test the hypothesis $H_0 : \lambda_1 = \lambda_2 \ (= \lambda)$. If H_0 is true,

$$P(X_1 = k, X_2 = l) = \frac{(2\lambda)^{k+l}}{(k+l)!} e^{-2\lambda} \binom{k+l}{k} \left(\frac{1}{2}\right)^{k+l}.$$

If we observe X_1 and X_2 and if $X_1 + X_2 = n$ then testing H_0 is exactly the sign test, with $T = X_1$. Indeed,

$$P(X_1 = k \mid X_1 + X_2 = n) = \binom{n}{k} \left(\frac{1}{2}\right)^n.$$

For instance, if $X_1 = 10$ and $X_2 = 20$ are observed, then the p-value for the two-sided alternative $H_1 : \lambda_1 \neq \lambda_2$ is $2 \sum_{i=0}^{10} \binom{30}{i} \left(\frac{1}{2}\right)^{30} = 2 \cdot 0.0494 = 0.0987$.

Example 7.3 Hogmanay Celebration[1] Roger van Gompel and Shona Falconer at the University of Dundee conducted an experiment to examine the

[1]Hogmanay is the Scottish New Year, celebrated on 31st December every year. The night involves a celebratory drink or two, fireworks and kissing complete strangers (not necessarily in that order).

drinking patterns of Members of the Scottish Parliament over the festive holiday season.

Being elected to the Scottish Parliament is likely to have created in members a sense of stereotypical conformity so that they appear to fit in with the traditional ways of Scotland, pleasing the tabloid newspapers and ensuring popular support. One stereotype of the Scottish people is that they drink a lot of whisky, and that they enjoy celebrating both Christmas and Hogmanay. However, it is possible that members of parliment tend to drink more whisky at one of these times compared to the other, and an investigation into this was carried out.

The measure used to investigate any such bias was the number of units of single malt scotch whisky ("drams") consumed over two 48-hour periods: Christmas Eve/Christmas Day and Hogmanay/New Year's Day. The hypothesis is that Members of the Scottish Parliament drink a significantly different amount of whisky over Christmas than over Hogmanay (either consistently more or consistently less). The following data were collected.

MSP	1	2	3	4	5	6	7	8	9
Drams at Christmas	2	3	3	2	4	0	3	6	2
Drams at Hogmanay	5	1	5	6	4	7	5	9	0

MSP	10	11	12	13	14	15	16	17	18
Drams at Christmas	2	5	4	3	6	0	3	3	0
Drams at Hogmanay	4	15	6	8	9	0	6	5	12

The MATLAB function `sign_test1` lists five summary statistics from the data for the sign test. The first is a p-value based on randomly assigning a '+' or '−' to tied values (see next subsection), and the second is the p-value based on the normal approximation, where ties are counted as half. n is the number of non-tied observations, $plus$ are the number of plusses in $y - x$, and tie is the number of tied observations.

```
>> x=[2 3 3 2 4 0 3 6 2 2 5  4 3 6 0 3 3 0];
>> y=[5 1 5 6 4 7 5 9 0 4 15 6 8 9 0 6 5 12];
>> [p1 p2 n plus tie] = sign_test1(x',y')

p1 =
   0.0021

p2 =
   0.0030

n =
   16
```

```
plus =
     2

tie =
     2
```

7.2.2 Treatments of Ties

Tied data present numerous problems in derivations of nonparametric methods, and are frequently encountered in real-world data. Even when observations are generated from a continuous distribution, due to limited precision on measurement and application, ties may appear. To deal with ties, MATLAB does one of three things via the third input in `sign_test1`:

R Randomly assigns '+' or '-' to tied values

C Uses least favorable assignment in terms of H_0

I Ignores tied values in test statistic computation

The preferable way to deal with ties is the first option (to randomize). Another equivalent way to deal with ties is to add a slight bit of "noise" to the data. That is, complete the sign test after modifying D by adding a small enough random variable that will not affect the ranking of the differences; i.e., $\tilde{D}_i = D_i + \epsilon_i$, where $\epsilon_i \sim \mathcal{N}(0, 0.0001)$. Using the second or third options in `sign_test1` will lead to biased or misleading results, in general.

7.3 SPEARMAN COEFFICIENT OF RANK CORRELATION

Charles Edward Spearman (Figure 7.2) was a late bloomer, academically. He received his Ph.D. at the age of 48, after serving as an officer in the British army for 15 years. He is most famous in the field of psychology, where he theorized that "general intelligence" was a function of a comprehensive mental competence rather than a collection of multi-faceted mental abilities. His theories eventually led to the development of factor analysis.

Spearman (1904) proposed the rank correlation coefficient long before statistics became a scientific discipline. For bivariate data, an observation has two coupled components (X, Y) that may or may not be related to each other. Let $\rho = \mathbb{C}\text{orr}(X, Y)$ represent the unknown correlation between the two components. In a sample of n, let $R_1, \dots, R_n$ denote the ranks for the first component X and $S_1, \dots, S_n$ denote the ranks for Y. For example, if $x_1 = x_{n:n}$ is the largest value from $x_1, ..., x_n$ and $y_1 = y_{1:n}$ is the smallest

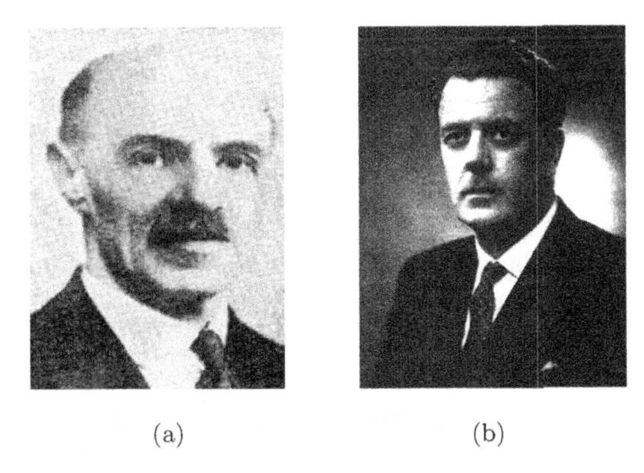

(a) (b)

Fig. 7.2 Charles Edward Spearman (1863–1945) and Maurice George Kendall (1907–1983)

value from $y_1, ..., y_n$, then $(r_1, s_1) = (n, 1)$. Corresponding to Pearson's (parametric) coefficient of correlation, the Spearman coefficient of correlation is defined as

$$\hat{\rho} = \frac{\sum_{i=1}^{n}(R_i - \bar{R})(S_i - \bar{S})}{\sqrt{\sum_{i=1}^{n}(R_i - \bar{R})^2 \cdot \sum_{i=1}^{n}(S_i - \bar{S})^2}}. \tag{7.1}$$

This expression can be simplified. From (7.1), $\bar{R} = \bar{S} = (n+1)/2$, and $\sum(R_i - \bar{R})^2 = \sum(S_i - \bar{S})^2 = n\mathbb{V}\mathrm{ar}(R_i) = n(n^2-1)/12$. Define D as the difference between ranks, i.e., $D_i = R_i - S_i$. With $\bar{R} = \bar{S}$, we can see that

$$D_i = (R_i - \bar{R}) - (S_i - \bar{S}),$$

and

$$\sum_{i=1}^{n} D_i^2 = \sum_{i=1}^{n}(R_i - \bar{R})^2 + \sum_{i=1}^{n}(S_i - \bar{S})^2 - 2\sum_{i=1}^{n}(R_i - \bar{R})(S_i - \bar{S}),$$

that is,

$$\sum_{i=1}^{n}(R_i - \bar{R})(S_i - \bar{S}) = \frac{n(n^2-1)}{12} - \frac{1}{2}\sum_{i=1}^{n} D_i^2.$$

By dividing both sides of the equation with $\sqrt{\sum_{i=1}^{n}(R_i - \bar{R})^2 \cdot \sum_{i=1}^{n}(S_i - \bar{S})^2} =$

$\sum_{i=1}^{n}(R_i - \bar{R})^2 = n(n^2-1)/12$, we obtain

$$\hat{\rho} = 1 - \frac{6\sum_{i=1}^{n} D_i^2}{n(n^2-1)}. \tag{7.2}$$

Consistent with Pearson's coefficient of correlation (the standard parametric measure of covariance), the Spearman coefficient of correlation ranges between -1 and 1. If there is perfect agreement, that is, all the differences are 0, then $\hat{\rho} = 1$. The scenario that maximizes $\sum D_i^2$ occurs when ranks are perfectly opposite: $r_i = n - s_i + 1$.

If the sample is large enough, the Spearman statistic can be approximated using the normal distribution. It was shown that if $n > 10$,

$$Z = (\hat{\rho} - \rho)\sqrt{n-1} \sim \mathcal{N}(0,1).$$

Assumptions: Actually, no assumptions are necessary for testing ρ other than the data are at least ordinal.

Example 7.4 Stichler, Richey, and Mandel (1953) list tread wear for tires (see table below), each tire measured by two methods based on (a) weight loss and (b) groove wear. In MATLAB, the function

```
spear(x,y)
```

computes the Spearman coefficient. For this example, $\hat{\rho} = 0.9265$. Note that if we opt for the parametric measure of correlation, the Pearson coefficient is 0.948.

Weight	Groove	Weight	Groove
45.9	35.7	41.9	39.2
37.5	31.1	33.4	28.1
31.0	24.0	30.5	28.7
30.9	25.9	31.9	23.3
30.4	23.1	27.3	23.7
20.4	20.9	24.5	16.1
20.9	19.9	18.9	15.2
13.7	11.5	11.4	11.2

Ties in the data: The statistics in (7.1) and (7.2) are not designed for paired data that include tied measurements. If ties exist in the data, a simple adjustment should be made. Define $u' = \sum u(u^2-1)/12$ and $v' = \sum v(v^2-1)/12$ where the u's and v's are the ranks for X and Y adjusted (e.g. averaged) for ties. Then,

$$\hat{\rho}' = \frac{n(n^2-1) - 6\sum_{i=1}^{n} D_i^2 - 6(u'+v')}{\{[n(n^2-1) - 12u'][n(n^2-1) - 12v']\}^{1/2}}$$

and it holds that, for large n,

$$Z = (\hat{\rho}' - \rho)\sqrt{n-1} \sim \mathcal{N}(0,1).$$

7.3.1 Kendall's Tau

Kendall (1938) derived an alternative measure of bivariate dependence by finding out how many pairs in the sample are "concordant", which means the signs between X and Y agree in the pairs. That is, out of $\binom{n}{2}$ pairs such as (X_i, Y_i) and (X_j, Y_j), we compare the sign of $(X_i - Y_i)$ to that of $(X_j - Y_j)$. Pairs for which one sign is plus and the other is minus are "discordant".

The Kendall's τ statistic is defined as

$$\tau = \frac{2S_\tau}{n(n-1)}, \quad S_\tau = \sum_{i=1}^{n} \sum_{j=i+1}^{n} \text{sign}\{r_i - r_j\},$$

where r_is are defined via ranks of the second sample corresponding to the ordered ranks of the first sample, $\{1, 2, \ldots, n\}$, that is,

$$\begin{pmatrix} 1 & 2 & \ldots & n \\ r_1 & r_2 & \ldots & r_n \end{pmatrix}$$

In this notation $\sum_{i=1}^{n} D_i^2$ from the Spearman's coefficient of correlation becomes $\sum_{i=1}^{n}(r_i - i)^2$. In terms of the number of concordant (n_c) and discordant ($n_D = \binom{n}{2} - n_c$) pairs,

$$\tau = \frac{n_c - n_D}{\binom{n}{2}}$$

and in the case of ties, use

$$\tau = \frac{n_c - n_D}{n_c + n_D}.$$

Example 7.5 Trends in Indiana's water use from 1986 to 1996 were reported by Arvin and Spaeth (1997) for Indiana Department of Natural Resources. About 95% of the surface water taken annually is accounted for by two categories: surface water withdrawal and ground-water withdrawal. Kendall's tau statistic showed no apparent trend in total surface water withdrawal over time (p-value ≈ 0.59), but ground-water withdrawal increased slightly over the 10 year span (p-value ≈ 0.13).

```
>> x=(1986:1996);
>> y1=[2.96,3.00,3.12,3.22,3.21,2.96,2.89,3.04,2.99,3.08,3.12];
>> y2=[0.175,0.173,0.197,0.182,0.176,0.205,0.188,0.186,0.202,...
       0.208,0.213];
```

```
>> y1_rank=ranks(y1); y2_rank=ranks(y2);
>> n=length(x); S1=0; S2=0;
>> for i=1:n-1
    for j=i+1:n
       S1=S1+sign(y1_rank(i)-y1_rank(j));
       S2=S2+sign(y2_rank(i)-y2_rank(j));
    end
  end
>> ktau1=2*S1/(n*(n-1))

ktau1 =
  -0.0909

>> ktau2=2*S2/(n*(n-1))

ktau2 =
  -0.6364
```

With large sample size n, we can use the following z-statistic as a normal approximation:

$$z_\tau = \frac{3\tau\sqrt{n(n-1)}}{\sqrt{2(2n+5)}}.$$

This can be used to test the null hypothesis of zero correlation between the populations. Kendall's tau is natural measure of the relationship between X and Y. We can describe it as an odds-ratio by noting that

$$\frac{1+\tau}{1-\tau} = \frac{P(C)}{P(D)},$$

where C is the event that any pair in the population is concordant, and D is the event any pair is discordant. Spearman's coefficient, on the other hand, cannot be explained this way. For example, in a population with $\tau = 1/3$, any two sets of observations are twice as likely to be concordant than discordant. On the other hand, computations for τ grow as $O(n^2)$, compared to the Spearman coefficient, that grows as $O(n \ln n)$

7.4 WILCOXON SIGNED RANK TEST

Recall that the sign test can be used to test differences in medians for two independent samples. A major shortcoming of the sign test is that only the sign of $D_i = X_i - m_0$, or $D_i = X_i - Y_i$, (depending if we have a one- or two-sample problem) contributes to the test statistics. Frank Wilcoxon suggested that, in addition to the sign, the absolute value of the discrepancy between the pairs should matter as well, and it could increase the efficiency of the sign test.

Suppose that, as in the sign test, we are interested in testing the hypothesis that a median of the unknown distribution is m_0. We make an important assumption of the data.

> **Assumption:** The differences $D_i,\ i = 1, \ldots, n$ are symmetrically distributed about 0.

This implies that positive and negative differences are equally likely. For this test, the absolute values of the differences $(|D_1|, |D_2|, \ldots, |D_n|)$ are ranked. The idea is to use $(|D_1|, |D_2|, \ldots, |D_n|)$ as a set of weights for comparing the differences between $(S_1, \ldots, S_n)$.

Under H_0 (the median of distribution is m_0), the expectation of the sum of positive differences should be equal to the expectation of the sum of the negative differences. Define

$$T^+ = \sum_{i=1}^{n} S_i \ r(|D_i|)$$

and

$$T^- = \sum_{i=1}^{n} (1 - S_i) \ r(|D_i|),$$

where $S_i \equiv S(D_i) = I(D_i > 0)$. Thus $T^+ + T^- = \sum_{i=1}^{n} i = n(n+1)/2$ and

$$T = T^+ - T^- = 2 \sum_{i=1}^{n} r(|D_i|) S_i - n(n+1)/2. \tag{7.3}$$

Under H_0, $(S_1, \ldots, S_n)$ are i.i.d. Bernoulli random variables with $p = 1/2$, independent of the corresponding magnitudes. Thus, when H_0 is true, $\mathbb{E}(T^+) = n(n+1)/4$ and $\mathbb{V}\mathrm{ar}(T^+) = n(n+1)(2n+1)/24$. Quantiles for T^+ are listed in Table 7.9. In MATLAB, the signed rank test based on T^+ is

`wilcoxon_signed2`.

Large sample tests are typically based on a normal approximation of the test statistic, which is even more effective if there are ties in the data.

> **Rule:** For the Wilcoxon signed-rank test, it is suggested to use T from (7.3) instead of T^+ in the case of large-sample approximation.

In this case, $\mathbb{E}(T) = 0$ and $\mathbb{V}\mathrm{ar}(T) = \sum_i (R(|D_i|)^2) = n(n+1)(2n+1)/6$ under H_0. Normal quantiles

$$P\left(\frac{T}{\sqrt{\mathbb{V}\mathrm{ar}(T)}} \le t \right) = \Phi(t).$$

Table 7.9 Quantiles of T^+ for the Wilcoxon signed rank test.

n	0.01	0.025	0.05	n	0.01	0.025	0.05
8	2	4	6	24	70	82	92
9	4	6	9	25	77	90	101
10	6	9	11	26	85	99	111
11	8	11	14	27	94	108	120
12	10	14	18	28	102	117	131
13	13	18	22	29	111	127	141
14	16	22	26	30	121	138	152
15	16	20	26	31	131	148	164
16	24	30	36	32	141	160	176
17	28	35	42	33	152	171	188
18	33	41	48	34	163	183	201
19	38	47	54	35	175	196	214
20	44	53	61	36	187	209	228
21	50	59	68	37	199	222	242
22	56	67	76	38	212	236	257
23	63	74	84	39	225	250	272

can be used to evaluate p-values of the observed statistics T with respect to a particular alternative (see the m-file `wilcoxon_signed`)

Example 7.6 Twelve sets of identical twins underwent psychological tests to measure the amount of aggressiveness in each person's personality. We are interested in comparing the twins to each other to see if the first born twin tends to be more aggressive than the other. The results are as follows, the higher score indicates more aggressiveness.

first born X_i:	86	71	77	68	91	72	77	91	70	71	88	87
second twin Y_i:	88	77	76	64	96	72	65	90	65	80	81	72

The hypotheses are: H_0 : the first twin does not tend to be more aggressive than the other, that is, $\mathbb{E}(X_i) \leq \mathbb{E}(Y_i)$, and H_1 : the first twin tends to be more aggressive than the other, i.e., $\mathbb{E}(X_i) > \mathbb{E}(Y_i)$. The Wilcoxon signed-rank test is appropriate if we assume that $D_i = X_i - Y_i$ are independent, symmetric, and have the same mean. Below is the output of `wilcoxon_signed`, where T statistics have been used.

```
>> fb = [86 71 77 68 91 72 77 91 70 71 88 87];
>> sb = [88 77 76 64 96 72 65 90 65 80 81 72];
>> [t1, z1, p] = wilcoxon_signed(fb, sb, 1)

 t1 =    17             %value of T


 z1 =    0.7565         %value of Z
```

```
    p =    0.2382          %p-value of the test
```

The following is the output of `wilcoxon_signed2` where T^+ statistics have been used. The p-values are identical, and there is insufficient evidence to conclude the first twin is more aggressive than the next.

```
>> [t2, z2, p] = wilcoxon_signed2(fb, sb, 1)

    t2 =41.5000            %value of T^+

    z2 = 0.7565

    p =0.2382
```

7.5 WILCOXON (TWO-SAMPLE) SUM RANK TEST

The Wilcoxon Sum Rank Test (WSuRT) is often used in place of a two sample t-test when the populations being compared are not normally distributed. It requires independent random samples of sizes n_1 and n_2.

> **Assumption:** Actually, no additional assumptions are needed for the Wilcoxon two-sample test.

An example of the sort of data for which this test could be used is responses on a Likert scale (e.g., 1 = much worse, 2 = worse, 3 = no change, 4 = better, 5 = much better). It would be inappropriate to use the t-test for such data because it is only of an ordinal nature. The Wilcoxon rank sum test tells us more generally whether the groups are homogeneous or one group is "better' than the other. More generally, the basic null hypothesis of the Wilcoxon sum rank test is that the two populations are equal. That is $H_0 : F_X(x) = F_Y(x)$. This test assumes that the shapes of the distributions are similar.

Let $\mathbf{X} = X_1, \dots, X_{n_1}$ and $\mathbf{Y} = Y_1, \dots, Y_{n_2}$ be two samples from populations that we want to compare. The $n = n_1 + n_2$ ranks are assigned as they were in the sign test. The test statistic W_n is the sum of ranks (1 to n) for $\mathbf{X}$. For example, if $X_1 = 1, X_2 = 13, X_3 = 7, X_4 = 9$, and $Y_1 = 2, Y_2 = 0, Y_3 = 18$, then the value of W_n is $2 + 4 + 5 + 6 = 17$.

If the two populations have the same distribution then the sum of the ranks of the first sample and those in the second sample should be the same relative to their sample sizes. Our test statistic is

$$W_n = \sum_{i=1}^{n} i S_i(\mathbf{X}, \mathbf{Y}),$$

where $S_i(\mathbf{X}, \mathbf{Y})$ is an indicator function defined as 1 if the ith ranked observation is from the first sample and as 0 if the observation is from the second sample. If there are no ties, then under H_0,

$$\mathbb{E}(W_n) = \frac{n_1(n+1)}{2} \quad \text{and} \quad \mathbb{V}\text{ar}(W_n) = \frac{n_1 n_2(n+1)}{12}.$$

The statistic W_n achieves its minimum when the first sample is entirely smaller than the second, and its maximum when the opposite occurs:

$$\min W_n = \sum_{i=1}^{n_1} i = \frac{n_1(n_1+1)}{2}, \qquad \max W_n = \sum_{i=n-n_1+1}^{n} i = \frac{n_1(2n-n_1+1)}{2}.$$

The exact distribution of W_n is computed in a tedious but straightforward manner. The probabilities for W_n are symmetric about the value of $\mathbb{E}(W_n) = n_1(n+1)/2$.

Example 7.7 Suppose $n_1 = 2, n_2 = 3$, and of course $n = 5$. There are $\binom{5}{2} = \binom{5}{3} = 10$ distinguishable configurations of the vector $(S_1, S_2, \ldots, S_5)$. The minimum of W_5 is 3 and the maximum is 9. Table 7.10 gives the values for W_5 in this example, along with the configurations of ones in the vector $(S_1, S_2, \ldots, S_5)$ and the probability under H_0. Notice the symmetry in probabilities about $\mathbb{E}(W_5)$.

Table 7.10 Distribution of W_5 when $n_1 = 2$ and $n_2 = 3$.

W_5	configuration	probability
3	(1,2)	1/10
4	(1,3)	1/10
5	(1,4), (2,3)	2/10
6	(1,5), (2,4)	2/10
7	(2,5), (3,4)	2/10
8	(3,5)	1/10
9	(4,5)	1/10

Let $k_{n_1,n_2}(m)$ be the number of all arrangements of zeroes and ones in $(S_1(\mathbf{X},\mathbf{Y}), \ldots, S_n(\mathbf{X},\mathbf{Y}))$ such that $W_n = \sum_{i=1}^{n} iS_i(\mathbf{X},\mathbf{Y}) = m$. Then the probability distribution

$$P(W_n = m) = \frac{k_{n_1,n_2}(m)}{\binom{n}{n_1}}, \qquad \frac{n_1(n_1+1)}{2} \le m \le \frac{n_1(2n-n_1+1)}{2},$$

can be used to perform an exact test. Deriving this distribution is no trivial matter, mind you. When n is large, the calculation of exact distribution of W_n is cumbersome.

The statistic W_n in WSuRT is an example of a *linear rank statistic* (see section on Properties of Ranks) for which the normal approximation holds,

$$W_n \sim \mathcal{N}\left(\frac{n_1(n+1)}{2}, \frac{n_1 n_2(n+1)}{12}\right).$$

A better approximation is

$$P(W_n \leq w) \approx \Phi(x) + \phi(x)(x^3 - 3x)\frac{n_1^2 + n_2^2 + n_1 n_2 + n}{20 n_1 n_2 (n+1)},$$

where $\phi(x)$ and $\Phi(x)$ are the PDF and CDF of a standard normal distribution and $x = (w - \mathbb{E}(W) + 0.5)/\sqrt{\mathbb{V}\mathrm{ar}(W_n)}$. This approximation is satisfactory for $n_1 > 5$ and $n_2 > 5$ if there are no ties.

Ties in the Data: If ties are present, let $t_1, \dots, t_k$ be the number of different observations among all the observations in the combined sample. The adjustment for ties is needed only in $\mathbb{V}\mathrm{ar}(W_n)$, because $\mathbb{E}(W_n)$ does not change. The variance decreases to

$$\mathbb{V}\mathrm{ar}(W_n) = \frac{n_1 n_2 (n+1)}{12} - \frac{n_1 n_2 \sum_{i=1}^{k}(t_i^3 - t_i)}{12 n (n+1)}. \tag{7.4}$$

For a proof of (7.4) and more details, see Lehmann (1998).

Example 7.8 Let the combined sample be $\{\ 2\ \boxed{2}\ \boxed{3}\ \boxed{4}\ 4\ 4\ 5\ \}$, where the boxed numbers are observations from the first sample. Then $n = 7$, $n_1 = 3$, $n_2 = 4$, and the ranks are $\{1.5\ \ 1.5\ \ 3\ \ 5\ \ 5\ \ 5\ \ 7\}$. The statistic $w = 1.5 + 3 + 5 = 9.5$ has mean $\mathbb{E}(W_n) = n_1(n+1)/2 = 12$. To adjust the variance for the ties first note that there are $k = 4$ different groups of observations, with $t_1 = 2, t_2 = 1, t_3 = 3$, and $t_4 = 1$. With $t_i = 1$, $t_i^3 - t_i = 0$, only the values of $t_i > 1$ (genuine ties) contribute to the adjusting factor in the variance. In this case,

$$\mathbb{V}\mathrm{ar}(W_7) = \frac{3 \cdot 4 \cdot 8}{12} - \frac{3 \cdot 4 \cdot ((8-2) + (27-3))}{12 \cdot 7 \cdot 8} = 8 - 0.5357 = 7.4643.$$

7.6 MANN-WHITNEY U TEST

Like the Wilcoxon test above, the Mann-Whitney test is applied to find differences in two populations, and does not assume that the populations are normally distributed. However, if we extend the method to tests involving population means (instead of just $\mathbb{E}(D_{ij}) = P(Y < X)$), we need an additional assumption.

Assumption: The shapes of the two distributions are identical.

This is satisfied if we have $F_X(t) = F_Y(t+\delta)$ for some $\delta \in \mathbb{R}$. Let $X_1, \dots, X_{n_1}$ and $Y_1, \dots, Y_{n_2}$ represent two independent samples. Define $D_{ij} = I(Y_j < X_i)$, $i = 1, \dots, n_1$ and $j = 1, \dots, n_2$. The Mann-Whitney statistic for testing the equality of distributions for X and Y is the linear rank statistic

$$U = \sum_{i=1}^{n_1} \sum_{j=1}^{n_2} D_{ij}.$$

It turns out that the test using U is equivalent to the test using W_n in the last section.

Equivalence of Mann-Whitney and Wilcoxon Sum Rank Test. Fix i and consider

$$\sum_{j=1}^{n_2} D_{ij} = D_{i1} + D_{i2} + \cdots + D_{i,n_2}. \tag{7.5}$$

The sum in (7.5) is exactly the number of index values j for which $Y_j < X_i$. Apparently, this sum is equal to the rank of the X_i in the combined sample, $r(X_i)$, minus the number of Xs which are $\leq X_i$. Denote the number of Xs which are $\leq X_i$ by k_i. Then,

$$\begin{aligned} U &= \sum_{i=1}^{n_1} (r(X_i) - k_i) = \sum_{i=1}^{n_1} r(X_i) - (k_1 + k_2 + \cdots + k_{n_1}) \\ &= \sum_{i=1}^{n_1} i S_i(\mathbf{X}, \mathbf{Y}) - \frac{n_1(n_1+1)}{2} = W_n - \frac{n_1(n_1+1)}{2}, \end{aligned}$$

because $k_1 + k_2 + \cdots + k_{n_1} = 1 + 2 + \cdots + n_1$. After all this, the Mann-Whitney (U) statistic and the Wicoxon sum rank statistic (W_n) are equivalent. As a result, the Wilcoxon Sum rank test and Mann-Whitney test are often referred simply as the *Wilcoxon-Mann-Whitney* test.

Example 7.9 Let the combined sample be $\{\,\boxed{7}\ 12\ 13\ \boxed{15}\ \boxed{15}\ 18\ 28\}$, where boxed observations come from sample 1. The statistic U is $0+2+2 = 4$. On the other hand, $W_7 - 3 \cdot 4/2 = (1 + 4.5 + 4.5) - 6 = 4$.

The MATLAB function `wmw` computes the Wilcoxon-Mann-Whitney test using the same arguments from tests listed above. In the example below, w is the sum of ranks for the first sample, and z is the standardized rank statistic for the case of ties.

```
>> [w,z,p]=wmw([1 2 3 4 5], [2 4 2 11 1], 0)
```

```
w = 27

z = -0.1057

p = 0.8740
```

7.7 TEST OF VARIANCES

Compared to parametric tests of the mean, statistical tests on population variances based on the assumption of normal distributed populations are less robust. That is, the parametric tests for variances are known to perform quite poorly if the normal assumptions are wrong.

Suppose we have two populations with CDFs F and G, and we collect random samples $X_1, ..., X_{n_1} \sim F$ and $Y_1, ..., Y_{n_2} \sim G$ (the same set-up used in the Mann-Whitney test). This time, our null hypothesis is

$$H_0 : {\sigma_X}^2 = {\sigma_Y}^2$$

versus one of three alternative hypotheses (H_1): ${\sigma_X}^2 \neq {\sigma_Y}^2$, ${\sigma_X}^2 < {\sigma_Y}^2$, ${\sigma_X}^2 > {\sigma_Y}^2$. If $\bar{x}$ and $\bar{y}$ are the respective sample means, the test statistic is based on

$$\tilde{R}(x_i) = \text{rank of } (x_i - \bar{x})^2 \text{ among all } n = n_1 + n_2 \text{ squared differences}$$
$$\tilde{R}(y_i) = \text{rank of } (y_i - \bar{y})^2 \text{ among all } n = n_1 + n_2 \text{ squared differences}$$

with test statistic

$$T = \sum_{i=1}^{n_1} \tilde{R}(x_i).$$

Assumption: The measurement scale needs to be interval (at least).

Ties in the Data: If there are ties in the data, it is better to use

$$T^* = \frac{T - n_1 V_R}{\sqrt{\frac{n_1 n_2}{n(n-1)} W_R - \frac{n_1 n_2}{n-1} V^2{}_R}}$$

where

$$V_R = n^{-1} \left[\sum^{n_1} \tilde{R}(x_i)^2 + \sum^{n_2} \tilde{R}(y_i)^2 \right] \quad \text{and}$$

$$W_R = \sum^{n_1} \tilde{R}(x_i)^4 + \sum^{n_2} \tilde{R}(y_i)^4.$$

The critical region for the test corresponds to the direction of the alternative hypothesis. This is called the *Conover test of equal variances*, and tabled quantiles for the null distribution of T are be found in Conover and Iman (1978). If we have larger samples ($n_1 \geq 10$, $n_2 \geq 10$), the following normal approximation for T can be used:

$$T \sim \mathcal{N}(\mu_T, \sigma_T^2), \quad \text{with } \mu_T = \frac{n_1(n+1)(2n+1)}{6},$$
$$\sigma_T^2 = \frac{n_1 n_2 (n+1)(2n+1)(8n+11)}{180}.$$

For example, with an α-level test, if H_1 : $\sigma_X{}^2 > \sigma_Y{}^2$, we reject H_0 if $z_0 = (T - \mu_T)/\sigma_T > z_\alpha$, where z_α is the $1 - \alpha$ quantile of the normal distribution. The test for three or more variances is discussed in Chapter 8, after the Kruskal-Wallis test for testing differences in three or more population medians.

Use the MATLAB function `SquaredRanksTest(x,y,p,side,data)` for the test of two variances, where x and y are the samples, p is the sought-after quantile from the null distribution of T, `side` $= 1$ for the test of $H_1 : \sigma_X{}^2 > \sigma_Y{}^2$ (use $p/2$ for the two-sided test), `side` $= -1$ for the test of $H_1 : \sigma_X{}^2 < \sigma_Y{}^2$ and `side` $= 0$ for the test of $H_1 : \sigma_X{}^2 \neq \sigma_Y{}^2$. The last argument, `data`, is optional; if you are using small samples, the procedure will look for the Excel file (`squared ranks critical values.xl`) containing the table values for a test with small samples. In the simple example below, the test statistic $T = -1.5253$ is inside the region the interval $(-1.6449, 1.6449)$ and we do not reject $H_0 : \sigma_X{}^2 = \sigma_Y{}^2$ at level $\alpha = 0.10$.

```
>>  x=[1,2,3,4,5]; y=[1,3,5,7,9];
>>  [T,T1,dec,ties,p,side,Tp1,Tp2]=SquaredRanksTest(x,y,.1,0)

 T=111.25              %T statistic in case of no ties

 T1=-1.5253            %T1 is the z-statistic in case of ties

 dec=0                 %do not reject H0 at the level specified

 ties=1                %1 indicates ties were found

 p=0.1000              %set type I error rate

 side=0                %chosen alternative hypothesis

 Tp1=-1.6449           %lower critical value

 Tp2=1.6449            %upper critical value
```

7.8 EXERCISES

7.1. With the Spearman correlation statistic, show that when the ranks are opposite, $\hat{\rho} = -1$.

7.2. Diet A was given to a group of 10 overweight boys between the ages of 8 and 10. Diet B was given to another independent group of 8 similar overweight boys. The weight loss is given in the table below. Using WMW test, test the hypothesis that the diets are of comparable effectiveness against the two-sided alternative. Use $\alpha = 5\%$ and normal approximation.

Diet A	7	2	3	-1	4	6	0	1	4	6
Diet B	5	6	4	7	8	9	7	2		

7.3. A psychological study involved the rating of rats along a dominance-submissiveness continuum. In order to determine the reliability of the ratings, the ranks given by two different observers were tabulated below. Are the ratings agreeable? Explain your answer.

Animal	Rank observer A	Rank observer B	Animal	Rank observer A	Rank observer B
A	12	15	I	6	5
B	2	1	J	9	9
C	3	7	K	7	6
D	1	4	L	10	12
E	4	2	M	15	13
F	5	3	N	8	8
G	14	11	O	13	14
H	11	10	P	16	16

7.4. Two vinophiles, X and Y, were asked to rank $N = 8$ tasted wines from best to worst (rank #1=highest, rank #8=lowest). Find the Spearman Coefficient of Correlation between the experts. If the sample size increased to $N = 80$ and we find $\hat{\rho}$ is ten times smaller than what you found above, what would the p-value be for the two-sided test of hypothesis?

Wine brand	a	b	c	d	e	f	g	h
Expert X	1	2	3	4	5	6	7	8
Expert Y	2	3	1	4	7	8	5	6

7.5. Use the link below to see the results of an experiment on the effect of prior information on the time to fuse random dot stereograms. One group (NV) was given either no information or just verbal information about the shape of the embedded object. A second group (group VV) received both verbal information and visual information (e.g., a drawing of the object). Does the median time prove to be greater for the NV group? Compare your results to those from a two-sample t-test.

`http://lib.stat.cmu.edu/DASL/Datafiles/FusionTime.html`

7.6. Derive the exact distribution of the Mann-Whitney U statistic in the case that $n_1 = 4$ and $n_2 = 2$.

7.7. A number of Vietnam combat veterans were discovered to have dangerously high levels of the dioxin 2,3,7,8-TCDD in blood and fat tissue as a result of their exposure to the defoliant Agent Orange. A study published in *Chemosphere (Vol. 20, 1990)* reported on the TCDD levels of 20 Massachusetts Vietnam veterans who were possibly exposed to Agent Orange. The amounts of TCDD (measured in parts per trillion) in blood plasma and fat tissue drawn from each veteran are shown in the table. Is there sufficient evidence of a difference between the distributions of TCDD levels in plasma and fat tissue for Vietnam veterans exposed to Agent Orange?

TCDD Levels in Plasma			TCDD Levels in Fat Tissue		
2.5	3.1	2.1	4.9	5.9	4.4
3.5	3.1	1.8	6.9	7.0	4.2
6.8	3.0	36.0	10.0	5.5	41.0
4.7	6.9	3.3	4.4	7.0	2.9
4.6	1.6	7.2	4.6	1.4	7.7
1.8	20.0	2.0	1.1	11.0	2.5
2.5	4.1		2.3	2.5	

7.8. For the two samples in Exercise 7.5, test for equal variances.

7.9. The following two data sets are part of a larger data set from Scanlon, T.J., Luben, R.N., Scanlon, F.L., Singleton, N. (1993), "Is Friday the 13th Bad For Your Health?," *BMJ*, 307, 1584–1586. The data analysis in this paper addresses the issues of how superstitions regarding Friday the 13th affect human behavior. Scanlon, et al. collected data on shopping patterns and traffic accidents for Fridays the 6th and the 13th between October of 1989 and November of 1992.

(i) The first data set is found on line at

`http://lib.stat.cmu.edu/DASL/Datafiles/Fridaythe13th.html`

The data set lists the number of shoppers in nine different supermarkets in southeast England. At the level $\alpha = 10\%$, test the hypothesis that "Friday 13th" affects spending patterns among South Englanders.

Year, Month	# of accidents Friday 6th	# of accidents Friday 13th	Sign	Hospital
1989, October	9	13	-	SWTRHA hospital
1990, July	6	12	-	
1991, September	11	14	-	
1991, December	11	10	+	
1992, March	3	4	-	
1992, November	5	12	-	

(ii) The second data set is the number of patients accepted in SWTRHA hospital on dates of Friday 6th and Friday 13th. At the level $\alpha = 10\%$, test the hypothesis that the "Friday 13th" effect is present.

7.10. Professor Inarb claims that 50% of his students in a large class achieve a final score 90 points or and higher. A suspicious student asks 17 randomly selected students from Professor Inarb's class and they report the following scores.

80 81 87 94 79 78 89 90 92 88 81 79 82 79 77 89 90

Test the hypothesis that the Professor Inarb's claim is not consistent with the evidence, i.e., that the 50%-tile (0.5-quantile, median) is not equal to 90. Use $\alpha = 0.05$.

7.11. Why does the moon look bigger on the horizon? Kaufman and Rock (1962) tested 10 subjects in an experimental room with moons on a horizon and straight above. The ratios of the perceived size of the horizon moon and the perceived size of the zenith moon were recorded for each person. Does the horizon moon seem bigger?

Subject	Zenith	Horizon	Subject	Zenith	Horizon
1	1.65	1.73	2	1	1.06
3	2.03	2.03	4	1.25	1.4
5	1.05	0.95	6	1.02	1.13
7	1.67	1.41	8	1.86	1.73
9	1.56	1.63	10	1.73	1.56

7.12. To compare the t-test with the WSuRT, set up the following simulation in MATLAB: (1) Generate $n = 10$ observations from $\mathcal{N}(0,1)$; (2) For the test of $H_0 : \mu = 1$ versus $H_1 : \mu < 1$, perform a t-test at $\alpha = 0.05$; (3) Run an analogous nonparametric test; (4) Repeat this simulation 1000 times and compare the power of each test by counting the number of times H_0 is rejected; (5) Repeat the entire experiment using a non-normal distribution and comment on your result.

Year, Month	# Shoppers Friday 6th	# Shoppers Friday 13th	Sign	Supermarket
1990, July	4942	4882	+	Epsom
1991, September	4895	4736	+	
1991, December	4805	4784	+	
1992, March	4570	4603	-	
1992, November	4506	4629	-	
1990, July	6754	6998	-	Guildford
1991, September	6704	6707	-	
1991, December	5871	5662	+	
1992, March	6026	6162	-	
1992, November	5676	5665	+	
1990, July	3685	3848	-	Dorking
1991, September	3799	3680	+	
1991, December	3563	3554	+	
1992, March	3673	3676	-	
1992, November	3558	3613	-	
1990, July	5751	5993	-	Chichester
1991, September	5367	5320	+	
1991, December	4949	4960	-	
1992, March	5298	5467	-	
1992, November	5199	5092	+	
1990, July	4141	4389	-	Horsham
1991, September	3674	3660	+	
1991, December	3707	3822	-	
1992, March	3633	3730	-	
1992, November	3688	3615	+	
1990, July	4266	4532	-	East Grinstead
1991, September	3954	3964	-	
1991, December	4028	3926	+	
1992, March	3689	3692	-	
1992, November	3920	3853	+	
1990, July	7138	6836	+	Lewisham
1991, September	6568	6363	+	
1991, December	6514	6555	-	
1992, March	6115	6412	-	
1992, November	5325	6099	-	
1990, July	6502	6648	-	Nine Elms
1991, September	6416	6398	+	
1991, December	6422	6503	-	
1992, March	6748	6716	+	
1992, November	7023	7057	-	
1990, July	4083	4277	-	Crystal Palace
1991, September	4107	4334	-	
1991, December	4168	4050	+	
1992, March	4174	4198	-	
1992, November	4079	4105	-	

REFERENCES

Arvin, D. V., and Spaeth, R. (1997), "Trends in Indiana's water use 1986–1996 special report," Technical report by State of Indiana Department of Datural Resources, Division of Water.

Conover, W. J., and Iman, R. L. (1978), "Some Exact Tables for the Squared Ranks Test," *Communications in Statistics*, 5, 491-513.

Hotelling, H., and Pabst, M. (1936), "Rank Correlation and Tests of Significance Involving the Assumption of Normality," *Annals of Mathematical Statistics*, 7, 29-43.

Kendall, M. G. (1938), "A New Measure of Rank Correlation," *Biometrika*, 30, 81-93.

Lehmann, E. L. (1998), *Nonparametrics: Statistical Methods Based on Ranks,* New Jersey: Prentice Hall.

Stichler, R.D., Richey, G.G. and Mandel, J. (1953), "Measurement of Treadware of Commercial Tires," *Rubber Age*, 2, 73.

Spearman, C. (1904), "The Proof and Measurement for Association Between Two Things," *American Journal of Psychology*, 15, 72-101.

Kaufman, L., and Rock, I. (1962), "The Moon Illusion," *Science*, 136, 953-961.

8

Designed Experiments

Luck is the residue of design.

Branch Rickey, former owner of the Brooklyn Dodgers (1881-1965)

This chapter deals with the nonparametric statistical analysis of designed experiments. The classical parametric methods in analysis of variance, from one-way to multi-way tables, often suffer from a sensitivity to the effects of non-normal data. The nonparametric methods discussed here are much more robust. In most cases, they mimic their parametric counterparts but focus on analyzing ranks instead of response measurements in the experimental outcome. In this way, the chapter represents a continuation of the rank tests presented in the last chapter.

We cover the *Kruskal-Wallis test* to compare three or more samples in an analysis of variance, the *Friedman test* to analyze two-way analysis of variance (ANOVA) in a "randomized block" design, and nonparametric tests of variances for three or more samples.

8.1 KRUSKAL-WALLIS TEST

The Kruskal-Wallis (KW) test is a logical extension of the Wilcoxon-Mann-Whitney test. It is a nonparametric test used to compare three or more samples. It is used to test the null hypothesis that all populations have identical distribution functions against the alternative hypothesis that at least two

(a) (b)

Fig. 8.1 William Henry Kruskal (1919 –); Wilson Allen Wallis (1912–1998)

of the samples differ only with respect to location (median), if at all.

The KW test is the analogue to the F-test used in the one-way ANOVA. While analysis of variance tests depend on the assumption that all populations under comparison are independent and normally distributed, the Kruskal-Wallis test places no such restriction on the comparison. Suppose the data consist of k independent random samples with sample sizes $n_1, \dots, n_k$. Let $n = n_1 + \cdots + n_k$.

sample 1	X_{11},	X_{12},	...	X_{1,n_1}
sample 2	X_{21},	X_{22},	...	X_{2,n_2}
⋮	⋮			
sample $k-1$	$X_{k-1,1}$,	$X_{k-1,2}$,	...	$X_{k-1,n_{k-1}}$
sample k	X_{k1},	X_{k2},	...	X_{k,n_k}

Under the null hypothesis, we can claim that all of the k samples are from a common population. The expected sum of ranks for the sample i, $\mathbb{E}(R_i)$, would be n_i times the expected rank for a single observation. That is, $n_i(n+1)/2$, and the variance can be calculated as $\mathbb{V}\mathrm{ar}(R_i) = n_i(n+1)(n-n_i)/12$. One way to test H_0 is to calculate $R_i = \sum_{j=1}^{n_i} r(X_{ij})$ – the total sum of ranks in sample i. The statistic

$$\sum_{i=1}^{k} \left[R_i - \frac{n_i(n+1)}{2} \right]^2, \tag{8.1}$$

will be large if the samples differ, so the idea is to reject H_0 if (8.1) is "too large". However, its distribution is a jumbled mess, even for small samples, so there is little use in pursuing a direct test. Alternatively we can use the normal approximation

$$\frac{R_i - \mathbb{E}(R_i)}{\sqrt{\mathbb{V}\text{ar}(R_i)}} \overset{\text{appr}}{\sim} \mathcal{N}(0,1) \Rightarrow \sum_{i=1}^{k} \frac{(R_i - \mathbb{E}(R_i))^2}{\mathbb{V}\text{ar}(R_i)} \overset{\text{appr}}{\sim} \chi^2_{k-1},$$

where the χ^2 statistic has only $k-1$ degrees of freedom due to the fact that only $k-1$ ranks are unique.

Based on this idea, Kruskal and Wallis (1952) proposed the test statistic

$$H' = \frac{1}{S^2}\left[\sum_{i=1}^{k} \frac{R_i^2}{n_i} - \frac{n(n+1)^2}{4}\right], \tag{8.2}$$

where

$$S^2 = \frac{1}{n-1}\left[\sum_{i=1}^{k}\sum_{j=1}^{n_i} r(X_{ij})^2 - \frac{n(n+1)^2}{4}\right].$$

If there are no ties in the data, (8.2) simplifies to

$$H = \frac{12}{n(n+1)} \sum_{i=1}^{k} \frac{1}{n_i}\left[R_i - \frac{n_i(n+1)}{2}\right]^2. \tag{8.3}$$

They showed that this statistic has an approximate χ^2 distribution with $k-1$ degrees of freedom.

The MATLAB routine

```
kruskal_wallis
```

implements the KW test using a vector to represent the responses and another to identify the population from which the response came. Suppose we have the following responses from three treatment groups:

$$(1,3,4), \quad (3,4,5), \quad (4,4,4,6,5)$$

be a sample from 3 populations. The MATLAB code for testing the equality of locations of the three populations computes a p-value of 0.1428.

```
> data   =  [ 1 3 4  3 4 5  4 4 4 6 5 ];
> belong =  [ 1 1 1  2 2 2  3 3 3 3 3 ];
> [H, p] = kruskal_wallis(data, belong)

 [H, p] =
         3.8923       0.1428
```

Example 8.1 The following data are from a classic agricultural experiment measuring crop yield in four different plots. For simplicity, we identify the treatment (plot) using the integers {1,2,3,4}. The third treatment mean measures far above the rest, and the null hypothesis (the treatment means are equal) is rejected with a p-value less than 0.0002.

```
> data= [83 91 94 89 89 96 91 92 90 84 91 90 81 83 84 83 ...
      88 91 89 101 100 91 93 96 95 94 81 78  82  81 77 79 81 80];
> belong = [1 1 1 1 1 1 1 1 1 1 2 2 2 2 2 2 2 2 2 ...
             3 3 3 3 3 3 3 3 4 4 4 4 4 4 4 4];
>> [H, p] = kruskal_wallis(data, belong)

 H =
     20.3371

 p =
     1.4451e-004
```

Kruskal-Wallis Pairwise Comparisons. If the KW test detects treatment differences, we can determine if two particular treatment groups (say i and j) are different at level α if

$$\left| \frac{R_i}{n_i} - \frac{R_j}{n_j} \right| > t_{n-k,1-\alpha/2} \sqrt{\frac{S^2(n-1-H')}{n-k} \cdot \left(\frac{1}{n_i} + \frac{1}{n_j} \right)}. \tag{8.4}$$

Example 8.2 We decided the four crop treatments were statistically different, and it would be natural to find out which ones seem better and which ones seem worse. In the table below, we compute the statistic

$$T = \frac{\left| \frac{R_i}{n_i} - \frac{R_j}{n_j} \right|}{\sqrt{\frac{S^2(n-1-H')}{n-k} \left(\frac{1}{n_i} + \frac{1}{n_j} \right)}}$$

for every combination of $1 \le i \ne j \le 4$, and compare it to $t_{30,0.975} = 2.042$.

(i,j)	1	2	3	4
1	0	1.856	1.859	5.169
2	1.856	0	3.570	3.363
3	1.859	3.570	0	6.626
4	5.169	3.363	6.626	0

This shows that the third treatment is the best, but not significantly different from the first treatment, which is second best. Treatment 2, which is third best is not significantly different from Treatment 1, but is different from Treatment 4 and Treatment 3.

Fig. 8.2 Milton Friedman (1912–2006)

8.2 FRIEDMAN TEST

The *Friedman Test* is a nonparametric alternative to the randomized block design (RBD) in regular ANOVA. It replaces the RBD when the assumptions of normality are in question or when variances are possibly different from population to population. This test uses the ranks of the data rather than their raw values to calculate the test statistic. Because the Friedman test does not make distribution assumptions, it is not as powerful as the standard test if the populations are indeed normal.

Milton Friedman published the first results for this test, which was eventually named after him. He received the Nobel Prize for Economics in 1976 and one of the listed breakthrough publications was his article "The Use of Ranks to Avoid the Assumption of Normality Implicit in the Analysis of Variance", published in 1937.

Recall that the RBD design requires repeated measures for each block at each level of treatment. Let X_{ij} represent the experimental outcome of subject (or "block") i with treatment j, where $i = 1, \dots, b$, and $j = 1, \dots, k$.

	Treatments			
Blocks	1	2	$\cdots$	k
1	X_{11}	X_{12}	$\cdots$	X_{1k}
2	X_{21}	X_{22}	$\cdots$	X_{2k}
$\vdots$	$\vdots$	$\vdots$		$\vdots$
b	X_{b1}	X_{b2}	$\cdots$	X_{bk}

To form the test statistic, we assign ranks $\{1, 2, \dots, k\}$ to each *row* in the table of observations. Thus the expected rank of any observation under H_0 is $(k+1)/2$. We next sum all the ranks by columns (by treatments) to obtain $R_j = \sum_{i=1}^{b} r(X_{ij})$, $1 \le j \le k$. If H_0 is true, the expected value for R_j is

$\mathbb{E}(R_j) = b(k+1)/2$. The statistic

$$\sum_{j=1}^{k} \left(R_j - \frac{b(k+1)}{2} \right)^2,$$

is an intuitive formula to reveal treatment differences. It has expectation $bk(k^2-1)/12$ and variance $k^2 b(b-1)(k-1)(k+1)^2/72$. Once normalized to

$$S = \frac{12}{bk(k+1)} \sum_{j=1}^{k} \left(R_j - \frac{b(k+1)}{2} \right)^2, \tag{8.5}$$

it has moments $\mathbb{E}(S) = k-1$ and $\mathbb{V}\text{ar}(S) = 2(k-1)(b-1)/b \approx 2(k-1)$, which coincide with the first two moments of χ^2_{k-1}. Higher moments of S also approximate well those of χ^2_{k-1} when b is large.

In the case of ties, a modification to S is needed. Let $C = bk(k+1)^2/4$ and $R^* = \sum_{i=1}^{b} \sum_{j=1}^{k} r(X_{ij})^2$. Then,

$$S' = \frac{k-1}{R^* - bk(k+1)^2/4} \left(\sum_{j=1}^{k} R_j^2 - bC \right), \tag{8.6}$$

is also approximately distributed as χ^2_{k-1}.

Although the Friedman statistic makes for a sensible, intuitive test, it turns out there is a better one to use. As an alternative to S (or S'), the test statistic

$$F = \frac{(b-1)S}{b(k-1) - S}$$

is approximately distributed as $F_{k-1,(b-1)(k-1)}$, and tests based on this approximation are generally superior to those based on chi-square tests that use S. For details on the comparison between S and F, see Iman and Davenport (1980).

Example 8.3 In an evaluation of vehicle performance, six professional drivers, (labelled I,II,III,IV,V,VI) evaluated three cars (A, B, and C) in a randomized order. Their grades concern only the performance of the vehicles and supposedly are not influenced by the vehicle brand name or similar exogenous information. Here are their rankings on the scale 1–10:

Car	I	II	III	IV	V	VI
A	7	6	6	7	7	8
B	8	10	8	9	10	8
C	9	7	8	8	9	9

To use the MATLAB procedure

```
friedman(data),
```

the first input vector represents blocks (drivers) and the second represents treatments (cars).

```
> data = [7  8  9;    6  10  7;    6  8  8; ...
          7  9  8;    7  10  9;    8  8  9];
> [S,F,pS,pF] = friedman(data)

S =
    8.2727
F =
    11.0976
pS =
    0.0160
pF =
    0.0029    % this p-value is more reliable
```

Friedman Pairwise Comparisons. If the p-value is small enough to warrant multiple comparisons of treatments, we consider two treatments i and j to be different at level α if

$$|R_i - R_j| > t_{(b-1)(k-1),1-\alpha/2}\sqrt{2 \cdot \frac{bR^* - \sum_{j=1}^{k} R_j^2}{(b-1)(k-1)}}. \tag{8.7}$$

Example 8.4 From Example 8.3, the three cars (A,B,C) are considered significantly different at test level $\alpha = 0.01$ (if we use the F-statistic). We can use the MATLAB procedure

```
friedman_pairwise_comparison(x,i,j,a)
```

to make a pairwise comparison between treatment i and treatment j at level a. The output $= 1$ if the treatments i and j are different, otherwise it is 0. The Friedman pairwise comparison reveals that car A is rated significantly lower than both car B and car C, but car B and car C are not considered to be different.

An alternative test for k matched populations is the test by Quade (1966), which is an extension of the Wilcoxon signed-rank test. In general, the *Quade test* performs no better than Friedman's test, but slightly better in the case $k = 3$. For that reason, we reference it but will not go over it in any detail.

8.3 VARIANCE TEST FOR SEVERAL POPULATIONS

In the last chapter, the test for variances from two populations was achieved with the nonparametric *Conover Test.* In this section, the test is extended to three or more populations using a set-up similar to that of the Kruskal-Wallis test. For the hypotheses H_0 : k variances are equal versus H_1 : some of the variances are different, let n_i = the number of observations sampled from each population and X_{ij} is the jth observation from population i. We denote the following:

- $n = n_1 + \cdots + n_k$
- $\bar{x}_i$ = sample average for i^{th} population
- $R(x_{ij})$ = rank of $(x_{ij} - \bar{x}_i)^2$ among n items
- $T_i = \sum_{j=1}^{n_i} R(x_{ij})^2$
- $\bar{T} = n^{-1} \sum_{j=1}^{k} T_j$
- $V_T = (n-1)^{-1} \left(\sum_i \sum_j R(x_{ij})^4 - n\bar{T}^2 \right)$

Then the test statistic is

$$T = \frac{\sum_{j=1}^{k} (T_j^2/n_j) - n\bar{T}^2}{V_T}. \tag{8.8}$$

Under H_0, T has an approximate χ^2 distribution with $k-1$ degrees of freedom, so we can test for equal variances at level α by rejecting H_0 if $T > \chi_{k-1}^2(1-\alpha)$. Conover (1999) notes that the asymptotic relative efficiency, relative to the regular test for different variances is 0.76 (when the data are actually distributed normally). If the data are distributed as *double-exponential,* the A.R.E. is over 1.08.

Example 8.5 For the crop data in the Example 8.1, we can apply the variance test and obtain $n = 34$, $T_1 = 3845$, $T_2 = 4631$, $T_3 = 4032$, $T_4 = 1174.5$, and $\bar{T} = 402.51$. The variance term $V_T = \left(\sum_i \sum_j R(x_{ij})^4 - 34(402.51)^2 \right) /33 = 129{,}090$ leads to the test statistic

$$T = \frac{\sum_{j=1}^{k} (T_j^2/n_j) - 34(402.51)^2}{V_T} = 4.5086.$$

Using the approximation that $T \sim {\chi^2}_3$ under the null hypothesis of equal variances, the p-value associated with this test is $P(T > 4.5086) = 0.2115$. There is no strong evidence to conclude the underlying variances for crop yields are significantly different.

Multiple Comparisons. If H_0 is rejected, we can determine which populations have unequal variances using the following paired comparisons:

$$\left|\frac{T_i}{n_i} - \frac{T_j}{n_j}\right| > \sqrt{\left(\frac{1}{n_i} + \frac{1}{n_j}\right) V_T \left(\frac{n-1-\bar{T}}{n-k}\right)}\; t_{n-k}(1-\alpha/2),$$

where $t_{n-k}(\alpha)$ is the α quantile of the t distribution with $n-k$ degrees of freedom. If there are no ties, $\bar{T}$ and V_T are simple constants: $\bar{T} = (n+1)(2n+1)/6$ and $V_T = n(n+1)(2n+1)(8n+11)/180$.

8.4 EXERCISES

8.1. Show, that when ties are not present, the Kruskal-Wallis statistic H' in (8.2) coincides with H in (8.3).

8.2. Generate three samples of size 10 from an exponential distribution with $\lambda = 0.10$. Perform both the F-test and the Kruskal-Wallis test to see if there are treatment differences in the three groups. Repeat this 1000 times, recording the p-value for both tests. Compare the simulation results by comparing the two histograms made from these p-values. What do the results mean?

8.3. The data set Hypnosis contains data from a study investigating whether hypnosis has the same effect on skin potential (measured in millivolts) for four emotions (Lehmann, p. 264). Eight subjects are asked to display fear, joy, sadness, and calmness under hypnosis. The data are recorded as one observation per subject for each emotion.

```
1 fear 23.1  1 joy 22.7  1 sadness 22.5  1 calmness 22.6
2 fear 57.6  2 joy 53.2  2 sadness 53.7  2 calmness 53.1
3 fear 10.5  3 joy  9.7  3 sadness 10.8  3 calmness  8.3
4 fear 23.6  4 joy 19.6  4 sadness 21.1  4 calmness 21.6
5 fear 11.9  5 joy 13.8  5 sadness 13.7  5 calmness 13.3
6 fear 54.6  6 joy 47.1  6 sadness 39.2  6 calmness 37.0
7 fear 21.0  7 joy 13.6  7 sadness 13.7  7 calmness 14.8
8 fear 20.3  8 joy 23.6  8 sadness 16.3  8 calmness 14.8
```

8.4. The points-per-game statistics from the 1993 NBA season were analyzed for basketball players who went to college in four particular ACC schools: Duke, North Carolina, North Carolina State, and Georgia Tech. We want to find out if scoring is different for the players from different schools. Can this be analyzed with a parametric procedure? Why or why not? The classical F-test that assumes normality of the populations yields $F = 0.41$ and H_0 is not rejected. What about the nonparametric procedure?

Duke	UNC	NCSU	GT
7.5	5.5	16.9	7.9
8.7	6.2	4.5	7.8
7.1	13.0	10.5	14.5
18.2	9.7	4.4	6.1
	12.9	4.6	4.0
	5.9	18.7	14.0
	1.9	8.7	
		15.8	

8.5. Some varieties of nematodes (roundworms that live in the soil and are frequently so small they are invisible to the naked eye) feed on the roots of lawn grasses and crops such as strawberries and tomatoes. This pest, which is particularly troublesome in warm climates, can be treated by the application of nematocides. However, because of size of the worms, it is difficult to measure the effectiveness of these pesticides directly. To compare four nematocides, the yields of equal-size plots of one variety of tomatoes were collected. The data (yields in pounds per plot) are shown in the table. Use a nonparametric test to find out which nematocides are different.

Nematocide A	Nematocide B	Nematocide C	Nematocide D
18.6	18.7	19.4	19.0
18.4	19.0	18.9	18.8
18.4	18.9	19.5	18.6
18.5	18.5	19.1	18.7
17.9		18.5	

8.6. An experiment was run to determine whether four specific firing temperatures affect the density of a certain type of brick. The experiment led to the following data. Does the firing temperature affect the density of the bricks?

Temperature	Density					
100	21.8	21.9	21.7	21.7	21.6	21.7
125	21.7	21.4	21.5	21.4		
150	21.9	21.8	21.8	21.8	21.6	21.5
175	21.9	21.7	21.8	21.4		

8.7. A chemist wishes to test the effect of four chemical agents on the strength of a particular type of cloth. Because there might be variability from one bolt to another, the chemist decides to use a randomized block design, with the bolts of cloth considered as blocks. She selects five bolts and applies all four chemicals in random order to each bolt. The resulting tensile strengths follow. How do the effects of the chemical agents differ?

Chemical	Bolt No. 1	Bolt No. 2	Bolt No. 3	Bolt No. 4	Bolt No. 5
1	73	68	74	71	67
2	73	67	75	72	70
3	75	68	78	73	68
4	73	71	75	75	69

8.8. The venerable auction house of Snootly & Snobs will soon be putting three fine 17th-and 18th-century violins, A, B, and C, up for bidding. A certain musical arts foundation, wishing to determine which of these instruments to add to its collection, arranges to have them played by each of 10 concert violinists. The players are blindfolded, so that they cannot tell which violin is which; and each plays the violins in a randomly determined sequence (BCA, ACB, etc.)

The violinists are not informed that the instruments are classic masterworks; all they know is that they are playing three different violins. After each violin is played, the player rates the instrument on a 10-point scale of overall excellence (1 = lowest, 10 = highest). The players are told that they can also give fractional ratings, such as 6.2 or 4.5, if they wish. The results are shown in the table below. For the sake of consistency, the $n = 10$ players are listed as "subjects."

	Subject									
Violin	1	2	3	4	5	6	7	8	9	10
A	9	9.5	5	7.5	9.5	7.5	8	7	8.5	6
B	7	6.5	7	7.5	5	8	6	6.5	7	7
C	6	8	4	6	7	6.5	6	4	6.5	3

8.9. From Exercise 8.5, test to see if the underlying variances for the four plot yields are the same. Use a test level of $\alpha = 0.05$.

REFERENCES

Friedman, M. (1937), "The Use of Ranks to Avoid the Assumption of Normality Implicit in the Analysis of Variance," *Journal of the American Statistical Association*, 32, 675-701.

Iman, R. L., and Davenport, J. M. (1980), "Approximations of the Critical Region of the Friedman Statistic," *Communications in Statistics A: Theory and Methods*, 9, 571-595.

Kruskal, W. H. (1952), "A Nonparametric Test for the Several Sample Problem," *Annals of Mathematical Statistics,* **23**, 525-540.

Kruskal W. H., and Wallis W. A. (1952), "Use of Ranks in One-Criterion Variance Analysis," *Journal of the American Statistical Association*, 47, 583-621.

Lehmann, E. L. (1975), *Testing Statistical Hypotheses*, New York: Wiley.

Quade, D. (1966), "On the Analysis of Variance for the k-sample Population," *Annals of Mathematical Statistics,* 37, 1747-1785.

9

Categorical Data

> Statistically speaking, U.S. soldiers have less of a chance of dying from all causes in Iraq than citizens have of being murdered in California, which is roughly the same geographical size. California has more than 2300 homicides each year, which means about 6.6 murders each day. Meanwhile, U.S. troops have been in Iraq for 160 days, which means they're incurring about 1.7 deaths, including illness and accidents each day.[1]
>
> Brit Hume, Fox News, August 2003.

A *categorical* variable is a variable which is nominal or ordinal in scale. Ordinal variables have more information than nominal ones because their levels can be ordered. For example, an automobile could be categorized in an ordinal scale (compact, mid-size, large) or a nominal scale (Honda, Buick, Audi). Opposed to interval data, which are quantitative, nominal data are *qualitative*, so comparisons between the variables cannot be described mathematically. Ordinal variables are more useful than nominal ones because they can possibly be ranked, yet they are not quite quantitative. Categorical data analysis is seemingly ubiquitous in statistical practice, and we encourage readers who are interested in a more comprehensive coverage to consult monographs by

[1]By not taking the total population of each group into account, Hume failed to note the relative risk of death (Section 9.2) to a soldier in Iraq was 65 times higher than the murder rate in California.

Agresti (1996) and Simonoff (2003).

At the turn of the 19th century, while probabilists in Russia, France and other parts of the world were hastening the development of statistical theory through probability, British academics made great methodological developments in statistics through applications in the biological sciences. This was due in part from the gush of research following Charles Darwin's publication of *The Origin of Species* in 1859. Darwin's theories helped to catalyze research in the variations of traits within species, and this strongly affected the growth of applied statistics and biometrics. Soon after, Gregor Mendel's previous findings in genetics (from over a generation before Darwin) were "rediscovered" in light of these new theories of evolution.

(a) (b)

Fig. 9.1 Charles Darwin (1843–1927), Gregor Mendel (1780–1880)

When it comes to the development of statistical methods, two individuals are dominant from this era: Karl Pearson and R. A. Fisher. Both were cantankerous researchers influenced by William S. Gosset, the man who derived the (Student's) t distribution. Karl Pearson, in particular, contributed seminal results to the study of categorical data, including the chi-square test of statistical significance (Pearson, 1900). Fisher used Mendel's theories as a framework for the research of biological inheritance[2]. Both researchers were motivated by problems in heredity, and both played an interesting role in its promotion.

Fisher, an upper-class British conservative and intellectual, theorized the decline of western civilization due to the diminished fertility of the upper classes. Pearson, his rival, was a staunch socialist, yet ironically advocated a "war on inferior races", which he often associated with the working class. Pearson said, "no degenerate and feeble stock will ever be converted into

[2]Actually, Fisher showed statistically that Mendel's data were probably fudged a little in order to support the theory for his new genetic model. See Section 9.2.

(a)

(b)

(c)

Fig. 9.2 Karl Pearson (1857–1936), William Sealy Gosset (a.k.a. Student) (1876–1937), and Ronald Fisher (1890–1962)

healthy and sound stock by the accumulated effects of education, good laws and sanitary surroundings." Although their research was undoubtedly brilliant, racial bigotry strongly prevailed in western society during this colonial period, and scientists were hardly exceptional in this regard.

9.1 CHI-SQUARE AND GOODNESS-OF-FIT

Pearson's chi-square statistic found immediate applications in biometry, genetics and other life sciences. It is introduced in the most rudimentary science courses. For instance, if you are at a party and you meet a college graduate of the social sciences, it's likely one of the few things they remember about the required statistics class they suffered through in college is the term "chi-square".

To motivate the chi-square statistic, let $X_1, X_2, \dots, X_n$ be a sample from any distribution. As in Chapter 6, we would like to test the goodness-of-fit hypothesis $H_0 : F_X(x) = F_0(x)$. Let the domain of the distribution $D = (a, b)$ be split into r non-overlapping intervals, $I_1 = (a, x_1]$, $I_2 = (x_1, x_2]$ ... $I_r = (x_{r-1}, b)$. Such intervals have (theoretical) probabilities $p_1 = F_0(x_1) - F_0(a)$, $p_2 = F_0(x_2) - F_0(x_1)$, ..., $p_r = F_0(b) - F_0(x_{r-1})$, under H_0.

Let $n_1, n_2, \dots, n_r$ be observed frequencies of intervals $I_1, I_2, \dots, I_r$. In this notation, n_1 is the number of elements of the sample $X_1, \dots, X_n$ that falls into the interval I_1. Of course, $n_1 + \cdots + n_r = n$ because the intervals are a partition of the domain of the sample. The discrepancy between observed frequencies n_i and theoretical frequencies np_i is the rationale for forming the statistic

$$X^2 = \sum_{i=1}^{r} \frac{(n_i - np_i)^2}{np_i}, \tag{9.1}$$

that has a chi-square (χ^2) distribution with $r-1$ degrees of freedom. Large values of X^2 are critical for H_0. Alternative representations include

$$X^2 = \sum_{i=1}^{r} \frac{n_i^2}{np_i} - n \quad \text{and} \quad X^2 = n\left[\sum_{i=1}^{r} \left(\frac{\hat{p}_i}{p_i}\right) \hat{p}_i - 1\right],$$

where $\hat{p}_i = n_i/n$.

In some experiments, the distribution under H_0 cannot be *fully* specified; for example, one might conjecture the data are generated from a normal distribution without knowing the exact values of μ or σ^2. In this case, the unknown parameters are estimated using the sample.

Suppose that k parameters are estimated in order to fully specify F_0. Then, the resulting statistic in (9.1) has a χ^2 distribution with $r-k-1$ degrees of freedom. A degree of freedom is lost with the estimation of a parameter. In fairness, if we estimated a parameter and then inserted it into the hypothesis without further acknowledgment, the hypothesis will undoubtedly fit the data at least as well as any alternative hypothesis we could construct with a known parameter. So the lost degree of freedom represents a form of handicapping.

There is no orthodoxy in selecting the categories or even the number of categories to use. If possible, make the categories approximately equal in probability. Practitioners may want to arrange interval selection so that all $np_i > 1$ and that at least 80% of the np_i's exceed 5. The rule-of-thumb is: $n \geq 10$, $r \geq 3$, $n^2/r \geq 10$, $np_i \geq 0.25$.

As mentioned in Chapter 6, the chi-square test is not altogether efficient for testing known continuous distributions, especially compared to individualized tests such as Shapiro-Wilk or Anderson-Darling. Its advantage is manifest with discrete data and special distributions that cannot be fit in a Kolmogorov-type statistical test.

Example 9.1 Mendel's Data. In 1865, Mendel discovered a basic genetic code by breeding green and yellow peas in an experiment. Because the yellow pea gene is dominant, the first generation hybrids all appeared yellow, but the second generation hybrids were about 75% yellow and 25% green. The green color reappears in the second generation because there is a 25% chance that two peas, both having a yellow and green gene, will contribute the green gene to the next hybrid seed. In another pea experiment[3] that considered both color and texture traits, the outcomes from repeated experiments came out as in Table 9.11

[3] See Section 16.1 for more detail on probability models in basic genetics.

Table 9.11 Mendel's Data

Type of Pea	Observed Number	Expected Number
Smooth Yellow	315	313
Wrinkled Yellow	101	104
Smooth Green	108	104
Wrinkled Green	32	35

The statistical analysis shows a strong agreement with the hypothesized outcome with a p-value of 0.9166. While this, by itself, is not sufficient proof to consider foul play, Fisher noted this kind of result in a sequence of several experiments. His "meta-analysis" (see Chapter 6) revealed a p-value around 0.00013.

```
>> o=[315 101 108 32];
>> th=[313 104 104 35];
>> sum( (o-th).^2 ./ th )

ans =
   0.5103

>> 1-chi2cdf( 0.5103, 4 - 1)

ans =
   0.9166
```

Example 9.2 Horse-Kick Fatalities. During the latter part of the nineteenth century, Prussian officials collected information on the hazards that horses posed to cavalry soldiers. A total of 10 cavalry corps were monitored over a period of 20 years. Recorded for each year and each corps was X, the number of fatalities due to kicks. Table 9.12 shows the distribution of X for these 200 "corps-years".

Altogether there were 122 fatalities (109(0) + 65 (1) + 22(2) + 3(3) + 1(4)), meaning that the observed fatality *rate* was $122/200 = 0.61$ fatalities per corps-year. A Poisson model for X with a mean of $\mu = .61$ was proposed by von Bortkiewicz (1898). Table 9.12 shows the expected frequency corresponding to $x = 0, 1, \dots,$ etc., assuming the Poisson model for X was correct. The agreement between the observed and the expected frequencies is remarkable. The MATLAB procedure below shows that the resulting X^2 statistic = 0.6104. If the Poisson distribution is correct, the statistic is distributed χ^2 with 3 degrees of freedom, so the p-value is computed $P(W > 0.6104) = 0.8940$.

```
>> o=[109 65 22 3 1];
```

Table 9.12 Horse-kick fatalities data

x	Observed Number of Corps-Years	Expected Number of Corps-Years
0	109	108.7
1	65	66.3
2	22	20.2
3	3	4.1
4	1	0.7
	200	200

```
>> th=[108.7 66.3 20.2 4.1 0.7];
>> sum( (o-th).^2 ./ th )
   ans = 0.6104

>> 1-chi2cdf( 0.6104, 5 - 1 - 1)
   ans = 0.8940
```

Example 9.3 Benford's Law. Benford's law (Benford, 1938; Hill, 1998) concerns relative frequencies of leading digits of various data sets, numerical tables, accounting data, etc. Benford's law, also called *the first digit law*, states that in numbers from many sources, the leading digit 1 occurs much more often than the others (namely about 30% of the time). Furthermore, the higher the digit, the less likely it is to occur as the leading digit of a number. This applies to figures related to the natural world or of social significance, be it numbers taken from electricity bills, newspaper articles, street addresses, stock prices, population numbers, death rates, areas or lengths of rivers or physical and mathematical constants.

To be precise, Benford's law states that the leading digit n, $(n = 1, \ldots, 9)$ occurs with probability $P(n) = \log_{10}(n+1) - \log_{10}(n)$, approximated to three digits in the table below.

Digit n	1	2	3	4	5	6	7	8	9
$P(n)$	0.301	0.176	0.125	0.097	0.079	0.067	0.058	0.051	0.046

The table below lists the distribution of the leading digit for all 307 numbers appearing in a particular issue of *Reader's Digest*. With p-value of 0.8719, the support for H_0 (The first digits in *Reader's Digest* are distributed according to Benford's Law) is strong.

Digit	1	2	3	4	5	6	7	8	9
count	103	57	38	23	20	21	17	15	13

The agreement between the observed digit frequencies and Benford's distribution is good. The MATLAB calculation shows that the resulting X^2 statistic is 3.8322. Under H_0, X^2 is distributed as χ^2_8 and more extreme values of X^2 are quite likely. The p-value is almost 90%.

```
>> x = [103  57  38  23  20  21  17   15  13];
>> e = 307*[0.301  0.176  0.125  0.097 0.079 ...
     0.067  0.058  0.051 0.046];
>> sum((x-e).^2 ./ e)
      ans = 3.8322
>> 1 - chi2cdf(3.8322, 8)
      ans = 0.8719
```

9.2 CONTINGENCY TABLES: TESTING FOR HOMOGENEITY AND INDEPENDENCE

Suppose there are m populations (more specifically, m levels of factor A: $(R_1, \ldots, R_m)$ under consideration. Furthermore, each observation can be classified in a different ways, according to another factor B, which has k levels $(C_1, \ldots, C_k)$. Let n_{ij} be the number of all observations at the ith level of A and j^{th} level of B. We seek to find out if the populations (from A) and treatments (from B) are independent. If we treat the levels of A as population groups and the levels of B as treatment groups, there are

$$n_{i\cdot} = \sum_{j=1}^{k} n_{ij}$$

observations in population i, where $i = 1, \ldots, m$. Each of the treatment groups is represented

$$n_{\cdot j} = \sum_{i=1}^{n} n_{ij},$$

times, and the total number of observations is

$$n_{1\cdot} + \cdots + n_{m\cdot} = n_{\cdot\cdot}.$$

The following table summarizes the above description.

	C_1	C_2	$\cdots$	C_k	Total
R_1	n_{11}	n_{12}		n_{1k}	$n_{1\cdot}$
R_2	n_{21}	n_{22}		n_{2k}	$n_{2\cdot}$
R_m	n_{m1}	n_{m2}		n_{mk}	$n_{m\cdot}$
Total	$n_{\cdot 1}$	$n_{\cdot 2}$		$n_{\cdot k}$	$n_{\cdot\cdot}$

We are interested in testing independence of factors A and B, represented by their respective levels $R_1, \ldots, R_m$ and $C_1, \ldots, C_k$, on the basis of observed frequencies n_{ij}. Recall the definition of independence of component random variables X and Y in the random vector (X, Y),

$$P(X = x_i, Y = y_j) = P(X = x_i) \cdot P(Y = y_j).$$

Assume that the random variable ξ is to be classified. Under the hypothesis of independence, the cell probabilities $P(\xi \in R_i \cap C_j)$ should be equal to the product of probabilities $P(\xi \in R_i) \cdot P(\xi \in C_j)$. Thus, to test the independence of factors A and B, we should evaluate how different the sample counterparts of cell probabilities

$$\frac{n_{ij}}{n_{\cdot\cdot}}$$

are from the product of marginal probability estimators:

$$\frac{n_{i\cdot}}{n_{\cdot\cdot}} \cdot \frac{n_{\cdot j}}{n_{\cdot\cdot}},$$

or equivalently, how different the observed frequencies, n_{ij}, are from the expected (under the hypothesis of independence) frequencies

$$\hat{n}_{ij} = n_{\cdot\cdot} \frac{n_{i\cdot}}{n_{\cdot\cdot}} \frac{n_{\cdot j}}{n_{\cdot\cdot}} = \frac{n_{i\cdot} \cdot n_{\cdot j}}{n_{\cdot\cdot}}.$$

The measure of discrepancy, defined as

$$X^2 = \sum_{i=1}^{m} \sum_{j=1}^{k} \frac{(n_{ij} - \hat{n}_{ij})^2}{\hat{n}_{ij}}, \tag{9.2}$$

and under the assumption of independence, (9.2) has a χ^2 distribution with $(m-1)(k-1)$ degrees of freedom. Here is the rationale: the observed frequencies n_{ij} are distributed as multinomial $\mathcal{M}n(n_{\cdot\cdot}; \theta_{11}, \ldots, \theta_{mk})$, where $\theta_{ij} = P(\xi \in R_i \cap C_j)$.

	C_1	C_2	$\cdots$	C_k	Total
R_1	θ_{11}	θ_{12}		θ_{1k}	$\theta_{1\cdot}$
R_2	θ_{21}	θ_{22}		θ_{2k}	$\theta_{2\cdot}$
R_m	θ_{m1}	θ_{m2}		θ_{mk}	$\theta_{m\cdot}$
Total	$\theta_{\cdot 1}$	$\theta_{\cdot 2}$		$\theta_{\cdot k}$	1

The corresponding likelihood is $L = \prod_{i=1}^{m} \prod_{j=1}^{k} (\theta_{ij})^{n_{ij}}, \sum_{i,j} \theta_{ij} = 1$. The null hypothesis of independence states that for any pair i, j, the cell probability is the product of marginal probabilities, $\theta_{ij} = \theta_{i\cdot} \cdot \theta_{\cdot j}$. Under H_0 the likelihood becomes

$$L = \prod_{i=1}^{m} \prod_{j=1}^{n} (\theta_{i\cdot} \cdot \theta_{\cdot j})^{n_{ij}}, \ \sum_i \theta_{i\cdot} = \sum_j \theta_{\cdot j} = 1.$$

If the estimators of $\theta_{i\cdot}$ and $\theta_{\cdot j}$ are $\hat{\theta}_{i\cdot} = n_{i\cdot}/n_{\cdot\cdot}$ and $\hat{\theta}_{\cdot j} = n_{\cdot j}/n_{\cdot\cdot}$, respectively, then, under H_0, the observed frequency n_{ij} should be compared to its theoretical counterpart,

$$\hat{n}_{ij} = \hat{\theta}_{ij} n_{\cdot\cdot} = \frac{n_{i\cdot} n_{\cdot j}}{n_{\cdot\cdot}}.$$

As the n_{ij} are binomially distributed, they can be approximated by the normal distribution, and the χ^2 forms when they are squared. The statistic is based on $(m-1)+(k-1)$ estimated parameters, $\theta_{i\cdot}$, $i = 1, \dots, m-1$, and $\theta_{\cdot j}$, $j = 1, \dots, k-1$. The remaining parameters are determined: $\theta_{m\cdot} = 1 - \sum_{i=1}^{m-1} \theta_{i\cdot}$, $\theta_{\cdot n} = 1 - \sum_{n=1}^{k-1} \theta_{\cdot j}$. Thus, the chi-square statistic

$$\chi^2 = \sum_{i=1}^{m} \sum_{j=1}^{k} \frac{(n_{ij} - \hat{n}_{ij})^2}{\hat{n}_{ij}}$$

has $mk - 1 - (m-1) - (k-1) = (m-1)(k-1)$ degrees of freedom.

Pearson first developed this test but mistakenly used $mk - 1$ degrees of freedom. It was Fisher (1922), who later deduced the correct degrees of freedom, $(m-1)(k-1)$. This probably did not help to mitigate the antagonism in their professional relationship!

Example 9.4 Icelandic Dolphins. From Rasmussen and Miller (2004), groups of dolphins were observed off the coast in Iceland, and their frequency of observation was recorded along with the time of day and the perceived activity of the dolphins at that time. Table 9.13 provides the data. To see if the activity is independent of the time of day, the MATLAB procedure

```
tablerxc(X)
```

takes the input table X and computes the χ^2 statistic, its associated p-value, and a table of expected values under the assumption of independence. In this example, the activity and time of day appear to be dependent.

Table 9.13 Observed Groups of Dolphins, Including *Time of Day* and *Activity*

Time-of-Day	Traveling	Feeding	Socializing
Morning	6	28	38
Noon	6	4	5
Afternoon	14	0	9
Evening	13	56	10

```
>>[chi2,pvalue,exp]=tablerxc([6 28 38; 6 4 5; 14 0 9; 13 56 10])

chi2 =
  68.4646

pvalue =
 8.4388e-013

exp =
  14.8571   33.5238   23.6190
   3.0952    6.9841    4.9206
   4.7460   10.7090    7.5450
  16.3016   36.7831   25.9153
```

Relative Risk. In simple 2×2 tables, the comparison of two proportions might be more important if those proportions veer toward zero or one. For example, a procedure that decreases production errors from 5% to 2% could be much more valuable than one that decreases errors in another process from 45% to 42%. For example, if we revisit the example introduced at the start of the chapter, the rate of murder in California is compared to the death rate of U.S. military personnel in Iraq in 2003. The relative risk, in this case, is rather easy to understand (even to the writers at Fox News), if overly simplified.

	Killed	Not Killed	Total
California	6.6	37,999,993.4	38,000,000
Iraq	1.7	149,998.3	150,000
Total	8.3	38,149,981.7	

Here we define the *relative risk* as the risk of death in Iraq (for U.S. soldiers) divided by the risk of murder for citizens of California. For example, McWilliams and Piotrowski (2005) determined the rate of 6.6 Californian homicide victims (out of 38,000,000 at risk) per day. On the other hand, there were 1.7 average daily military related deaths in Iraq (with 150,000 solders at risk).

$$\frac{\theta_{11}}{\theta_{11}+\theta_{12}}\left(\frac{\theta_{21}}{\theta_{21}+\theta_{22}}\right)^{-1} = \frac{1.7}{150,000}\left(\frac{6.6}{38,000,000}\right)^{-1} = 65.25.$$

Fixed Marginal Totals. The categorical analysis above was developed based on assuming that each observation is to be classified according to the stochastic nature of the two factors. It is actually common, however, to have either row or column totals fixed. If row totals are fixed, for example, we are observing $n_{j.}$ observations distributed into k bins, and essentially comparing multinomial observations. In this case we are testing differences in the multinomial parameter sets. However, if we look at the experiment this way (where $n_{.j}$ is fixed) the test statistic and rejection region remain the same. This is also true if *both* row and column totals are fixed. This is less common; for example, if $m = k = 2$, this is essentially Fisher's exact test.

9.3 FISHER EXACT TEST

Along with Pearson, R. A. Fisher contributed important new methods for analyzing categorical data. Pearson and Fisher both recognized that the statistical methods of their time were not adequate for small categorized samples, but their disagreements are more well known. In 1922, Pearson, used his position as editor of *Biometrika* to attack Fisher's use of the chi-squared test. Fisher attacked Pearson with equal fierceness. While at University College, Fisher continued to criticize Pearson's ideas even after his passing. With Pearson's son Egon also holding a chair there, the departmental politics were awkward, to say the least.

Along with his original concept of maximum likelihood estimation, Fisher pioneered research in small sample analysis, including a simple categorical data test that bears his name (*Fisher Exact Test*). Fisher (1966) described a test based on the claims of a British woman who said she could taste a cup of tea, with milk added, and identify whether the milk or tea was added to the cup first. She was tested with eight cups, of which she knew four had the tea added first, and four had the milk added first. The results are listed below.

	Lady's Guess		
First Poured	Tea	Milk	Total
Tea	3	1	4
Milk	1	3	4
Total	4	4	

Both *marginal totals* are fixed at four, so if X is the number of times the woman guessed tea was poured first when, in truth, tea *was* poured first, then X determines the whole table, and under the null hypothesis (that she is just guessing), X has a hypergeometric distribution with PMF

$$p_X(x) = \frac{\binom{4}{x}\binom{4}{4-x}}{\binom{8}{4}}.$$

To see this more easily, count the number of ways to choose x cups from the correct 4, and the remaining $4 - x$ cups from the incorrect 4 and divide by the total number of ways to choose 4 cups from the 8 total. The lady guessed correctly $x = 3$ times. In this case, because the only better guess is all four, the p-value is $P(X = 3) + P(X = 4) = 0.229 + 0.014 = 0.243$. Because the sample is so small, not much can be said of the experimental results.

In general, the Fisher exact test is based on the null hypothesis that two factors, each with two factor levels, are independent, conditional on fixing marginal frequencies for *both* factors (e.g., the number of times tea was poured first and the number of times the lady guesses that tea was poured first).

9.4 MC NEMAR TEST

Quinn McNemar's expertise in statistics and psychometrics led to an influential textbook titled *Psychological Statistics*. The McNemar test (McNemar, 1947b) is a simple way to test *marginal homogeneity* in 2×2 tables. This is not a regular contingency table, so the usual analysis of contingency tables would not be applicable.

Consider such a table that, for instance, summarizes agreement between 2 evaluators choosing only two grades 0 and 1, so in the table below, a represents the number of times that both evaluators graded an outcome with 0. The marginal totals, unlike the Fisher Exact Test, are not fixed.

	0	1	total
0	a	b	$a+b$
1	c	d	$c+d$
total	$a+c$	$b+d$	$a+b+c+d$

Marginal homogeneity (i.e., the graders give the same proportion of zeros and ones, on average) implies that row totals should be close to the corresponding column totals, or

$$\begin{aligned} a+b &\approx a+c \\ c+d &\approx b+d. \end{aligned} \tag{9.3}$$

More formally, suppose that a matched pair of Bernoulli random variables (X, Y) is to be classified into a table,

	0	1	marginal
0	θ_{00}	θ_{01}	$\theta_{0\cdot}$
1	θ_{10}	θ_{11}	$\theta_{1\cdot}$
marginal	$\theta_{\cdot 0}$	$\theta_{\cdot 1}$	1

in which $\theta_{ij} = P(X = i, Y = j)$, $\theta_{i\cdot} = P(X = i)$ and $\theta_{\cdot j} = P(Y = j)$, for $i, j \in \{0, 1\}$. The null hypothesis H_0 can be expressed as a hypothesis of symmetry

$$H_0 : \quad \theta_{01} = P(X = 0, Y = 1) = P(X = 1, Y = 0) = \theta_{10}, \tag{9.4}$$

but after adding $\theta_{00} = P(X = 0, Y = 0)$ or $\theta_{11} = P(X = 1, Y = 1)$ to the both sides in (9.4), we get H_0 in the form of marginal homogeneity,

$$\begin{aligned} H_0 : \quad \theta_{0\cdot} = P(X = 0) &= P(Y = 0) = \theta_{\cdot 0}, \quad \text{or equivalently} \\ H_0 : \quad \theta_{1\cdot} = P(X = 1) &= P(Y = 1) = \theta_{\cdot 1}\ . \end{aligned}$$

As a and d on both sides of (9.3) cancel out, implying $b \approx c$. A sensible test statistic for testing H_0 might depend on how much b and c differ. The values of a and d are called ties and do not contribute to the testing of H_0.

When, $b + c > 20$, the McNemar statistic is calculated as

$$X^2 = \frac{(b-c)^2}{b+c},$$

which has a χ^2 distribution with 1 degree of freedom. Some authors recommend a version of the McNemar test with a correction for discontinuity, calculated as $X^2 = (|b - c| - 1)^2/(b + c)$, but there is no consensus among experts that this statistic is better.

If $b + c < 20$, a simple statistics

$$T = b$$

can be used. If H_0 is true, $T \sim \mathcal{B}in(b + c, 1/2)$ and testing is as in the sign-test. In some sense, what the standard two-sample paired t-test is for normally distributed responses, the McNemar test is for paired binary responses.

Example 9.5 A study by Johnson and Johnson (1972) involved 85 patients with Hodgkin's disease. Hodgkin's disease is a cancer of the lymphatic system; it is known also as a lymphoma. Each patient in the study had a sibling who did not have the disease. In 26 of these pairs, both individuals had a tonsillectomy (T). In 37 pairs, neither of the siblings had a tonsillectomy (N). In 15 pairs, only the individual with Hodgkin's had a tonsillectomy and in 7 pairs, only the non-Hodgkin's disease sibling had a tonsillectomy.

	Sibling/T	Sibling/N	Total
Patient/T	26	15	41
Patient/N	7	37	44
Total	33	52	85

The pairs (X_i, Y_i), $i = 1, \ldots, 85$ represent siblings – one of which is a patient with Hodgkin's disease (X) and the second without the disease (Y). Each of the siblings is also classified (as $T = 1$ or $N = 0$) with respect to having a tonsillectomy.

	$Y = 1$	$Y = 0$
$X = 1$	26	15
$X = 0$	7	37

The test we are interested in is based on $H_0 : P(X = 1) = P(Y = 1)$, i.e., that the probabilities of siblings having a tonsillectomy are the same with and without the disease. Because $b + c > 20$, the statistic of choice is

$$\chi^2 = \frac{(b - c)^2}{b + c} = 8^2/(7 + 15) = 2.9091.$$

The p-value is $p = P(W \geq 2.9091) = 0.0881$, where $W \sim \chi^2_1$. Under H_0, $T = 15$ is a realization of a binomial $\mathcal{B}in(22, 0.5)$ random variable and the p value is $2 \cdot P(T \geq 15) = 2 \cdot P(T > 14) = 0.1338$, that is,

```
>> 2 * (1-binocdf(14, 22, 0.5))
ans =
    0.1338
```

With such a high p-value, there is scant evidence to reject the null hypothesis of homogeneity of the two groups of patients with respect to having a tonsillectomy.

9.5 COCHRAN'S TEST

Cochran's (1950) test is essentially a randomized block design (RBD), as described in Chapter 8, but the responses are dichotomous. That is, each treatment-block combination receives a 0 or 1 response.

If there are only two treatments, the experimental outcome is equivalent to McNemar's test with marginal totals equaling the number of blocks. To see this, consider the last example as a collection of dichotomous outcomes; each of the 85 patients are initially classified into two blocks depending on whether the patient had or had not received a tonsillectomy. The response is 0 if the patient's sibling did not have a tonsillectomy and 1 if they did.

Example 9.6 Consider the software debugging data in Table 9.14. Here the software reviewers (A,B,C,D,E) represent five blocks, and the 27 bugs are considered to be treatments. Let the column totals be denoted $\{C_1, \cdots, C_5\}$ and denote row totals as $\{R_1, \cdots, R_{27}\}$. We are essentially testing H_0 : treatments (software bugs) have an equal chance of being discovered, versus H_a : some software bugs are more prevalent (or easily found) than others. the test statistic is

$$T_C = \frac{\sum_{j=1}^{m} \left(C_j - \frac{n}{m}\right)^2}{\left(\frac{\sum_{i=1}^{k} R_i(m-R_i)}{m(m-1)}\right)}$$

where $n = \sum C_j = \sum R_i$, $m = 5$ (blocks) and $k = 27$ treatments (software bugs). Under H_0, T_C has an approximate chi-square distribution with $m - 1$ degrees of freedom. In this example, $T_C = 17.647$, corresponding to a test p-value of 0.00145.

9.6 MANTEL-HAENSZEL TEST

Suppose that k independent classifications into a 2x2 table are observed. We could denote the ith such table by

Table 9.14 Five Reviewers Found 27 Issues in Software Example as in Gilb and Graham (1993)

A	B	C	D	E	A	B	C	D	E
1	1	1	1	1	0	0	1	0	0
1	0	1	0	1	0	0	1	0	0
1	1	1	0	1	0	0	0	1	0
1	0	1	1	1	1	1	1	0	1
1	0	1	1	1	0	0	1	0	1
1	0	1	1	1	1	0	0	0	0
1	1	1	1	1	0	1	0	0	0
1	1	1	1	1	1	0	1	1	1
0	0	1	0	0	0	0	0	0	0
1	0	1	0	0	0	0	0	0	0
0	1	0	0	0	0	1	0	0	1
1	0	0	1	1	1	0	0	0	0
1	0	1	0	1	1	0	0	0	0
0	0	1	0	1					

x_i	$r_i - x_i$	r_i
$c_i - x_i$	$n_i - r_i - c_i + x_i$	$n_i - r_i$
c_i	$n_i - c_i$	n_i

Fig. 9.3 Quinn McNemar (1900-1986), William Gemmell Cochran (1909-1980), and Nathan Mantel (1919-2002)

It is assumed that the marginal totals (r_i, n_i or just n_i) are fixed in advance and that the sampling was carried out until such fixed marginal totals are satisfied. If each of the k tables represent an independent study of the same classifications, the Mantel-Haenszel Test essentially pools the studies together in a "meta-analysis" that combines all experimental outcomes into a single statistic. For more about non-parametric approaches to this kind of problem, see the section on meta-analysis in Chapter 6.

For the ith table, p_{1i} is the proportion of subjects from the first row falling in the first column, and likewise, p_{2i} is the proportion of subjects from the 2nd row falling in the first column. The hypothesis of interest here is if the population proportions p_{1i} and p_{2i} coincide over all k experiments.

Suppose that in experiment i there are n_i observations. All items can be categorized as type 1 (r_i of them) or type 2 ($n_i - r_i$ of them). If c_i items are selected from the total of n_i items, the probability that exactly x_i of the selected items are of the type 1 is

$$\frac{\binom{r_i}{x_i} \cdot \binom{n_i - r_i}{c_i - x_i}}{\binom{n_i}{c_i}}. \tag{9.5}$$

Likewise, all items can be categorized as type A (c_i of them) or type B ($n_i - c_i$ of them). If r_i items are selected from the total of n_i items, the probability that exactly x_i of the selected are of the type A is

$$\frac{\binom{c_i}{x_i} \cdot \binom{n_i - c_i}{r_i - x_i}}{\binom{n_i}{r_i}}. \tag{9.6}$$

Of course these two probabilities are equal, i.e,

$$\frac{\binom{r_i}{x_i} \cdot \binom{n_i - r_i}{c_i - x_i}}{\binom{n_i}{c_i}} = \frac{\binom{c_i}{x_i} \cdot \binom{n_i - c_i}{r_i - x_i}}{\binom{n_i}{r_i}}.$$

These are hypergeometric probabilities with mean and variance

$$\frac{r_i \cdot c_i}{n_i}, \quad \text{and} \quad \frac{r_i \cdot c_i \cdot (n_i - r_i) \cdot (n_i - c_i)}{n_i^2 (n_i - 1)},$$

respectively. The k experiments are independent and the statistic

$$T = \frac{\sum_{i=1}^{k} x_i - \sum_{i=1}^{k} \frac{r_i c_i}{n_i}}{\sqrt{\sum_{i=1}^{k} \frac{r_i \cdot c_i \cdot (n_i - r_i) \cdot (n_i - c_i)}{n_i^2 (n_i - 1)}}} \tag{9.7}$$

is approximately normal (if n_i is large, the distributions of the x_i's are close to binomial and thus the normal approximation holds. In addition, summing over k independent experiments makes the normal approximation more accurate.) Large values of $|T|$ indicate that the proportions change across the k experiments.

Example 9.7 The three 2×2 tables provide classification of people from 3 Chinese cities, Zhengzhou, Taiyuan, and Nanchang with respect to smoking habits and incidence of lung cancer (Liu, 1992).

	Zhengzhou			Taiyuan			Nanchang		
Cancer Diagnosis:	yes	no	total	yes	no	total	yes	no	total
Smoker	182	156	338	60	99	159	104	89	193
Non-Smoker	72	98	170	11	43	54	21	36	57
Total	254	254	508	71	142	213	125	125	250

We can apply the Mantel-Haenszel Test to decide if the proportions of cancer incidence for smokers and non-smokers coincide for the three cities, i.e., $H_0 : p_{1i} = p_{2i}$ where p_{1i} is the proportion of incidence of cancer among smokers in the city i, and p_{2i} is the proportion of incidence of cancer among nonsmokers in the city i, $i = 1, 2, 3$. We use the two-sided alternative, $H_1 : p_{1i} \neq p_{2i}$ for some $i \in \{1, 2, 3\}$ and fix the type-I error rate at $\alpha = 0.10$.

From the tables, $\sum_i x_i = 182 + 60 + 104 = 346$. Also, $\sum_i r_i c_i / n_i = 338 \cdot 254/508 + 159 \cdot 71/213 + 193 \cdot 125/250 = 169 + 53 + 96.5 = 318.5$. To compute T in (9.7),

$$\begin{aligned}\sum_i \frac{r_i\, c_i\, (n_i - r_i)\, (n_i - c_i)}{n_i^2\, (n_i - 1)} &= \frac{338 \cdot 254 \cdot 170 \cdot 254}{508^2 \cdot 507} + \frac{159 \cdot 71 \cdot 54 \cdot 142}{213^2 \cdot 212} \\ &+ \frac{193 \cdot 125 \cdot 57 \cdot 125}{250^2 \cdot 249} \\ &= 28.33333 + 9 + 11.04518 = 48.37851.\end{aligned}$$

Therefore,

$$T = \frac{\sum_i x_i - \sum_i \frac{r_i\, c_i}{n_i}}{\sqrt{\sum_i \frac{r_i\, c_i\, (n_i - r_i)\, (n_i - c_i)}{n_i^2\, (n_i - 1)}}} = \frac{346 - 318.5}{\sqrt{48.37851}} \approx 3.95.$$

Because T is approximately $\mathcal{N}(0, 1)$, the p-value (via MATLAB) is

```
>> [st, p] = mantel_haenszel([182 156; 72 98; 60 99; 11 43; 104 89; 21 36])
   st = 3.9537
   p  = 7.6944e-005
```

In this case, there is clear evidence that the differences in cancer rates is not constant across the three cities.

9.7 CENTRAL LIMIT THEOREM FOR MULTINOMIAL PROBABILITIES

Let $E_1, E_2, \ldots, E_r$ be events that have probabilities $p_1, p_2, \ldots, p_r$; $\sum_i p_i = 1$. Suppose that in n independent trials the event E_i appears n_i times ($n_1 + \cdots + n_r = n$). Consider

$$\zeta^{(n)} = \left(\sqrt{\frac{n}{p_1}}\left(\frac{n_1}{n} - p_1\right), \ldots, \sqrt{\frac{n}{p_r}}\left(\frac{n_r}{n} - p_r\right)\right).$$

The vector $\zeta^{(n)}$ can be represented as

$$\zeta^{(n)} = \frac{1}{\sqrt{n}}\sum_{j=1}^{n} \xi^{(j)},$$

where components $\xi^{(j)}$ are given by ${p_i}^{-1/2}[\mathbf{1}(E_i) - p_i]$, $i = 1, \ldots, r$. Vectors $\xi^{(j)}$ are i.i.d., with $E(\xi_i^{(j)}) = {p_i}^{-1}(E\mathbf{1}(E_i) - p_i) = 0$, $E(\xi_i^{(j)})^2 = ({p_i}^{-1})p_i(1-p_i) = 1 - p_i$, and $E(\xi_i^{(j)}\xi_\ell^{(j)}) = (p_i p_\ell)^{-1/2}(E\mathbf{1}(E_i)\mathbf{1}(E_\ell) - p_i p_\ell) = -\sqrt{p_i p_\ell}$, $i \neq \ell$.

Result. When $n \to \infty$, the random vector $\zeta^{(n)}$ is asymptotically normal with mean 0 and the covariance matrix,

$$\Sigma = \begin{bmatrix} 1-p_1 & -\sqrt{p_1 p_2} & \cdots & -\sqrt{p_1 p_r} \\ -\sqrt{p_2 p_1} & 1-p_2 & \cdots & -\sqrt{p_2 p_r} \\ & & \cdots & \\ -\sqrt{p_r p_1} & -\sqrt{p_r p_2} & \cdots & 1-p_r \end{bmatrix} = I - zz',$$

where I is the $r \times r$ identity matrix and $z = (\sqrt{p_1} \;\; \sqrt{p_2} \;\;, \ldots, \sqrt{p_r})'$. The matrix Σ is singular. Indeed, $\Sigma z = z - z(z'z) = 0$, due to $z'z = 1$.

As a consequence, $\lambda = 0$ is characteristic value of Σ corresponding to a characteristic vector z. Because $|\zeta^{(n)}|^2$ is a continuous function of $\zeta^{(n)}$, its limiting distribution is the same as $|\zeta|^2$, where $|\zeta|^2$ is distributed as χ^2 with $r-1$ degrees of freedom.

This is more clear if we consider the following argument. Let Ξ be an orthogonal matrix with the first row equal to $(\sqrt{p_1}, \ldots, \sqrt{p_r})$, and the rest being arbitrary, but subject to orthogonality of Ξ. Let $\eta = \Xi\zeta$. Then $E\eta = 0$ and $\Sigma\eta = E\eta\eta' = E(\Xi\zeta)(\Xi\zeta)' = \Xi E\zeta\zeta'\Xi' = \Xi\Sigma\Xi' = I - (\Xi z)(\Xi z)'$, because $\Xi' = \Xi^{-1}$ It follows that $\Xi z = (1, 0, 0, \ldots, 0)$ and $(\Xi z)(\Xi z)'$ is a matrix with element at the position (1,1) as the only nonzero element. Thus,

$$\Sigma_\eta = I - (\Xi z)(\Xi z)' = \begin{bmatrix} 0 & 0 & 0 & \dots & 0 \\ 0 & 1 & 0 & \dots & 0 \\ 0 & 0 & 1 & \dots & 0 \\ & & & \dots & \\ 0 & 0 & 0 & \dots & 1 \end{bmatrix},$$

and $\eta_1 = 0, \quad w.p.1;\ \eta_2, \dots, \eta_r$ are i.i.d. $\mathcal{N}(0,1)$. The orthogonal transformation preserves the L_2 norm,

$$|\zeta|^2 = |\eta|^2 = \sum_{i=2}^{r} \eta_i^2 \stackrel{d}{=} \chi^2_{r-1}.$$

But, $|\zeta^{(n)}|^2 = \sum_{i=1}^{r} \frac{(n_i - np_i)^2}{np_i} \stackrel{d}{\to} |\zeta|^2$.

9.8 SIMPSON'S PARADOX

Simpson's Paradox is an example of changing the favor-ability of marginal proportions in a set of contingency tables due to aggregation of classes. In this case the manner of classification can be thought as a "lurking variable" causing seemingly paradoxical reversal of the inequalities in the marginal proportions when they are aggregated. Mathematically, there is no paradox – the set of vectors can not be ordered in the traditional fashion.

As an example of Simpson's Paradox, Radelet (1981) investigated the relationship between race and whether criminals (convicted of homicide) receive the death penalty (versus a lesser sentence) for regional Florida court cases during 1976–1977. Out of 326 defendants who were Caucasian or African-American, the table below shows that a higher percentage of Caucasian defendants (11.88%) received a death sentence than for African-American defendants (10.24%).

Race of Defendant	Death Penalty	Lesser Sentence
Caucasian	19	141
African-American	17	149
Total	36	290

What the table doesn't show you is the real story behind these statistics. The next $2 \times 2 \times 2$ table lists the death sentence frequencies categorized by the defendant's race and the (murder) victim's race. The table above is constructed by aggregating over this new category. Once the full table is shown, we see the importance of the victim's race in death penalty decisions. African-

Americans were sentenced to death more often if the victim was Caucasian (17.5% versus 12.6%) or African-American (5.8% to 0.0%). Why is this so? Because of the dramatic difference in marginal frequencies (i.e., 9 Caucasians defendants with African-American victims versus 103 African-American defendants with African-American victims). When both marginal associations point to a single conclusion (as in the table below) but that conclusion is contradicted when aggregating over a category, this is Simpson's paradox.[4]

Race of Defendant	Race of Victim	Death Penalty	Lesser Sentence
Caucasian	Caucasian	19	132
	African-American	0	9
African-American	Caucasian	11	52
	African-American	6	97

9.9 EXERCISES

9.1. Duke University has always been known for its great school spirit, especially when it comes to Men's basketball. One way that school enthusiasm is shown is by donning Duke paraphernalia including shirts, hats, shorts and sweat-shirts. A class of Duke students explored possible links between school spirit (measured by the number of students wearing paraphernalia) and some other attributes. It was hypothesized that males would wear Duke clothes more frequently than females. The data were collected on the Bryan Center walkway starting at 12:00 pm on ten different days. Each day 50 men and 50 women were tallied. Do the data bear out this claim?

	Duke Paraphernalia	No Duke Paraphernalia	Total
Male	131	369	500
Female	52	448	500
Total	183	817	1000

9.2. Gene Siskel and Roger Ebert hosted the most famous movie review shows in history. Below are their respective judgments on 43 films that were released in 1995. Each critic gives his judgment with a "thumbs

[4]Note that other covariate information about the defendant and victim, such as income or wealth, might have led to similar results

up" or "thumbs down." Do they have the same likelihood of giving a movie a positive rating?

		Ebert's Review	
		Thumbs Up	Thumbs Down
Siskel's	Thumbs Up	18	6
Review	Thumbs Down	9	10

9.3. Bickel, Hammel, and OConnell (1975) investigated whether there was any evidence of gender bias in graduate admissions at the University of California at Berkeley. The table below comes from their cross-classification of 4,526 applications to graduate programs in 1973 by gender (male or female), admission (whether or not the applicant was admitted to the program) and program (A, B, C, D, E or F). What does the data reveal?

A: Admit	Male	Female
Admitted	512	89
Rejected	313	19

B: Admit	Male	Female
Admitted	353	17
Rejected	207	8

C: Admit	Male	Female
Admitted	120	202
Rejected	205	391

D: Admit	Male	Female
Admitted	138	131
Rejected	279	244

E: Admit	Male	Female
Admitted	53	94
Rejected	138	299

F: Admit	Male	Female
Admitted	22	24
Rejected	351	317

9.4. When an epidemic of severe intestinal disease occurred among workers in a plant in South Bend, Indiana, doctors said that the illness resulted from infection with the amoeba *Entamoeba histolytica*[5]. There are actually two races of these amoebas, large and small, and the large ones were

[5]Source: J. E. Cohen (1973). Independence of Amoebas. In *Statistics by Example: Weighing Chances*, edited by F. Mosteller, R. S. Pieters, W. H. Kruskal, G. R. Rising, and R. F. Link, with the assistance of R. Carlson and M. Zelinka, p. 72. Addison-Wesley: Reading, MA.

believed to be causing the disease. Doctors suspected that the presence of the small ones might help people resist infection by the large ones. To check on this, public health officials chose a random sample of 138 apparently healthy workers and determined if they were infected with either the large or small amoebas. The table below gives the resulting data. Is the presence of the large race independent of the presence of the small one?

	Large Race		
Small Race	Present	Absent	Total
Present	12	23	35
Absent	35	68	103
Total	47	91	138

9.5. A study was designed to test whether or not aggression is a function of anonymity. The study was conducted as a field experiment on Halloween; 300 children were observed unobtrusively as they made their rounds. Of these 300 children, 173 wore masks that completely covered their faces, while 127 wore no masks. It was found that 101 children in the masked group displayed aggressive or antisocial behavior versus 36 children in unmasked group. What conclusion can be drawn? State your conclusion in terminology of the problem, using $\alpha = 0.01$.

9.6. Deathbed scenes in which a dying mother or father holds to life until after the long-absent son returns home and dies immediately after are all too familiar in movies. Do such things happen in everyday life? Are some people able to postpone their death until after an anticipated event takes place? It is believed that famous people do so with respect to their birthdays to which they attach some importance. A study by David P. Phillips (in Tanur, 1972, pp. 52-65) seems to be consistent with the notion. Phillips obtained data[6] on months of birth and death of 1251 famous Americans; the deaths were classified by the time period between the birth dates and death dates as shown in the table below. What do the data suggest?

b 6	e 5	f 4	o 3	r 2	e 1	Birth Month	a 1	f 2	t 3	e 4	r 5
90	100	87	96	101	86	119	118	121	114	113	106

[6]348 were people listed in *Four Hundred Notable Americans* and 903 are listed as foremost families in three volumes of *Who Was Who* for the years 1951–60, 1943–50 and 1897–1942.

9.7. Using a calculator mimic the MATLAB results for X^2 from Benford's law example (from p. 158). Here are some theoretical frequencies rounded to 2 decimal places:

92.41	54.06	•	29.75	24.31	•	•	15.72	14.06

Use χ^2 tables and compare X^2 with the critical χ^2 quantile at $\alpha = 0.05$.

9.8. Assume that a contingency table has two rows and two columns with frequencies of a and b in the first row and frequencies of c and d in the second row.

(a) Verify that the χ^2 test statistic can be expressed as

$$\chi^2 = \frac{(a+b+c+d)(ad-bc)^2}{(a+b)(c+d)(b+d)(a+c)}.$$

(b) Let $\hat{p}_1 = a/(a+c)$ and $\hat{p}_2 = b/(b+d)$. Show that the test statistic

$$z = \frac{(\hat{p}_1 - \hat{p}_2) - 0}{\sqrt{\frac{\bar{p}\bar{q}}{n_{\cdot 1}} + \frac{\bar{p}\bar{q}}{n_{\cdot 2}}}}, \quad \text{where } \bar{p} = \frac{a+b}{a+b+c+d}$$

and $\bar{q} = 1 - \bar{p}$, coincides with χ^2 from (a).

9.9. Generate a sample of size $n = 216$ from $\mathcal{N}(0,1)$. Select intervals by partitioning $\mathbb{R}$ at points -2.7, -2.2, -2, -1.7, -1.5, -1.2, -1, -0.8, -0.5, -0.3, 0, 0.2, 0.4, 0.9, 1, 1.4, 1.6, 1.9, 2, 2.5, and 2.8. Using a χ^2-test, confirm the normality of the sample. Repeat this procedure using sample contaminated by the Cauchy distribution in the following way: `0.95*normal_sample + 0.05*cauchy_sample`.

9.10. It is well known that when the arrival times of customers constitute a Poisson process with the rate λt, the inter-arrival times follow an exponential distribution with density $f(t) = \lambda e^{-\lambda t}$, $t \geq 0, \lambda > 0$. It is often of interest to establish that the process is Poisson because many theoretical results are available for such processes, ubiquitous in the domain of Industrial Engineering.

In the following example, $n = 109$ inter-arrival times of an arrival process were recorded, averaged ($\bar{x} = 2.5$) and categorized into time intervals as follows:

Interval	[0,1)	[1,2)	[2,3)	[3,4)	[4,5)	[5,6)	[6,∞)
Frequency	34	20	16	15	9	7	8

Test the hypothesis that the process described with the above inter-arrival times is Poisson, at level $\alpha = 0.05$. You must first estimate λ from the data.

9.11. In a long study of heart disease, the day of the week on which 63 seemingly healthy men died was recorded. These men had no history of disease and died suddenly.

Day of Week	Mon.	Tues.	Weds.	Thurs.	Fri.	Sat.	Sun.
No. of Deaths	22	7	6	13	5	4	6

(i) Test the hypothesis that these men were just as likely to die on one day as on any other. Use $\alpha = 0.05$. (ii) Explain in words what constitutes Type II error in the above testing.

9.12. Write a MATLAB function `mcnemar.m`. If $b + c \geq 20$, use the χ^2 approximation. If $b + c < 20$ use exact binomial p-values. You will need `chi2cdf` and `bincdf`. Use your program to solve exercise 9.4.

9.13. Doucet et al. (1999) compared applications to different primary care programs at Tulane University. The "Medicine/Pediatrics" program students are trained in both primary care specialties. The results for 148 survey responses, in the table below, are broken down by race. Does ethnicity seem to be a factor in program choice?

	Medical School Applicants		
Ethnicity	Medicine	Pediatrics	Medicine/Pediatrics
White	30	35	19
Black	11	6	9
Hispanic	3	9	6
Asian	9	3	8

9.14. The Donner party is the name given to a group of emigrants, including the families of George Donner and his brother Jacob, who became trapped in the Sierra Nevada mountains during the winter of 1846–47. Nearly half of the party died. The experience has become legendary as one of the most spectacular episodes in the record of Western migration in the United States. In total, of the 89 men, women and children in the Donner party, 48 survived, 41 died. The following table are gives the numbers of males/famales according their survival status:

	Male	Female
Died	32	9
Survived	23	25

Test the hypothesis that in the population of consisting of members of Donner's Party the gender and survival status were independent. Use $\alpha = 0.05$. The following table are gives the numbers of males/famales who survived according to their age (children/adults). Test the hypothesis that in the population of consisting of surviving members of Donner's Party the gender and age were independent. Use $\alpha = 0.05$.

	Adult	Children
Male	7	16
Female	10	15

Fig. 9.4 Surviving daughters of George Donner, Georgia (4 y.o.) and Eliza (3 y.o.) with their adoptive mother Mary Brunner.

Interesting facts (not needed for the solution):

- Two-thirds of the women survived; two-thirds of the men died.
- Four girls aged three and under died; two survived. No girls between the ages of 4 and 16 died.
- Four boys aged three and under died; none survived. Six boys between the ages of 4 and 16 died.
- All the adult males who survived the entrapment (Breen, Eddy, Foster, Keseberg) were fathers.
- All the bachelors (single males over age 21) who were trapped in the Sierra died. Jean-Baptiste Trudeau and Noah James survived the entrapment, but were only about 16 years old and are not considered bachelors.

9.15. West of Tokyo lies a large alluvial plain, dotted by a network of farming villages. Matui (1968) analyzed the position of the 911 houses making up one of those villages. The area studied was a rectangle, 3 km by 4 km. A grid was superimposed over a map of the village, dividing its

12 square kilometers into 1200 plots, each 100 meters on a side. The number of houses on each of those plots was recorded in a 30 by 40 matrix of data. Test the hypothesis that the distribution of number of houses per plot is Poisson. Use $\alpha = 0.05$.

Number	0	1	2	3	4	≥ 5
Frequency	584	398	168	35	9	6

Hint: Assume that parameter $\lambda = 0.76$ (approximately the ratio 911/1200). Find theoretical frequencies first. For example, the theoretical frequency for Number = 2 is $np_2 = 1200 \times 0.76^2/2! \times \exp\{-0.76\} = 162.0745$, while the observed frequency is 168. Subtract an additional degree of freedom because λ is estimated from the data.

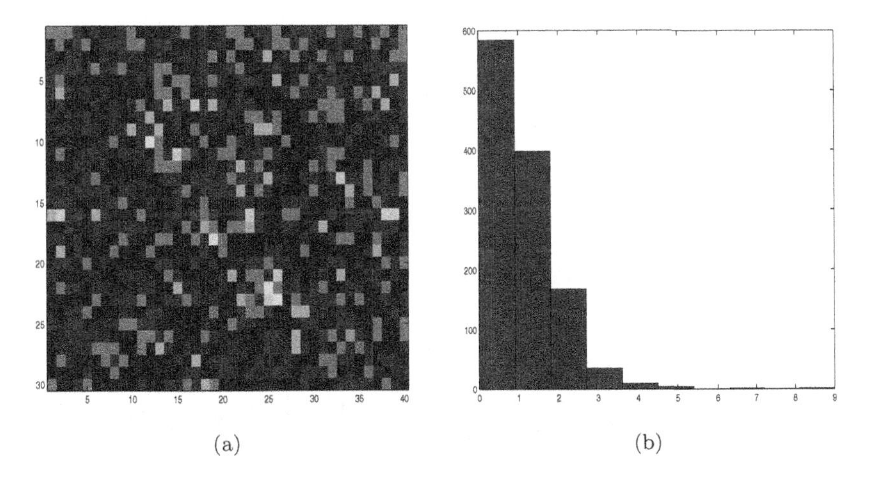

Fig. 9.5 (a) Matrix of 1200 plots (30 × 40). Lighter color corresponds to higher number of houses; (b) Histogram of number of houses per plot.

9.16. A poll was conducted to determine if perceptions of the hazards of smoking were dependent on whether or not the person smoked. One hundred people were randomly selected and surveyed. The results are given below.

	Very Dangerous [code 0]	Dangerous [code 1]	Somewhat Dangerous [code 2]	Not Dangerous [code 3]
Smokers	11 (18.13)	15 (15.19)	14 (9.80)	9 ( )
Nonsmokers	26 (18.87)	16 ( )	6 ( )	3 (6.12)

(a) Test the hypothesis that smoking status does not affect perception of the dangers of smoking at $\alpha = 0.05$ (Five theoretical/expected frequencies are given in the parentheses).

(b) Observed frequencies of perceptions of danger [codes] for smokers are

[code 0]	[code 1]	[code 2]	[code 3]
11	15	14	9

Are the codes coming from a discrete uniform distribution (i.e., each code is equally likely)? Use $\alpha = 0.01$.

REFERENCES

Agresti, A. (1992), *Categorical Data Analysis,* 2nd ed, New York: Wiley.

Benford, F. (1938), "The Law of Anomalous Numbers," *Proceedings of the American Philosophical Society*, 78, 551.

Bickel, P. J., Hammel, E. A., and O'Connell, J. W. (1975), "Sex Bias in Graduate Admissions: Data from Berkeley," *Science*, 187, 398-404.

Cochran, W. G. (1950), "The Comparison of Percentages in Matched Samples," *Biometrika*, 37, 256-266.

Darwin, C. (1859), *The Origin of Species by Means of Natural Selection*, 1st ed, London, UK: Murray.

Deonier, R. C., Tavare, S., and Waterman, M. S. (2005), *Computational Genome Analysis: An Introduction.* New York: Springer Verlag.

Doucet, H., Shah, M. K., Cummings, T. L., and Kahm, M. J. (1999), "Comparison of Internal Medicine, Pediatric and Medicine/Pediatrics Applicants and Factors Influencing Career Choices," *Southern Medical Journal*, 92, 296-299.

Fisher, R. A. (1918), "The Correlation Between Relatives on the Supposition of Mendelian Inheritance," *Philosophical Transactions of the Royal Society of Edinburgh*, 52, 399433.

_______ (1922), "On the Interpretation of Chi-Square from Contingency Tables, and the Calculation of P," *Journal of the Royal Statistical Society*, 85, 87-94.

_______ (1966), *The Design of Experiments*, 8th ed., Edinburgh, UK: Oliver and Boyd.

Gilb, T., and Graham, D. (1993), *Software Inspection*, Reading, MA: Addison-Wesley.

Hill, T. (1998), "The First Digit Phenomenon," *American Scientist*, 86, 358.

Johnson, S., and Johnson, R. (1972), "Tonsillectomy History in Hodgkin's Disease," *New England Journal of Medicine,* 287, 1122-1125.

Liu, Z. (1992), "Smoking and Lung Cancer in China: Combined Analysis of Eight Case-Control Studies," *International Journal of Epidemiology*, 21, 197-201.

Mantel, N., and Haenszel, W. (1959), "Statistical Aspects of the Analysis of Data from Retrospective Studies of Disease," *Journal of the National Cancer Institute*, 22, 719-729.

Matui, I. (1968), "Statistical Study of the Distribution of Scattered Villages in Two Regions of the Tonami Plain, Toyama Prefecture," in *Spatial Patterns,* Eds. Berry and Marble, Englewood Clifs, NJ: Prentice-Hall.

McNemar Q. (1947), "A Note on the Sampling Error of the Difference Between Correlated Proportions or Percentages," *Psychometrika*, 12, 153-157.

McWilliams, W. C. and Piotrowski , H. (2005) *The World Since 1945: A History Of International Relations*, Lynne Rienner Publishers.

_______ (1960), "At Random: Sense and Nonsense," *American Psychologist*, 15, 295-300.

_______ (1969), *Psychological Statistics,* 4th Edition, New York: Wiley.

Pearson, K. (1900), "On the Criterion that a Given System of Deviations from the Probable in the Case of a Correlated System of Variables is such that it can be Reasonably Supposed to have Arisen from Random Sampling," *Philosophical Magazine*, 50, 157 - 175.

Radelet, M. (1981), "Racial Characteristics and the Imposition of the Death Penalty," *American Sociological Review*, 46, 918-927.

Rasmussen, M. H., and Miller, L. A. (2004), "Echolocation and Social Signals from White-beaked Dolphins, Lagenorhyncus albirostris, recorded in Icelandic waters," in *Echolocation in Bats and Dolphins*, ed. J.A.Thomas, et al, Chicago: University of Chicago Press.

Simonoff, J. S. (2003), *Analyzing Categorical Data,* New York: Springer Verlag.

Tanur J. M. ed. (1972), *Statistics: A Guide to the Unknown,* San Francisco: Holden-Day.

von Bortkiewicz, L. (1898), "Das Gesetz der Kleinen Zahlen," Leipzig, Germany: Teubner.

10

Estimating Distribution Functions

The harder you fight to hold on to specific assumptions, the more likely there's gold in letting go of them.

John Seely Brown, former Chief Scientist at Xerox Corporation

10.1 INTRODUCTION

Let $X_1, X_2, \ldots, X_n$ be a sample from a population with continuous CDF F. In Chapter 3, we defined the *empirical (cumulative) distribution function* (EDF) based on a random sample as

$$F_n(x) = \frac{1}{n}\sum_{i=1}^{n} \mathbf{1}(X_i \le x).$$

Because $F_n(x)$, for a fixed x, has a sampling distribution directly related to the binomial distribution, its properties are readily apparent and it is easy to work with as an estimating function.

The EDF provides a sound estimator for the CDF, but not through any methodology that can be extended to general estimation problems in nonparametric statistics. For example, what if the sample is right truncated? Or censored? What if the sample observations are not independent or identically distributed? In standard statistical analysis, the method of *maximum likelihood* provides a general methodology for achieving inference procedures on unknown parameters, but in the nonparametric case, the unknown parameter is the function $F(x)$ (or, equivalently, the survival function $S(x) = 1 - F(x)$). Essentially, there are an infinite number of parameters. In the next section we develop a general formula for estimating the distribution function for non-i.i.d. samples. Specifically, the Kaplan-Meier estimator is constructed to estimate $F(x)$ when censoring is observed in the data.

This theme continues in Chapter 11 where we introduce *Density Estimation* as a practical alternative to estimating the CDF. Unlike the cumulative distribution, the density function provides a better visual summary of how the random variable is distributed. Corresponding to the EDF, the *empirical density function* is a discrete uniform probability distribution on the observed data, and its graph doesn't explain much about the distribution of the data. The properties of the more refined density estimators in Chapter 11 are not so easily discerned, but it will give the researcher a smoother and visually more interesting estimator to work with.

In medical research, survival analysis is the study of lifetime distributions along with associated factors that affect survival rates. The time event might be an organism's death, or perhaps the occurrence or recurrence of a disease or symptom.

10.2 NONPARAMETRIC MAXIMUM LIKELIHOOD

As a counterpart to the parametric likelihood, we define the nonparametric likelihood of the sample $X_1, \ldots, X_n$ as

$$L(F) = \prod_{i=1}^{n} \left(F(x_i) - F(x_i^-) \right), \tag{10.1}$$

where $F(x_i^-)$ is defined as $P(X < x_i)$. This framework was first introduced by Kiefer and Wolfowitz (1956).

One serious problem with this definition is that $L(F) = 0$ if F is continuous, which we might assume about the data. In order for L to be positive, the argument (F) must put positive weight (or probability mass) on every one of the observations in the sample. Even if we know F is continuous, the nonparametric maximum likelihood estimator (NPMLE) must be non-continuous at the points of the data.

For a reasonable class of estimators, we consider nondecreasing functions F that can have discrete and continuous components. Let $p_i = F(X_{i:n}) - F(X_{i-1:n})$, where $F(X_{0:n})$ is defined to be 0. We know that $p_j > 0$ is required, or else $L(F) = 0$. We also know that $p_1 + \cdots + p_n = 1$, because if the sum is less than one, there would be probability mass assigned outside the set $x_1, \ldots, x_n$. That would be impractical because if we reassigned that residual probability mass (say $q = 1 - p_1 - \cdots - p_n > 0$) to any one of the values x_i, the likelihood $L(F)$ would increase in the term $F(x_i) - F(x_i^-) = p_i + q$. So the NPMLE not only assigns probability mass to every observation, but *only* to that set, hence the likelihood can be equivalently expressed as

$$L(p_1, \ldots, p_n) = \prod_{i=1}^{n} p_i,$$

which, under the constraint that $\sum p_i = 1$, is the *multinomial* likelihood. The NPMLE is easily computed as $\hat{p}_i = 1/n, \quad i = 1, \ldots, n$. Note that this solution is quite intuitive – it places equal "importance" on all n of the observations, and it satisfies the constraint given above that $\sum p_i = 1$. This essentially proves the following theorem.

Theorem 10.1 *Let $X_1, \ldots, X_n$ be a random sample generated from F. For any distribution function F_0, the nonparametric likelihood $L(F_0) \leq L(F_n)$, so that the empirical distribution function is the nonparametric maximum likelihood estimator.*

10.3 KAPLAN-MEIER ESTIMATOR

The nonparametric likelihood can be generalized to all sorts of observed data sets beyond a simple i.i.d. sample. The most commonly observed phenomenon outside the i.i.d. case involves *censoring*. To describe censoring, we will consider $X \geq 0$, because most problems involving censoring consist of lifetime measurements (e.g., time until failure).

(a)

(b)

Fig. 10.1 Edward Kaplan (1920–2006) and Paul Meier (1924–).

Definition 10.1 *Suppose X is a lifetime measurement. X is* **right censored** *at time t if we know the failure time occurred after time t, but the actual time* *is unknown. X is* **left censored** *at time t if we know the failure time occurred before time t, but the actual time is unknown.*

Definition 10.2 Type-I censoring *occurs when n items on test are stopped at a fixed time t_0, at which time all surviving test items are taken off test and are right censored.*

Definition 10.3 Type-II censoring *occurs when n items $(X_1, \ldots, X_n)$ on test are stopped after a prefixed number of them (say, $k \leq n$) have failed, leaving the remaining items to be right censored at the random time $t = X_{k:n}$.*

Type I censoring is a common problem in drug treatment experiments based on human trials; if a patient receiving an experimental drug is known to survive up to a time t but leaves the study (and humans are known to leave such clinical trials much more frequently than lab mice) the lifetime is right censored.

Suppose we have a sample of possibly right-censored values. We will assume the random variables represent lifetimes (or "occurrence times"). The sample is summarized as $\{(X_i, \delta_i), \quad i = 1, \ldots, n\}$, where X_i is a time measurement, and δ_i equals 1 if the X_i represents the lifetime, and equals 0 if X_i is a (right) censoring time. If $\delta_i = 1$, X_i contributes $dF(x_i) \equiv F(x_i) - F(x_i^-)$ to the likelihood (as it does in the i.i.d. case). If $\delta_i = 0$, we know only that the lifetime surpassed time X_i, so this event contributes $1 - F(x_i)$ to the likelihood. Then

$$L(F) = \prod_{i=1}^{n} (1 - F(x_i))^{1-\delta_i} (dF(x_i))^{\delta_i}. \tag{10.2}$$

The argument about the NPMLE has changed from (10.1). In this case, no probability mass need be assigned to a value X_i for which $\delta_i = 0$, because in that case, $dF(X_i)$ does not appear in the likelihood. Furthermore, the accumulated probability mass of the NPMLE on the observed data does not necessarily sum to one, because if the largest value of X_i is a censored observation, the term $S(X_i) = 1 - F(X_i)$ will only be positive if probability mass is assigned to a point or interval to the right of X_i.

Let p_i be the probability mass assigned to $X_{i:n}$. This new notation allows for positive probability mass (call it p_{n+1}) that can be assigned to some arbitrary point or interval after the last observation $X_{n:n}$. Let $\tilde{\delta}_i$ be the censoring indicator associated with $X_{i:n}$. Note that even though $X_{1:n} < \cdots < X_{n:n}$ are ordered, the set $(\tilde{\delta}_1, \ldots, \tilde{\delta}_n)$ is not necessarily so ($\tilde{\delta}_i$ is called a *concomitant*).

If $\tilde{\delta}_i = 1$, the likelihood is clearly maximized by setting probability mass (say p_i) on $X_{i:n}$. If $\tilde{\delta}_i = 0$, some mass will be assigned to the right of $X_{i:n}$, which has interval probability $p_{i+1} + \cdots + p_{n+1}$. The likelihood based on censored data is expressed

$$L(p_1, \ldots, p_{n+1}) = \prod_{i=1}^{n} p_i^{\tilde{\delta}_i} \left(\sum_{j=i+1}^{n+1} p_j \right)^{1-\tilde{\delta}_i}.$$

Instead of maximizing the likelihood in terms of $(p_1, \ldots, p_{n+1})$, it will prove to be much easier using the transformation

$$\lambda_i = \frac{p_i}{\sum_{j=i}^{n+1} p_j}.$$

This is a convenient one-to-one mapping where

$$\sum_{j=i}^{n+1} p_j = \prod_{j=1}^{i-1}(1-\lambda_j), \quad p_i = \lambda_i \prod_{j=1}^{i-1}(1-\lambda_j).$$

The likelihood simplifies to

$$\begin{aligned} L(\lambda_1, \ldots, \lambda_{n+1}) &= \prod_{i=1}^{n} \left(\left(\lambda_i \prod_{j=1}^{i-1}(1-\lambda_j) \right)^{\tilde{\delta}_i} \left(\prod_{j=1}^{i}(1-\lambda_j) \right)^{1-\tilde{\delta}_i} \right) \\ &= \left(\prod_{i=1}^{n} \lambda_i^{\tilde{\delta}_i}(1-\lambda_i)^{1-\tilde{\delta}_i} \right) \left(\prod_{i=1}^{n-1}(1-\lambda_i)^{n-i} \right) \\ &= \prod_{i=1}^{n} \left(\frac{\lambda_i}{1-\lambda_i} \right)^{\tilde{\delta}_i} (1-\lambda_i)^{n-i+1}. \end{aligned}$$

As a function of $(\lambda_1, \ldots, \lambda_{n+1})$, L is maximized at $\hat{\lambda}_i = \tilde{\delta}_i/(n-i+1)$, $i = 1, \ldots, n+1$. Equivalently,

$$\hat{p_i} = \frac{\tilde{\delta}_i}{n-i+1} \prod_{j=1}^{i-1} \left(1 - \frac{\tilde{\delta}_j}{n-j+1} \right).$$

The NPMLE of the distribution function (denoted $F_{KM}(x)$) can be expressed as a sum in p_i. For example, at the observed order statistics, we see that

$$\begin{aligned} S_{KM}(x_{i:n}) &\equiv 1 - F_{KM}(x_{i:n}) = \prod_{j=1}^{i} \left(1 - \frac{1}{n-j+1} \right)^{\tilde{\delta}_j} \qquad (10.3) \\ &= \prod_{j=1}^{i} \left(1 - \frac{\tilde{\delta}_j}{n-j+1} \right). \end{aligned}$$

This is the *Kaplan–Meier* nonparametric estimator, developed by Kaplan and Meier (1958) for censored lifetime data analysis. It's been one of the most influential developments in the past century; their paper is the most cited paper in statistics (Stigler, 1994). E. L. Kaplan and Paul Meier never actually met during this time, but they both submitted their idea of the "product limit estimator" to the *Journal of the American Statistical Association* at approximately the same time, so their joint results were amalgamated through letter correspondence.

For non-censored observations, the Kaplan-Meier estimator is identical to the regular MLE. The difference occurs when there is a censored observation – then the Kaplan-Meier estimator takes the "weight" normally assigned to that observation and distributes it evenly among all observed values to the right of the observation. This is intuitive because we know that the true value of the censored observation must be somewhere to the right of the censored value, but we don't have any more information about what the exact value should be.

The estimator is easily extended to sets of data that have potential tied values. If we define d_j= number of failures at x_j, m_j = number of observations that had survived up to x_j^-, then

$$F_{KM}(t) = 1 - \prod_{x_j \le t} \left(1 - \frac{d_j}{m_j}\right). \tag{10.4}$$

Example 10.1 Muenchow (1986) tested whether male or female flowers (of *Western White Clematis*), were equally attractive to insects. The data in the Table 10.15 represent waiting times (in minutes), which includes censored data. In MATLAB, use the function

```
KMcdfSM(x,y,j)
```

where x is a vector of event times, y is a vector of zeros (indicating censor) and ones (indicating failure), and $j = 1$ indicates the vector values ordered ($j = 0$ means the data will be sorted first).

Example 10.2 Data from Crowder et al. (1991) lists strength measurements (in coded units) for 48 pieces of weathered cord. Seven of the pieces of cord were damaged and yielded strength measurements that are considered right censored. That is, because the damaged cord was taken off test, we know only the lower limit of its strength. In the MATLAB code below, vector `data` represents the strength measurements, and the vector `censor` indicates (with a zero) if the corresponding observation in `data` is censored.

```
>> data = [36.3,41.7,43.9,49.9,50.1,50.8,51.9,52.1,52.3,52.3,52.4,52.6,...
```

Table 10.15 Waiting Times for Insects to Visit Flowers

Male Flowers			Female Flowers		
1	9	27	1	19	57
1	9	27	2	23	59
2	9	30	4	23	67
2	11	31	4	26	71
4	11	35	5	28	75
4	14	36	6	29	75*
5	14	40	7	29	78*
5	14	43	7	29	81
6	16	54	8	30	90*
6	16	61	8	32	94*
6	17	68	8	35	96
7	17	69	9	35	96*
7	18	70	14	37	100*
8	19	83	15	39	102*
8	19	95	18	43	105*
8	19	102*	18	56	
		104*			

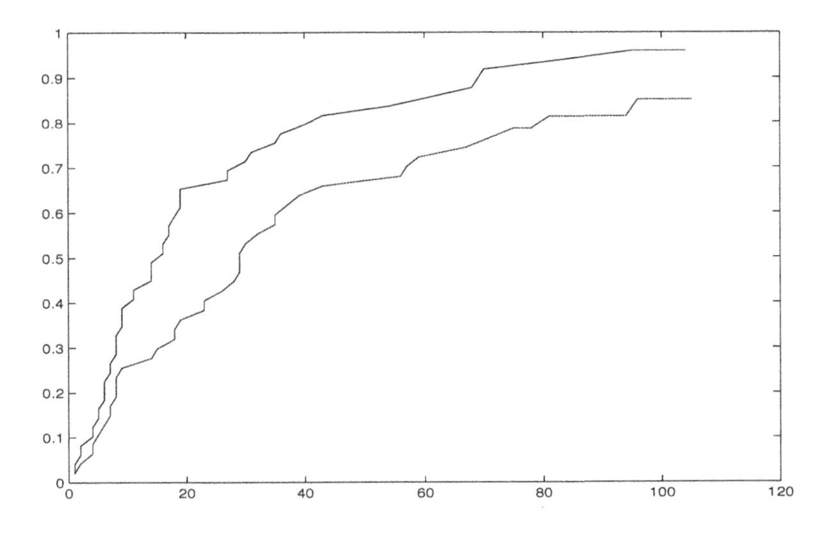

Fig. 10.2 Kaplan-Meier estimator for Waiting Times (solid line for male flowers, dashed line for female flowers).

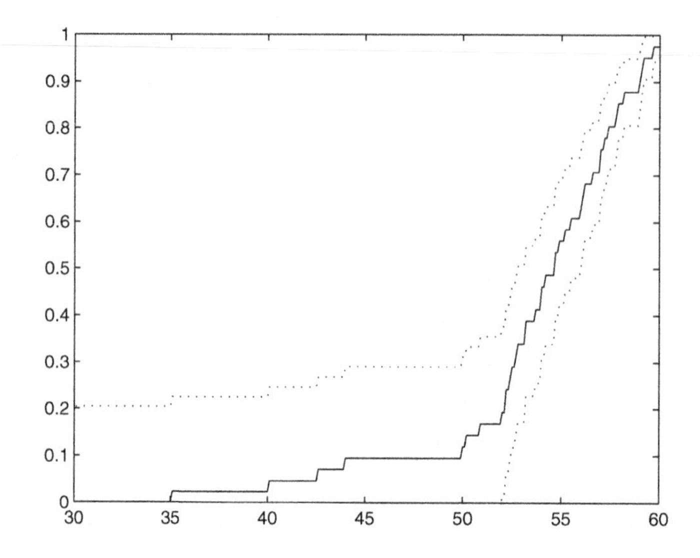

Fig. 10.3 Kaplan-Meier estimator cord strength (in coded units).

```
        52.7,53.1,53.6,53.6,53.9,53.9,54.1,54.6,54.8,54.8,55.1,55.4,55.9,...
        56.0,56.1,56.5,56.9,57.1,57.1,57.3,57.7,57.8,58.1,58.9,59.0,59.1,...
        59.6,60.4,60.7,26.8,29.6,33.4,35.0,40.0,41.9,42.5];
>> censor=[ones(1,41),zeros(1,7)];
>> [kmest,sortdat,sortcen]= kmcdfsm(data',censor',0);
>> plot(sortdat,kmest,'k');
```

The table below shows how the Kaplan-Meier estimator is calculated using the formula in (10.4) for the first 16 measurements, which includes seven censored observations. Figure 10.3 shows the estimated survival function for the cord strength data.

Uncensored	x_j	m_j	d_j	$\frac{m_j-d_j}{m_j}$	$1-F_{KM}(x_j)$
	26.8	48	0	1.000	1.000
	29.6	47	0	1.000	1.000
	33.4	46	0	1.000	1.000
	35.0	45	0	1.000	1.000
1	36.3	44	1	0.977	0.977
	40.0	43	0	1.000	0.977
2	41.7	42	1	0.976	0.954
	41.9	41	0	1.000	0.954
	42.5	40	0	1.000	0.954
3	43.9	39	1	0.974	0.930
4	49.9	38	1	0.974	0.905
5	50.1	37	1	0.973	0.881
6	50.8	36	1	0.972	0.856
7	51.9	35	1	0.971	0.832
8	52.1	34	1	0.971	0.807
9	52.3	33	2	0.939	0.758
$\vdots$	$\vdots$	$\vdots$	$\vdots$	$\vdots$	$\vdots$

Example 10.3 Consider observing the lifetime of a series system. Recall a series system is a system of $k \geq 1$ components that fails at the time the first component fails. Suppose we observe n different systems that are each made of k_i identical components ($i = 1, \ldots, n$) with lifetime distribution F. The lifetime data is denoted $(x_1, \ldots, x_n)$. Further suppose there is (random) right censoring, and $\delta_i = I(x_i$ represents a lifetime measurement). How do we estimate F?

If $F(x)$ is continuous with derivative $f(x)$, then the ith system's survival function is $S(x)^{k_i}$ and its corresponding likelihood is

$$\ell_i(F) = k_i\,(1 - F(x))^{k_i - 1}\,f(x).$$

It's easier to express the full likelihood in terms of $S(x) = 1 - F(x)$:

$$L(S) = \prod_{i=1}^{n} \left(k_i\,(S(x_i))^{k_i-1}\,f(x_i)\right)^{\delta_i} \left(S(x_i)^{k_i}\right)^{1-\delta_i},$$

where $1 - \delta$ indicates censoring.

To make the likelihood more easy to solve, let's examine the ordered sample $y_i = x_{i:n}$ so we observe $y_1 < y_2 < \cdots < y_n$. Let $\tilde{k}_i$ and $\tilde{\delta}_i$ represent the size of the series system and the censoring indicator for y_i. Note that $\tilde{k}_i$ and $\tilde{\delta}_i$ are concomitants of y_i.

The likelihood, now as a function of $(y_1, \ldots, y_n)$, is expressed

$$\begin{aligned}\tilde{L}(S) &= \prod_{i=1}^{n} \left(\tilde{k}_i \left(S(y_i)\right)^{\tilde{k}_i - 1} f(y_i)\right)^{\tilde{\delta}_i} \left(S(y_i)^{\tilde{k}_i}\right)^{1-\tilde{\delta}_i} \\ &\propto \prod_{i=1}^{n} f(y_i)^{\tilde{\delta}_i} S(y_i)^{\tilde{k}_i - \tilde{\delta}_i}.\end{aligned}$$

For estimating F nonparametrically, it is again clear that $\hat{F}$ (or $\hat{S}$) will be a step-function with jumps occurring only at points of observed system failure. With this in mind, let $S_i = S(y_i)$ and $\alpha_i = S_i/S_{i-1}$. Then $f_i = S_{i-1} - S_i = \prod_{r=1}^{i-1} \alpha_r(1-\alpha_i)$. If we let $\tau_j = \tilde{k}_j + \cdots + \tilde{k}_n$, the likelihood can be expressed simply (see Exercise 10.4) as

$$\tilde{L}(S) = \prod_{i=1}^{n} \alpha_i^{\tau_i - \tilde{\delta}_i} (1-\alpha_i)^{\tilde{\delta}_i},$$

and the nonparametric MLE for $S(x)$, in terms of the ordered system lifetimes, is

$$\hat{S}(y_i) = \prod_{r=1}^{i} \left(\frac{\tau_r - \tilde{\delta}_r}{\tau_r}\right).$$

Note the special case in which $k_i = 1$ for all i, we end up with the Kaplan-Meier estimator.

10.4 CONFIDENCE INTERVAL FOR F

Like all estimators, $\hat{F}(x)$ is only as good as its measurement of uncertainty. Confidence intervals can be constructed for $F(x)$ just as they are for regular parameters, but a typical inference procedure refers to a *pointwise* confidence interval about $F(x)$ where x is fixed.

A simple, approximate $1-\alpha$ confidence interval can be constructed using a normal approximation

$$\hat{F}(x) \pm z_{1-\alpha/2}\hat{\sigma}_{\hat{F}},$$

where $\hat{\sigma}_{\hat{F}}$ is our estimate of the standard deviation of $\hat{F}(x)$. If we have an i.i.d. sample, $\hat{F} = F_n$, and $\sigma^2_{F_n} = F(x)[1-F(x)]/n$, so that

$$\hat{\sigma}^2_{\hat{F}} = F_n(x)[1-F_n(x)]/n.$$

Recall that $nF_n(x)$ is distributed as binomial $\mathcal{B}in(n, F(x))$, and an exact interval for $F(x)$ can be constructed using the bounding procedure for the binomial parameter p in Chapter 3.

In the case of right censoring, a confidence interval can be based on the Kaplan-Meier estimator, but the variance of $F_{KM}(x)$ does not have a simple form. Greenwood's formula (Greenwood, 1926), originally concocted for grouped data, can be applied to construct a $1-\alpha$ confidence interval for the survival function ($S = 1 - F$) under right censoring:

$$S_{KM}(t_i) \pm z_{\alpha/2} \hat{\sigma}_{KM}(t_i),$$

where

$$\hat{\sigma}^2_{KM}(t_i) = \hat{\sigma}^2(S_{KM}(t_i)) = S_{KM}(t_i)^2 \sum_{t_j \le t_i} \frac{d_j}{m_j(m_j - d_j)}.$$

It is important to remember these are *pointwise* confidence intervals, based on fixed values of t in $F(t)$. Simultaneous confidence bands are a more recent phenomenon and apply as a confidence statement for F across all values of t for which $0 < F(t) < 1$. Nair (1984) showed that the confidence bands by Hall and Wellner (1980) work well in various settings, even though they are based on large-sample approximations. An approximate $1-\alpha$ confidence band for $S(t)$, for values of t less than the largest observed failure time, is

$$S_{KM}(t) \pm \sqrt{-\frac{1}{2n} \ln\left(\frac{\alpha}{2}\right)} S_{KM}(t) \left(1 + \hat{\sigma}^2_{KM}(t)\right).$$

This interval is based on rough approximation for an infinite series, and a slightly better approximation can be obtained using numerical procedures suggested in Nair (1984). Along with the Kaplan-Meier estimator of the distribution of cord strength, Figure (10.3) also shows a 95% simultaneous confidence band. The pointwise confidence interval at t=50 units is (0.8121, 0.9934). The confidence band, on the other hand, is (0.7078, 1.0000). Note that for small strength values, the band reflects a significant amount of uncertainty in $F_{KM}(x)$. See also the MATLAB procedure `survBand`.

10.5 PLUG-IN PRINCIPLE

With an i.i.d. sample, the EDF serves not only as an estimator for the underlying distribution of the data, but through the EDF, any particular parameter θ of the distribution can also be estimated. Suppose the parameter has a particular functional relationship with the distribution function F:

$$\theta = \theta(F).$$

Examples are easy to construct. The population mean, for example, can be expressed

$$\mu = \mu(F) = \int_{-\infty}^{\infty} x dF(x)$$

and variance is

$$\sigma^2 = \sigma^2(F) = \int_{-\infty}^{\infty} (x - \mu)^2 dF(x).$$

As F_n is the sample analog to F, so $\theta(F_n)$ can serve as a sample-based estimator for θ. This is the idea of the *plug-in principle.* The estimator for the population mean:

$$\hat{\mu} = \mu(F_n) = \int_{-\infty}^{\infty} x dF_n(x) = \sum_x x_i dF_n(x_i) = \bar{x}.$$

Obviously, the plug-in principle is not necessary for simply estimating the mean, but it is reassuring to see it produce a result that is consistent with standard estimating techniques.

Example 10.4 The quantile x_p can be expressed as a function of F: $x_p = \inf\{x : \int_x^{\infty} dF(x) \leq 1 - p\}$. The sample equivalent is the value $\hat{x}_p = \inf\{x : \int_x^{\infty} dF_n(x) \leq 1 - p\}$. If F is continuous, then we have $x_p = F^{-1}(p)$ and $F_n(\hat{x}_p) = p$ is solved uniquely. If F is discrete, $\hat{x}_p$ is the smallest value of x for which

$$n^{-1} \sum_{i=1}^{n} \mathbf{1}(x \leq x_i) \leq 1 - p,$$

or, equivalently, the smallest order statistic $x_{i:n}$ for which $i/n \leq p$, i.e., $(i+1)/n > p$. For example, with the flower data in Table 10.15, the median waiting times are easily estimated as the smallest values (x) for which $F_{KM}(x) \leq 1/2$, which are 16 (for the male flowers) and 29 (for the female flowers).

If the data are not i.i.d., the NPMLE $\hat{F}$ can be plugged in for F in $\theta(F)$. This is a key selling point to the plug-in principle; it can be used to formulate estimators where we might have no set rule to estimate them. Depending on the sample, $\hat{F}$ might be the EDF or the Kaplan-Meier estimator. The plug-in technique is simple, and it will form a basis for estimating uncertainty using re-sampling techniques in Chapter 15.

Example 10.5 To find the average cord strength from the censored data, for example, it would be imprudent to merely average the data, as the censored observations represent a lower bound on the data, hence the true mean will be underestimated. By using the plug in principle, we will get a more accurate estimate; the code below estimates the mean cord strength as 54.1946 (see also the MATLAB m-file pluginmu. The sample mean, ignoring the censoring indicator, is 51.4438.

```
>> [cdfy svdata svcensor ] = kmcdfsm(vdata,vcensor,ipresorted);
>> if min(svdata)>0;
      skm = 1-cdfy;    %survival function
      skm1 = [1, skm'];
      svdata2 = [0 svdata'];
      svdata3 = [svdata' svdata(end)];
      dx = svdata3 - svdata2;
      mu_hat = skm1 *dx';
  else;
      cdfy1 = [0, cdfy'];
      cdfy2 = [cdfy' 1];
      df = cdfy2 - cdfy1;
      svdata1 = [svdata', 0];
      mu_hat = svdata1 *df';
  end;
>> mu_hat

ans =
    154.1946
```

10.6 SEMI-PARAMETRIC INFERENCE

The *proportional hazards* model for lifetime data relates two populations according to a common underlying hazard rate. Suppose $r_0(t)$ is a baseline hazard rate, where $r(t) = f(t)/(1 - F(t))$. In reliability theory, $r(t)$ is called the *failure rate*. For some covariate x that is observed along with the lifetime, the positive function of $\Psi(x)$ describes how the level of x can change the failure rate (and thus the lifetime distribution):

$$r(t; x) = r_0(t)\Psi(x).$$

This is termed a *semi-parametric model* because $r_0(t)$ is usually left unspecified (and thus a candidate for nonparametric estimation) where as $\Psi(x)$ is a known positive function, at least up to some possibly unknown parameters. Recall that the CDF is related to the failure rate as

$$\int_{-\infty}^{x} r(u)du \equiv R(u) = -\ln S(x),$$

where $S(x) = 1 - F(x)$ is called the survivor function. $R(t)$ is called the *cumulative failure rate* in reliability and life testing. In this case, $S_0(t)$ is the baseline survivor function, and relates to the lifetime affected by $\Psi(x)$ as

$$S(t;x) = S_0(t)^{\Psi(x)}.$$

The most commonly used proportional hazards model used in survival analysis is called the *Cox Model* (named after Sir David Cox), which has the form

$$r(t;x) = r_0(t)e^{x'\beta}.$$

With this model, the (vector) parameter β is left unspecified and must be estimated. Suppose the baseline hazard function of two different populations are related by proportional hazards as $r_1(t) = r_0(t)\lambda$ and $r_2(t) = r_0(t)\theta$. Then if T_1 and T_2 represent lifetimes from these two populations,

$$P(T_1 < T_2) = \frac{\lambda}{\lambda + \theta}.$$

The probability does not depend at all on the underlying baseline hazard (or survivor) function. With this convenient set-up, nonparametric estimation of $S(t)$ is possible through maximizing the nonparametric likelihood. Suppose n possibly right-censored observations $(x_1, \ldots, x_n)$ from $F = 1 - S$ are observed. Let ξ_i represent the number of observations at risk just before time x_i. Then, if δ_i=1 indicates the lifetime was observed at x_i,

$$L(\beta) = \prod_{i=1}^{n} \left(\frac{e^{x'_i\beta}}{\sum_{j\in\xi_i} e^{x'_j\beta}} \right)^{\delta_i}.$$

In general, the likelihood must be solved numerically. For a thorough study of inference with a semi-parametric model, we suggest *Statistical Models and Methods for Lifetime Data* by Lawless. This area of research is paramount in survival analysis.

Related to the proportional hazard model, is the *accelerated lifetime model* used in engineering. In this case, the baseline survivor function $S_0(t)$ can represent the lifetime of a test product under usage conditions. In an accelerated life test, and additional stress is put on the test unit, such as high or low temperature, high voltage, high humidity, etc. This stress is characterized through the function $\Psi(x)$ and the survivor function of the stressed test item is

$$S(t;x) = S_0(t\Psi(x)).$$

Accelerated life testing is an important tool in product development, especially for electronics manufacturers who produce gadgets that are expected to last several years on test. By increasing the voltage in a particular way, as one example, the lifetimes can be shortened to hours. The key is how much faith the manufacturer has on the known acceleration function $\Psi(x)$.

In MATLAB, the Statistics Toolbox offers the routine `coxphfit`, which computes Cox proportional hazards estimator for input data, much in the same way the `kmcdfsm` computes the Kaplan-Meier estimator.

10.7 EMPIRICAL PROCESSES

If we express the sample as $X_1(\omega), \dots, X_n(\omega)$, we note that $F_n(x)$ is both a function of x and $\omega \in \Omega$. From this, the EDF can be treated as a random process. The Glivenko-Cantelli Theorem from Chapter 3 states that the EDF $F_n(x)$ converges to $F(x)$ (i) almost surely (as random variable, x fixed), and (ii) uniformly in x, (as a function of x with ω fixed). This can be expressed as:

$$P\left(\omega \;\middle|\; \lim_{n\to\infty} \sup_x |F_n(x) - F(x)| = 0\right) = 1.$$

Let $W(x)$ be a standard Brownian motion process. It is defined as a stochastic process for which $W(0) = 0$, $W(t) \sim \mathcal{N}(0, t)$, $W(t)$ has independent increments, and the paths of $W(t)$ are continuous. A Brownian Bridge is defined as $B(t) = W(t) - tW(1)$, $0 \le t \le 1$. Both ends of a Brownian Bridge, $B(0)$ and $B(1)$, are tied to 0, and this property motivates the name. A Brownian motion $W(x)$ has covariance function $\gamma(t,s) = t \wedge s = min(t,s)$. This is because $\mathbb{E}(W(t)) = 0, \mathbb{V}\text{ar}(W(t)) = s$, for $s < t, \mathbb{C}\text{ov}(W(t), W(s)) = \mathbb{C}\text{ov}(W(s), (W(t) - W(s)) + W(s))$ and W has independent increments.

Define the random process $B_n(x) = \sqrt{n}(F_n(x) - F(x))$. This process converges to a Brownian Bridge Process, $B(x)$, in the sense that all finite dimensional distributions of $B_n(x)$ (defined by a selection of $x_1, \dots, x_m$) converge to the corresponding finite dimensional distribution of a Brownian Bridge $B(x)$.

Using this, one can show that a Brownian Bridge has mean zero and covariance function $\gamma(t,s) = t \wedge s - ts$. If $s < t$, $\gamma(s,t) = s(1-t)$. For $s < t$, $\gamma(s,t) = \mathbb{E}(W(s) - sW(1))(W(t) - tW(1)) = \cdots = s - st$. Because the Brownian Bridge is a Gaussian process, it is uniquely determined by its second order properties. The covariance function $\gamma(t,s)$ for the process $\sqrt{n}(F_n(t) - F(t))$ is:

$$\begin{aligned}
\gamma(t,s) &= \mathbb{E}\left[\sqrt{n}(F_n(t) - F(t)) \cdot \sqrt{n}(F_n(s) - F(s))\right] \\
&= n\mathbb{E}(F_n(t) - F(t))(F_n(s) - F(s)) = \frac{1}{n}(F(t) \wedge F(s) - F(t)F(s)).
\end{aligned}$$

Proof:

$$
\begin{aligned}
\mathbb{E}\gamma(t,s) &= \mathbb{E}\left[\left(\frac{1}{n}\sum_i(\mathbf{1}(X_i<t)-F(t)\right)\cdot\left(\frac{1}{n}\sum_j(\mathbf{1}(X_j<s)-F(s)\right)\right] \\
&= \frac{1}{n^2}\mathbb{E}\left[\sum_{i,j}(\mathbf{1}(X_i<t)-F(t))(\mathbf{1}(X_j<s)-F(s))\right] \\
&= \frac{1}{n}\mathbb{E}(\mathbf{1}(X_1<t)-F(t))(\mathbf{1}(X_1<s)-F(s)) \\
&= \frac{1}{n}\mathbb{E}\left[\mathbf{1}(X_1<t\wedge s)-F(t)\mathbf{1}(X_1<s)-F(s)\mathbf{1}(X_1<t)+F(t)F(s)\right] \\
&= \frac{1}{n}(F(t\wedge s)-F(t)F(s)).
\end{aligned}
$$

This result is independent of F, as long as F is continuous, as the sample $X_1,\ldots,X_n$ could be transformed to uniform: $Y_1=F(X_1),\ldots,Y_n=F(X_n)$. Let $G_n(t)$ be the empirical distribution based on $Y_1,\ldots,Y_n$. For the uniform distribution the covariance is $\gamma(t,s)=t\wedge s-ts$, which is exactly the correlation function of the Brownian Bridge. This leads to the following result:

Theorem 10.2 *The random process $\sqrt{n}(F_n(x)-F(x))$ converges in distribution to the Brownian Bridge process.*

10.8 EMPIRICAL LIKELIHOOD

In Chapter 3 we defined the likelihood ratio based on the likelihood function $L(\theta)=\prod f(x_i;\theta)$, where $X_1,\ldots,X_n$ were i.i.d. with density function $f(x;\theta)$. The likelihood ratio function

$$
R(\theta_0)=\frac{L(\theta_0)}{\sup_\theta L(\theta)} \tag{10.5}
$$

allows us to construct efficient tests and confidence intervals for the parameter θ. In this chapter we extend the likelihood ratio to nonparametric inference, although it is assumed that the research interest lies in some parameter $\theta=\theta(F)$, where $F(x)$ is the unknown CDF.

The likelihood ratio extends naturally to nonparametric estimation. If we focus on the nonparametric likelihood from the beginning of this chapter, from an i.i.d. sample of $X_1,\ldots,X_n$ generated from $F(x)$,

$$
L(F)=\prod_{i=1}^n dF(x_i)=\prod_{i=1}^n\left(F(x_i)-F(x_i{}^-)\right).
$$

The likelihood ratio corresponding to this would be $R(F) = L(F)/L(F_n)$, where F_n is the empirical distribution function. $R(F)$ is called the *empirical likelihood ratio.* In terms of F, this ratio doesn't directly help us creating confidence intervals. All we know is that for any CDF F, $R(F) \leq 1$ and reaches its maximum only for $F = F_n$. This means we are considering only functions F that assign mass on the values $X_i = x_i$, $i = 1, \ldots, n$, and R is reduced to function of $n-1$ parameters $R(p_1, \ldots, p_{n-1})$ where $p_i = dF(x_i)$ and $\sum p_i = 1$.

It is more helpful to think of the problem in terms of an unknown parameter of interest $\theta = \theta(F)$. Recall the *plug-in principle* can be applied to estimate θ with $\hat{\theta} = \theta(F_n)$. For example, with $\mu = \int x dF(x)$ was merely the sample mean, i.e. $\int x dF_n(x) = \bar{x}$. We will focus on the mean as our first example to better understand the empirical likelihood.

Confidence Interval for the Mean. Suppose we have an i.i.d. sample $X_1, \ldots, X_n$ generated from an unknown distribution $F(x)$. In the case $\mu(F) = \int x dF(x)$, define the set $\mathcal{C}_{\mathbf{p}}(\mu)$ on $\mathbf{p} = (p_1, \ldots, p_n)$ as

$$\mathcal{C}_{\mathbf{p}}(\mu) = \left\{ \mathbf{p} : \sum_{i=1}^{n} p_i x_i = \mu, p_i \geq 0,\ i = 1, \ldots, n,\ \sum_{i=1}^{n} p_i = 1 \right\}.$$

The empirical likelihood associated with μ maximizes $L(\mu)$ over $\mathcal{C}_{\mathbf{p}}(\mu)$. The restriction $\sum p_i x_i = \mu$ is called the *structural constraint.* The empirical likelihood ratio (ELR) is this empirical likelihood divided by the unconstrained NPMLE, which is just $L(1/n, \ldots, 1/n) = n^{-n}$. If we can find a set of solutions to the empirical likelihood, Owen (1988) showed that

$$X^2 = -2 \log R(\mu) = -2 \log \left(\sup_{\mathbf{p} \in \mathcal{C}_{\mathbf{p}}} \prod_{i=1}^{n} n p_i \right)$$

is approximately distributed χ_1^2 if μ is correctly specified, so a nonparametric confidence interval for μ can be formed using the values of $-2 \log R(\mu)$.

MATLAB software is available to help: `elm.m` computes the empirical likelihood for a specific mean, allowing the user to iterate to make a curve for $R(\mu)$ and, in the process, construct confidence intervals for μ by solving $R(\mu) = r_0$ for specific values of r_0. Computing $R(\mu)$ is no simple matter; we can proceed with Lagrange multipliers to maximize $\sum p_i x_i$ subject to $\sum p_i = 1$ and $\sum \ln(n p_i) = \ln(r_0)$. The best numerical optimization methods are described in Chapter 2 of Owen (2001).

Example 10.6 Recall Exercise 6.2. Fuller et al. (1994) examined polished window strength data to estimate the lifetime for a glass airplane window. The units are ksi (or 1,000 psi). The MATLAB code below constructs the empirical likelihood for the mean glass strength, which is plotted in Figure

10.4 (a). In this case, a 90% confidence interval for μ is constructed by using the value of r_0 so that $-2\ln r_0 < \chi_1^2(0.90) = 2.7055$, or $r_0 > 0.2585$. The confidence interval is computed as (28.78 ksi, 33.02 ksi).

```
>> x = [18.83 20.8 21.657 23.03 23.23 24.05 24.321 25.5 25.52 25.8 ...
   26.69 26.77 26.78 27.05 27.67 29.9 31.11 33.2 33.73 33.76 33.89 ...
   34.76 35.75 35.91 36.98 37.08 37.09 39.58 44.045 45.29 45.381 ];
>> n=size(x); i=1;
>> for mu=min(x):0.1:max(x)
    R_mu=elm(x, mu,zeros(1,1), 100, 1e-7, 1e-9, 0 );
    ELR_mu(i)=R_mu; Mu(i)=mu; i=i+1;
  end
```

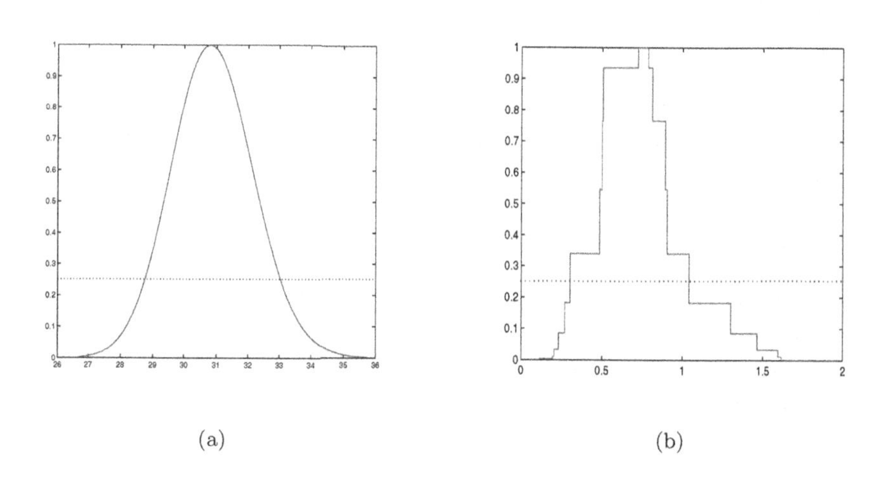

Fig. 10.4 Empirical likelihood ratio as a function of (a) the mean and (b) the median (for different samples).

Owen's extension of Wilk's theorem for parametric likelihood ratios is valid for other functions of F, including the variance, quantiles and more. To construct R for the median, we need only change the structural constraint from $\sum p_i x_i = \mu$ to $\sum p_i \operatorname{sign}(x_i - x_{0.50}) = 0$.

Confidence Interval for the Median. In general, computing $R(x)$ is difficult. For the case of estimating a population quantile, however, the optimizing becomes rather easy. For example, suppose that n_1 observations out of n are less than the population median $x_{0.50}$ and $n_2 = n - n_1$ observations are greater than $x_{0.50}$. Under the constraint $\hat{x}_{0.50} = x_{0.50}$, the nonparametric likelihood estimator assigns mass $(2n_1)^{-1}$ to each observation less than $x_{0.50}$ and assigns mass $(2n_2)^{-1}$ to each observation to the right of $x_{0.50}$, leaving us with

$$R(x_{0.50}; n_1, n_2) = \left(\frac{n}{2n_1}\right)^{n_1} \left(\frac{n}{2n_2}\right)^{n_2}.$$

Example 10.7 Figure 10.4(b), based on the MATLAB code below, shows the empirical likelihood for the median based on 30 randomly generated numbers from the exponential distribution (with μ=1 and $x_{0.50} = -\ln(0.5) = 0.6931$). A 90% confidence interval for $x_{0.50}$, again based on $r_0 > 0.2585$, is (0.3035, 0.9021).

```
>> n=30;
>> x=exprnd(1,1,n); y=sort(x);
>> n1=1:1:n; n2=n-n1;
>> R=((0.5*n./n1).^n1).*((0.5*n./n2).^n2);
>> stairs(y,R)
```

For general problems, computing the empirical likelihood is no easy matter, and to really utilize the method fully, more advanced study is needed. This section provides a modest introduction to let you know what is possible using the empirical likelihood. Students interested in further pursuing this method are recommended to read Owen's book.

10.9 EXERCISES

10.1. With an i.i.d. sample of n measurements, use the plug-in principle to derive an estimator for population variance.

10.2. Twelve people were interviewed and asked how many years they stayed at their first job. Three people are still employed at their first job and have been there for 1.5, 3.0 and 6.2 years. The others reported the following data for years at first job: 0.4, 0.9, 1.1, 1.9, 2.0, 3.3, 5.3, 5.8, 14.0. Using hand calculations, compute a nonparametric estimator for the distribution of T = time spent (in years) at first job. Verify your hand calculations using MATLAB. According to your estimator, what is the estimated probability that a person stays at their job for less than four years? Construct a 95% confidence interval for this estimate.

10.3. Using the estimator in Exercise 10.2, use the plug-in principle to compute the underlying mean number of years a person stays at their first job. Compare it to the faulty estimators based on using (a) only the noncensored items and (b) using the censored times but ignoring the censoring mechanism.

10.4. Consider Example 10.3, where we observe series-system lifetimes of a series system. We observe n different systems that are each made of k_i

identical components ($i = 1, \ldots, n$) with lifetime distribution F. The lifetime data is denoted $(\tilde{x}_1, \ldots, \tilde{x}_n)$ and are possibly right censored. Show that if we let $\tau_j = \tilde{k}_j + \cdots + \tilde{k}_n$, the likelihood can be expressed as (10.5) and solve for the nonparametric maximum likelihood estimator.

10.5. Suppose we observe m different k-out-of-n systems and each system contains i.i.d. components (with distribution F), and the ith system contains n_i components. Set up the nonparametric likelihood function for F based on the n system lifetimes (but do not solve the likelihood).

10.6. Go to the link below to download survival times for 87 people with lupus nephritis. They were followed for 15+ or more years after an initial renal biopsy. The *duration* variable indicates how long the patient had the disease before the biopsy; construct the Kaplan-Meier estimator for survival, ignoring the duration variable.

`http://lib.stat.cmu.edu/datasets/lupus`

10.7. Recall Exercise 6.3 based on 100 measurements of the speed of light in air. Use empirical likelihood to construct a 90% confidence interval for the mean and median.

`http://www.itl.nist.gov/div898/strd/univ/data/Michelso.dat`

10.8. Suppose the empirical likelihood ratio for the mean was equal to $R(\mu) = \mu \mathbf{1}(0 \leq \mu \leq 1) + (2 - \mu)\mathbf{1}(1 \leq \mu \leq 2)$. Find a 95% confidence interval for μ.

10.9. The *Receiver Operating Characteristic* (ROC) curve is a statistical tool to compare diagnostic tests. Suppose we have a sample of measurements (scores) $X_1, \ldots, X_n$ from a diseased population $F(x)$, and a sample of $Y_1, \ldots, Y_m$ from a healthy population $G(y)$. The healthy population has lower scores, so an observation is categorized as being diseased if it exceeds a given threshold value, e.g., if $X > c$. Then the rate of false-positive results would be $P(Y > c)$. The ROC curve is defined as the plot of $R(p) = F(G^{-1}(p))$. The ROC estimator can be computed using the plug-in principle:

$$\hat{R}(p) = F_n(G_m^{-1}(p)).$$

A common test to see if the diagnostic test is effective is to see if $R(p)$ remains well above 0.5 for $0 \leq p \leq 1$. The *Area Under the Curve* (AUC) is defined as

$$AUC = \int_0^1 R(p)dp.$$

Show that $AUC = P(X \leq Y)$ and show that by using the plug-in principle, the sample estimator of the AUC is equivalent to the Mann-Whitney two-sample test statistic.

REFERENCES

Brown, J. S. (1997), *What It Means to Lead*, Fast Company, 7. New York. Mansueto Ventures, LLC.

Cox, D. R. (1972), "Regression Models and Life Tables," *Journal of the Royal Statistical Society (B)*, 34, 187-220.

Crowder, M. J., Kimber, A. C., Smith, R. L., and Sweeting, T. J. (1991), *Statistical Analysis of Reliability Data*, London, Chapman & Hall.

Fuller Jr., E. R., Frieman, S. W., Quinn, J. B., Quinn, G. D., and Carter, W. C. (1994), "Fracture Mechanics Approach to the Design of Glass Aircraft Windows: A Case Study", *SPIE Proceedings*, Vol. 2286, (Society of Photo-Optical Instrumentation Engineers (SPIE), Bellingham, WA).

Greenwood, M. (1926), "The Natural Duration of Cancer," in *Reports on Public Health and Medical Subjects*, 33, London: H. M. Stationery Office.

Hall, W. J., and Wellner, J. A. (1980), "Confidence Bands for a Survival Curve," *Biometrika*, 67, 133-143.

Kaplan, E. L., and Meier, P. (1958), "Nonparametric Estimation from Incomplete Observations," *Journal of the American Statistical Association*, 53, 457-481.

Kiefer, J., and Wolfowitz, J. (1956), "Consistency of the Maximum Likelihood Estimator in the Presence of Infinitely Many Incidental Parameters," *Annals of Mathematical Statistics*, 27, 887-906.

Lawless, J. F. (1982), *Statistical Models and Methods for Lifetime Data*, New York: Wiley.

Muenchow, G. (1986), "Ecological Use of Failure Time Analysis," *Ecology* 67, 246250.

Nair, V. N. (1984), "Confidence Bands for Survival Functions with Censored Data: A Comparative Study," *Technometrics*, 26, 265-275.

Owen, A. B. (1988), "Empirical Likelihood Ratio Confidence Intervals for a Single Functional," *Biometrika*, 75, 237-249.

_______ (1990), "Empirical Likelihood Confidence Regions," *Annals of Statistics*, 18, 90-120.

_______ (2001), *Empirical Likelihood*, Boca Raton, FL: Chapman & Hall/CRC.

Stigler, S. M. (1994), "Citations Patterns in the Journals of Statistics and Probability," *Statistical Science*, 9, 94-108.

11

Density Estimation

George McFly: Lorraine, my density has brought me to you.
Lorraine Baines: What?
George McFly: Oh, what I meant to say was...
Lorraine Baines: Wait a minute, don't I know you from somewhere?
George McFly: Yes. Yes. I'm George, George McFly.
I'm your density. I mean... your destiny.

From the movie *Back to the Future*, 1985

Probability density estimation goes hand in hand with nonparametric estimation of the cumulative distribution function discussed in Chapter 10. There, we noted that the density function provides a better visual summary of how the random variable is distributed across its support. Symmetry, skewness, disperseness and unimodality are just a few of the properties that are ascertained when we visually scrutinize a probability density plot.

Recall, for continuous i.i.d. data, the *empirical density function* places probability mass $1/n$ on each of the observations. While the plot of the empirical *distribution* function (EDF) emulates the underlying distribution function, for continuous distributions the empirical density function takes no shape beside the changing frequency of discrete jumps of $1/n$ across the domain of the underlying distribution – see Figure 11.2(a).

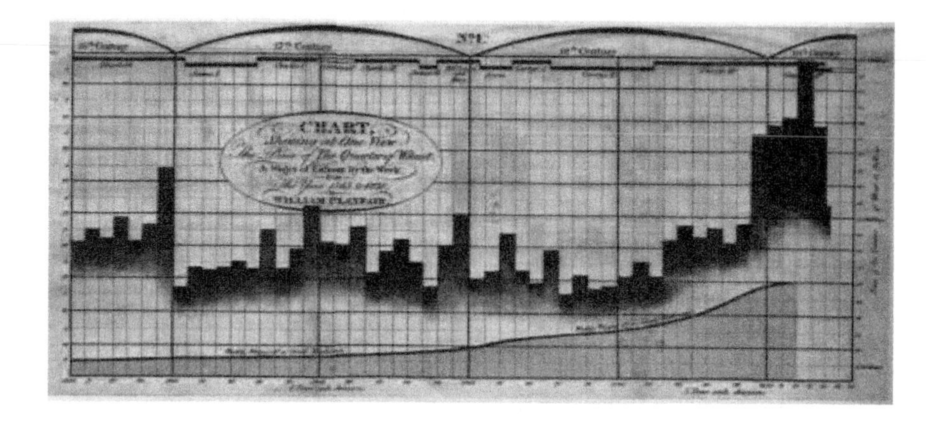

Fig. 11.1 Playfair's 1786 bar chart of wheat prices in England

11.1 HISTOGRAM

The histogram provides a quick picture of the underlying density by weighting fixed intervals according the their relative frequency in the data. Pearson (1895) coined the term for this empirical plot of the data, but its history goes as far back as the 18th century. William Playfair (1786) is credited with the first appearance of a bar chart (see Figure 11.1) that plotted the price of wheat in England through the 17th and 18th centuries.

In MATLAB, the procedure `hist(x)` will create a histogram with ten bins using the input vector `x`. Figure 11.2 shows (a) the empirical density function where vertical bars represent Dirac's point masses at the observations, and (b) a 10-bin histogram for a set of 30 generated $\mathcal{N}(0,1)$ random variables. Obviously, by aggregating observations within the disjoint intervals, we get a better, *smoother* visual construction of the frequency distribution of the sample.

```
>> x = rand_nor(0,1,30,1);
>> hist(x)
>> histfit(x,1000)
```

The histogram represents a rudimentary smoothing operation that provides the user a way of visualizing the true empirical density of the sample. Still, this simple plot is primitive, and depends on the subjective choices the user makes for bin widths and number of bins. With larger data sets, we can increase the number of bins while still keeping average bin frequency at a reasonable number, say 5 or more. If the underlying data are continuous, the histogram appears less discrete as the sample size (and number of bins) grow, but with smaller samples, the graph of binned frequency counts will not pick up the nuances of the underlying distribution.

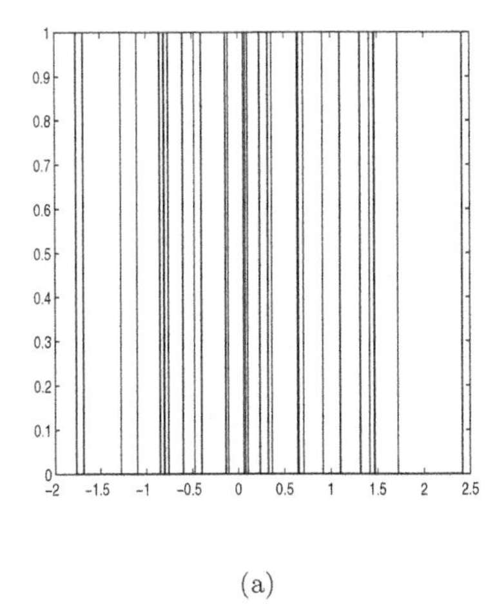

(a)

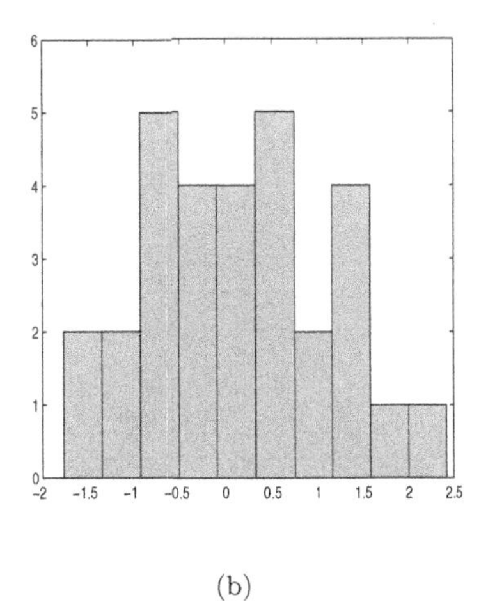

(b)

Fig. 11.2 Empirical "density" (a) and histogram (b) for 30 normal $\mathcal{N}(0, 1)$ variables.

The MATLAB function **histfit(x,n)** plots a histogram with **n** bins along with the best fitting normal density curve. Figure 11.3 shows how the appearance of continuity changes as the histogram becomes more refined (with more bins of smaller bin width). Of course, we do not have such luxury with smaller or medium sized data sets, and are more likely left to ponder the question of underlying normality with a sample of size 30, as in Figure 11.2(b).

```
>> x = rand_nor(0,1,5000,1);
>> histfit(x,10)
>> histfit(x,1000)
```

If you have no scruples, the histogram provides for you many opportunities to mislead your audience, as you can make the distribution of the data appear differently by choosing your own bin widths centered at a set of points arbitrarily left to your own choosing. If you are completely untrustworthy, you might even consider making bins of unequal length. That is sure to support a conjectured but otherwise unsupportable thesis with your data, and might jump-start a promising career for you in politics.

11.2 KERNEL AND BANDWIDTH

The idea of the *density estimator* is to spread out the weight of a single observation in a plot of the empirical density function. The histogram, then, is the picture of a density estimator that spreads the probability mass of each sample item *uniformly* throughout the interval (i.e., bin) it is observed in.

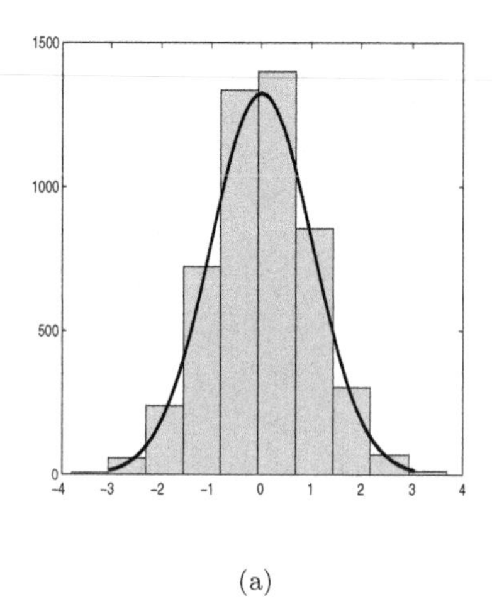

(a)

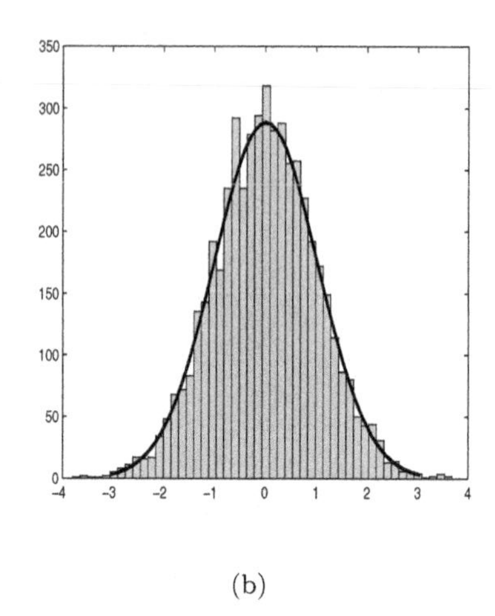

(b)

Fig. 11.3 Histograms with normal fit of 5000 generated variables using (a) 10 bins and (b) 50 bins.

Note that the observations are in no way expected to be uniformly spread out within any particular interval, so the mass is not spread equally around the observation unless it happens to fall exactly in the center of the interval.

In this chapter, we focus on the kernel density estimator that more fairly spreads out the probability mass of each observation, not arbitrarily in a fixed interval, but smoothly around the observation, typically in a symmetric way. With a sample $X_1, \ldots, X_n$, we write the density estimator

$$\hat{f}(x) = \frac{1}{nh_n} \sum_{i=1}^{n} K\left(\frac{x - x_i}{h_n}\right), \tag{11.1}$$

for $X_i = x_i$, $i = 1, \ldots, n$. The *kernel function* K represents how the probability mass is assigned, so for the histogram, it is just a constant in any particular interval. The smoothing function h_n is a positive sequence of bandwidths analogous to the bin width in a histogram.

The kernel function K has five important properties –

1. $K(x) \geq 0 \quad \forall x$
2. $K(x) = K(-x) \quad \text{for } x > 0$
3. $\int K(u)du = 1$
4. $\int uK(u)du = 0$
5. $\int u^2K(u)du = \sigma_K^2 < \infty.$

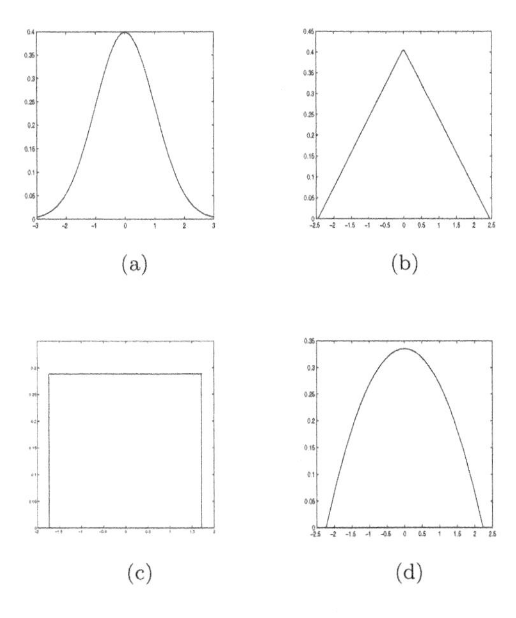

Fig. 11.4 (a) Normal, (b) Triangular, (c) Box, and (d) Epanechnickov kernel functions.

Figure 11.4 shows four basic kernel functions:

1. Normal (or Gaussian) kernel $K(x) = \phi(x)$,
2. Triangular kernel $K(x) = c^{-2}(c - |x|)\,\mathbf{1}(-c < x < c),\ c > 0.$
3. Epanechnickov kernel (described below).
4. Box kernel, $K(x) = \mathbf{1}(-c < x < c)/(2c),\ c > 0.$

While K controls the shape, h_n controls the spread of the kernel. The accuracy of a density estimator can be evaluated using the mean integrated squared error, defined as

$$\begin{aligned} \text{MISE} &= \mathbb{E}\left(\int (f(x) - \hat{f}(x))^2 dx\right) \\ &= \int \text{Bias}^2(\hat{f}(x))dx + \int \mathbb{V}\text{ar}(\hat{f}(x))dx. \end{aligned} \tag{11.2}$$

To find a density estimator that minimizes the MISE under the five mentioned constraints, we also will assume that $f(x)$ is continuous (and twice differentiable), $h_n \to 0$ and $nh_n \to \infty$ as $n \to \infty$. Under these conditions it can be shown that

$$\begin{aligned} \text{Bias}(\hat{f}(x)) &= \frac{\sigma_K^2}{2} f''(x) + O(h_n^4) \text{ and} \\ \mathbb{V}\text{ar}\left(\hat{f}(x)\right) &= \frac{f(x)R(K)}{nh_n} + O(n^{-1}), \end{aligned} \tag{11.3}$$

where $R(g) = \int g(u)^2 du$.

We determine (and minimize) the MISE by our choice of h_n. From the equations in (11.3), we see that there is a tradeoff. Choosing h_n to reduce bias will increase the variance, and vice versa. The choice of bandwidth is important in the construction of $\hat{f}(x)$. If h is chosen to be small, the subtle nuances in the main part of the density will be highlighted, but the tail of the distribution will be unseemly bumpy. If h is chosen large, the tails of the distribution are better handled, but we fail to see important characteristics in the middle quartiles of the data.

By substituting in the bias and variance in the formula for (11.2), we minimize MISE with

$$h_n^* = \left(\frac{R(K)}{\sigma_K^4 R(f')}\right)^{1/5} n^{-1/5}.$$

At this point, we can still choose $K(x)$ and insert a "representative" density for $f(x)$ to solve for the bandwidth. Epanechnickov (1969) showed that, upon substituting in $f(x) = \phi(x)$, the kernel that minimizes MISE is

$$K_E(x) = \begin{cases} \frac{3}{4}(1 - x^2) & |x| \le 1 \\ 0 & |x| > 1. \end{cases}$$

The resulting bandwidth becomes $h_n^* \approx 1.06\hat{\sigma}n^{-1/5}$, where $\hat{\sigma}$ is the sample standard deviation. This choice relies on the approximation of σ for $f(x)$. Alternative approaches, including cross-validation, lead to slightly different answers.

Adaptive kernels were derived to alleviate this problem. If we use a more general smoothing function tied to the density at x_j, we could generalize the density estimator as

$$\hat{f}(x) = \frac{1}{n}\sum_{i=1}^{n} \frac{1}{h_{n,i}} K\left(\frac{x - x_i}{h_{n,i}}\right). \tag{11.4}$$

This is an advanced topic in density estimation, and we will not further pursue learning more about optimal estimators based on adaptive kernels here. We will also leave out details about estimator limit properties, and instead point out that if h_n is a decreasing function of n, under some mild regularity conditions, $|\hat{f}(x) - f(x)| \xrightarrow{P} 0$. For details and more advanced topics in density

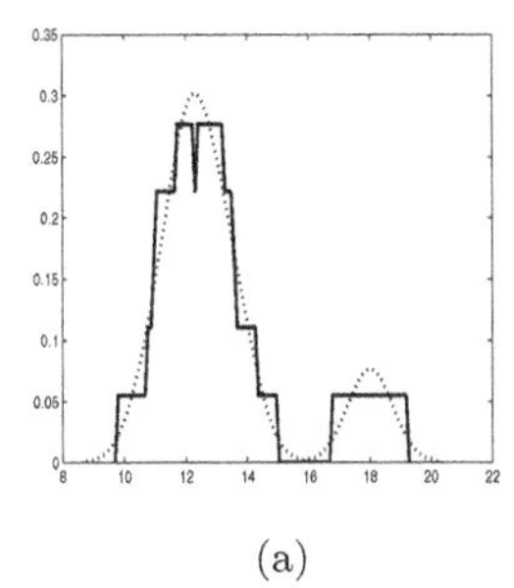

(a)

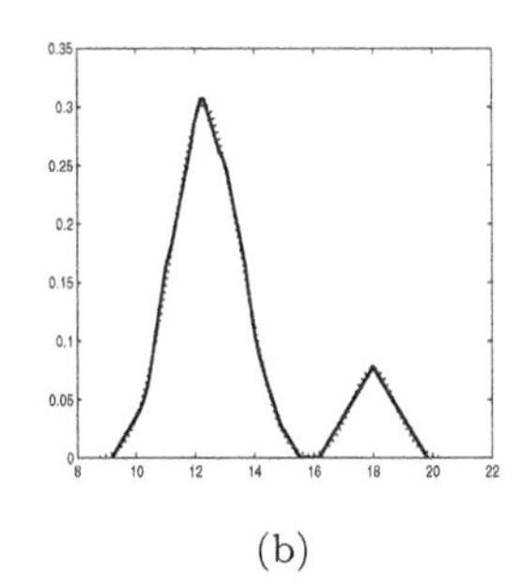

(b)

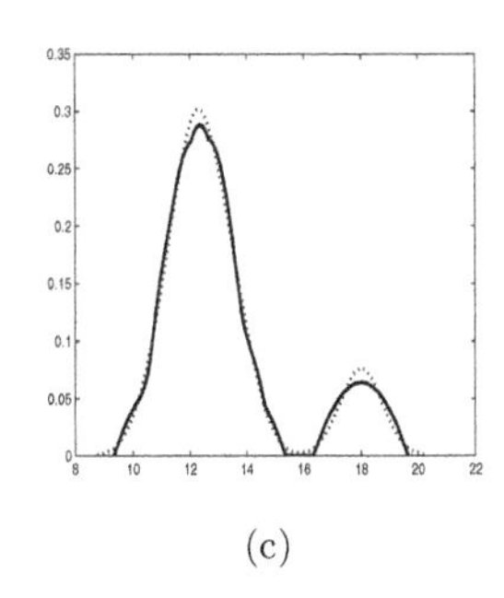

(c)

Fig. 11.5 Density estimation for sample of size n=7 using various kernels: (all) Normal, (a) Box, (b) Triangle, (c) Epanechnikov.

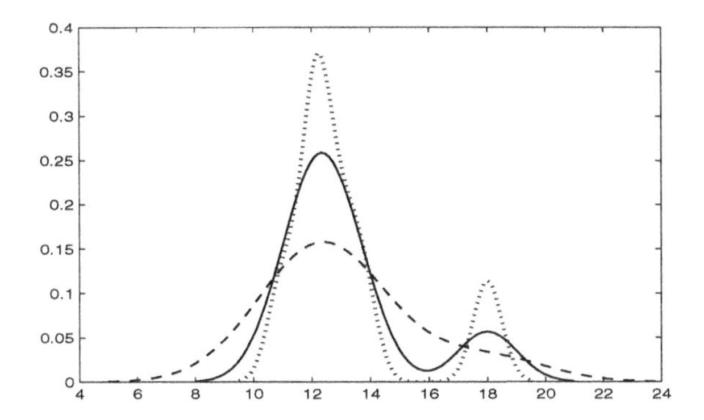

Fig. 11.6 Density estimation for sample of size $n = 7$ using various bandwidths.

estimation, see Silverman (1986) and Efromovich (1999).

The (univariate) density estimator from MATLAB, called

`ksdensity(data)`,

is illustrated in Figure 11.5 using a sample of seven observations. The default estimate is based on a normal kernel; to use another kernel, just enter 'box', 'triangle', or 'epanechnikov' (see code below). Figure 11.5 shows how the normal kernel compares to the (a) box, (2) triangle and (c) epanechnikov kernels. Figure 11.6 shows the density estimator using the same data based on the normal kernel, but using three different bandwidths. Note the optimal bandwidth (0.7449) can be found by allowing a third argument in the `ksdensity` output.

```
>> data1=[11,12,12.2,12.3,13,13.7,18];
```

```
>> data2=[50,21,25.5,40.0,41,47.6,39];
>> [f1,x1]=ksdensity(data1,'kernel','box');
>> plot(x1,f1,'-k')
>> hold on
>> [f2,x2,band]=ksdensity(data1);
>> plot(x2,f2,':k')
>> band

band =
    0.7449

>> [f1,x1]=ksdensity(data1,'width',2);
>> plot(x1,f1,'--k')
>> hold on
>> [f1,x1]=ksdensity(data1,'width',1);
>> plot(x1,f1,'-k')
>> [f1,x1]=ksdensity(data1,'width',.5);
>> plot(x1,f1,':k')
```

Censoring. The MATLAB function `ksdensity` also handles right-censored data by adding an optional vector designating censoring. Although we will not study the details about the way density estimators handle this problem, censored observations are treated in a way similar to nonparametric maximum likelihood, with the weight assigned to the censored observation x_c being distributed proportionally to non-censored observations $x_t \geq x_c$ (see the Kaplan-Meier estimator in Chapter 10). General weighting can also be included in the density estimation for `ksdensity` with an optional vector of weights.

Example 11.1 Radiation Measurements. In some situations, the experimenter might prefer to subjectively decide on a proper bandwidth instead of the objective choice of bandwidth that minimizes MISE. If outliers and subtle changes in the probability distribution are crucial in the model, a more jagged density estimator (with a smaller bandwidth) might be preferred to the optimal one. In Davies and Gather (1993), 2001 radiation measurements were taken from a balloon at a height of 100 feet. Outliers occur when the balloon rotates, causing the balloon's ropes to block direct radiation from the sun to the measuring device. Figure 11.7 shows two density estimates of the raw data, one based on a narrow bandwidth and the other more smooth density based on a bandwidth 10 times larger (0.01 to 0.1). Both densities are based upon a normal (Gaussian) kernel. While the more jagged estimator does show the mode and skew of the density as clearly as the smoother estimator, outliers are more easily discerned.

```
>> T=load('balloondata.txt');
>> T1=T(:,1); T2=T(:,2);
>> [f1,x1]=ksdensity(T1,'width',.01);
```

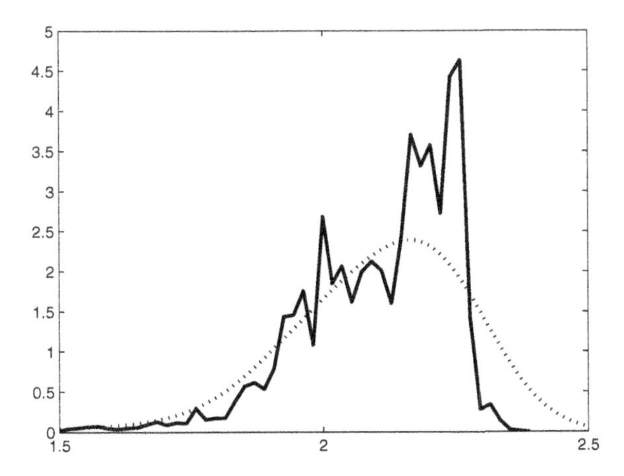

Fig. 11.7 Density estimation for 2001 radiation measurements using bandwidths `band = 0.5` and `band=0.05`.

```
>> plot(x1,f1,'-k')
>> hold on
>> [f2,x2]=ksdensity(T1,'width',.1);
>> plot(x2,f2,':k')
```

11.2.1 Bivariate Density Estimators

To plot density estimators for bivariate data, a three-dimensional plot can be constructed using MATLAB function `kdfft2`, noting that both `x` and `y`, the vectors designating plotting points for the density, must be of the same size.

In Figure 11.8, (univariate) density estimates are plotted for the seven observations `[data1, data2]`. In Figure 11.9, `kdfft2` is used to produce a two-dimensional density plot for the seven bivariate observations (coupled together).

11.3 EXERCISES

11.1. Which of the following serve as kernel functions for a density estimator? Prove your assertion one way or the other.

a. $K(x) = \mathbf{1}(-1 < x < 1)/2$,

b. $K(x) = \mathbf{1}(0 < x < 1)$,

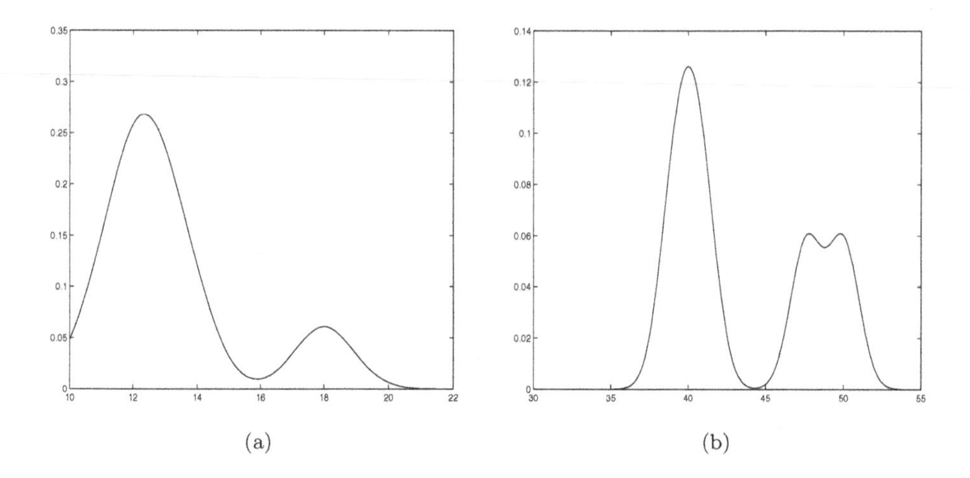

Fig. 11.8 (a) Univariate density estimator for first variable; (b) Univariate density estimator for second variable.

c. $K(x) = 1/x$,

d. $K(x) = \frac{3}{2}(2x+1)(1-2x)\,\mathbf{1}(-\frac{1}{2} < x < \frac{1}{2})$,

e. $K(x) = 0.75(1-x^2)\,\mathbf{1}(-1 < x < 1)$

11.2. With a data set of 12, 15, 16, 20, estimate $p^* = P$(observation is less than 15) using a density estimator based on a normal (Gaussian) kernel with $h_n = \sqrt{3/n}$. Use hand calculations instead of the MATLAB function.

11.3. Generate 12 observations from a mixture distribution, where half of the observations are from $\mathcal{N}(0,1)$ and the other half are from $\mathcal{N}(1, 0.64)$. Use the MATLAB function `ksdensity` to create a density estimator. Change bandwidth to see its effect on the estimator. Repeat this procedure using 24 observations instead of 12.

11.4. Suppose you have chosen kernel function $K(x)$ and smoothing function h_n to construct your density estimator, where $-\infty < K(x) < \infty$. What should you do if you encounter a right censored observation? For example, suppose the right-censored observation is ranked m lowest out of n, $m \leq n-1$.

11.5. Recall Exercise 6.3 based on 100 measurements of the speed of light in air. In that chapter we tested the data for normality. Use the same data to construct a density estimator that you feel gives the best visual display of the information provided by the data. What parameters did you choose? The data can be downloaded from

`http://www.itl.nist.gov/div898/strd/univ/data/Michelso.dat`

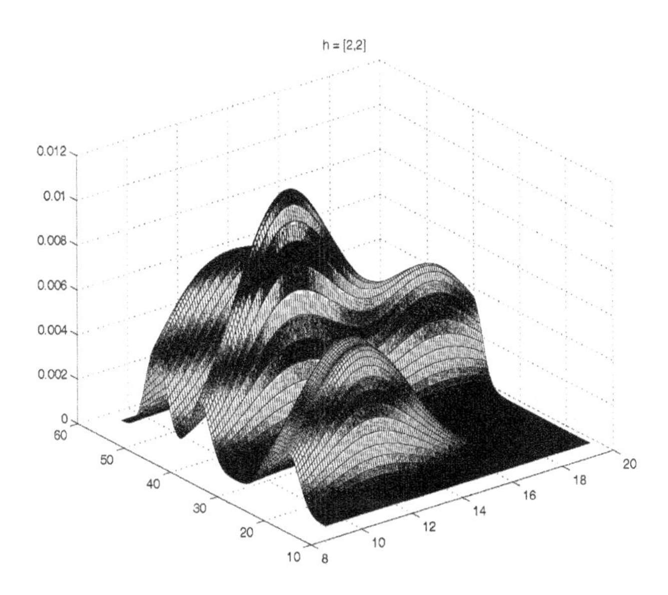

Fig. 11.9 Bivariate Density estimation for sample of size $n = 7$ using bandwidth = [2,2].

11.6. Go back to Exercise 10.6, where a link is provided to download right-censored survival times for 87 people with lupus nephritis. Construct a density estimator for the survival, ignoring the duration variable.

`http://lib.stat.cmu.edu/datasets/lupus`

REFERENCES

Davies, L., and Gather, U. (1993), "The Identification of Multiple Outliers" (discussion paper), *Journal of the American Statistical Association*, 88, 782-792.

Efromovich, S. (1999), *Nonparametric Curve Estimation: Methods, Theory and Applications,* New York: Springer Verlag.

Epanechnikov, V. A. (1969), "Nonparametric Estimation of a Multivariate Probability Density," *Theory of Probability and its Applications*, 14, 153-158.

Pearson, K. (1895), *Contributions to the Mathematical Theory of Evolution II, Philosophical Transactions of the Royal Society of London (A)*,186,

343-414

Playfair, W. (1786), *Commercial and Political Atlas: Representing, by Copper-Plate Charts, the Progress of the Commerce, Revenues, Expenditure, and Debts of England, during the Whole of the Eighteenth Century.* London: Corry.

Silverman, B. (1986), *Density Estimation for Statistics and Data Analysis*, New York: Chapman & Hall.

12

Beyond Linear Regression

Essentially, all models are wrong, but some models are useful.

George Box, from *Empirical Model-Building and Response Surfaces*

Statistical methods using linear regression are based on the assumptions that errors, and hence the regression responses, are normally distributed. Variable transformations increase the scope and applicability of linear regression toward real applications, but many modeling problems cannot fit in the confines of these model assumptions.

In some cases, the methods for linear regression are robust to minor violations of these assumptions. This has been shown in diagnostic methods and simulation. In examples where the assumptions are more seriously violated, however, estimation and prediction based on the regression model are biased. Some *residuals* (measured difference between the response and the model's estimate of the response) can be overly large in this case, and wield a large influence on the estimated model. The observations associated with large residuals are called outliers, which cause error variances to inflate and reduce the power of the inferences made.

In other applications, parametric regression techniques are inadequate in capturing the true relationship between the response and the set of predictors. General "curve fitting" techniques for such data problems are introduced in the next chapter, where the model of the regression is unspecified and not necessarily linear.

In this chapter, we look at simple alternatives to basic least-squares regression. These estimators are constructed to be less sensitive to the outliers that can affect regular regression estimators. *Robust* regression estimators are made specifically for this purpose. Nonparametric or *rank* regression relies more on the order relations in the paired data rather than the actual data measurements, and *isotonic* regression represents a nonparametric regression model with simple constraints built in, such as the response being monotone with respect to one or more inputs. Finally, we overview generalized linear models which although parametric, encompass some nonparametric methods, such as contingency tables, for example.

12.1 LEAST SQUARES REGRESSION

Before we introduce the less-familiar tools of nonparametric regression, we will first review basic linear regression that is taught in introductory statistics courses. Ordinary least-squares regression is synonymous with parametric regression only because of the way the errors in the model are treated. In the simple linear regression case, we observe n independent pairs (X_i, Y_i), where the linear regression of Y on X is the conditional expectation $\mathbb{E}(Y|X)$. A characterizing property of normally distributed X and Y is that the conditional expectation is linear, that is, $\mathbb{E}(Y|X) = \beta_0 + \beta_1 X$.

Standard least squares regression estimates are based on minimizing squared errors $\sum_i (Y_i - \hat{Y}_i)^2 = \sum_i (Y_i - [\beta_0 + \beta_1 X_i])^2$ with respect to the parameters β_1 and β_0. The least squares solutions are

$$\hat{\beta}_1 = \frac{\sum_{i=1}^n (X_i - \bar{X})(Y_i - \bar{Y})}{\sum_{i=1}^n (X_i - \bar{X})^2} \tag{12.1}$$

$$= \frac{\sum_{i=1}^n (X_i Y_i - n\bar{X}\bar{Y})}{\sum_{i=1}^n X_i^2 - n\bar{X}^2}.$$

$$\hat{\beta}_0 = \bar{Y} - \hat{\beta}_1 \bar{X}. \tag{12.2}$$

This solution is familiar from elementary parametric regression. In fact, $(\hat{\beta}_0, \hat{\beta}_1)$ are the MLEs of (β_0, β_1) in the case the errors are normally distributed. But with the minimized least squares approach (treating the sum of squares as a "loss function"), no such assumptions were needed, so the model is essentially nonparametric. However, in ordinary regression, the distributional properties of $\hat{\beta}_0$ and $\hat{\beta}_1$ that are used in constructing tests of hypothesis and confidence intervals are pinned to assuming these errors are homogenous and normal.

12.2 RANK REGRESSION

The truest nonparametric method for modeling bivariate data is Spearman's correlation coefficient which has no specified model (between X and Y) and no assumed distributions on the errors. Regression methods, by their nature, require additional model assumptions to relate a random variable X to Y via a function for the regression of $\mathbb{E}(Y|X)$. The technique discussed here is nonparametric except for the chosen regression model; error distributions are left to be arbitrary. Here we assume the linear model

$$Y_i = \beta_0 + \beta_1 x_i, \quad i = 1, \ldots, n$$

is appropriate and, using the squared errors as a loss function, we compute $\hat{\beta}_0$ and $\hat{\beta}_1$ as in (12.2) and (12.1) as the least-squares solution.

Suppose we are interested in testing H_0 that the population slope is equal to β_{10} against the three possible alternatives, $H_1 : \beta_1 > \beta_{10}$, $H_1 : \beta_1 < \beta_{10}$, $H_1 : \beta_1 \neq \beta_{10}$. Recall that in standard least-squares regression, the Pearson coefficient of linear correlation ($\hat{\rho}$) between the Xs and Ys is connected to β_1 via

$$\hat{\rho} = \hat{\beta}_1 \cdot \frac{\sqrt{\sum_i X_i^2 - n(\bar{X})^2}}{\sqrt{\sum_i Y_i^2 - n(\bar{Y})^2}}.$$

To test the hypothesis about the slope, first calculate $U_i = Y_i - \beta_{10} X_i$, and find the Spearman coefficient of rank correlation $\hat{\rho}$ between the X_is and the U_is. For the case in which $\beta_{10} = 0$, this is no more than the standard Spearman correlation statistic. In any case, under the assumption of independence, $(\hat{\rho} - \rho)\sqrt{n-1} \sim \mathcal{N}(0, 1)$ and the tests against alternatives H_1 are

Alternative	p-value
$H_1 : \beta_1 \neq \beta_{10}$	$p = 2P(Z \geq \lvert\hat{\rho}\rvert\sqrt{n-1})$
$H_1 : \beta_1 < \beta_{10}$	$p = P(Z \leq \hat{\rho}\sqrt{n-1})$
$H_1 : \beta_1 > \beta_{10}$	$p = P(Z \geq \hat{\rho}\sqrt{n-1})$

where $Z \sim \mathcal{N}(0, 1)$. The table represents a simple nonparametric regression test based only on Spearman's correlation statistic.

Example 12.1 Active Learning. Kvam (2000) examined the effect of active learning methods on student retention by examining students of an introductory statistics course eight months after the course finished. For a class taught using an emphasis on active learning techniques, scores were compared to equivalent final exam scores.

Exam 1	14	15	18	16	17	12	17	15	17	14	17	13	15	18	14
Exam 2	14	10	11	8	17	9	11	13	12	13	14	11	11	15	9

Scores for the first (x-axis) and second (y-axis) exam scores are plotted in Figure 12.1(a) for 15 active-learning students. In Figure 12.1(b), the solid line represents the computed Spearman correlation coefficient for X_i and $U_i = Y_i - \beta_{10} X_i$ with β_{10} varying from -1 to 1. The dashed line is the p-value corresponding to the test $H_1 : \beta_1 \neq \beta_{10}$. For the hypothesis $H_0 : \beta_1 \geq 0$ versus $H_1 : \beta_1 < 0$, the p-value is about 0.12 (the p-value for the two-sided test, from the graph, is about 0.24).

Note that at $\beta_{10} = 0.498$, $\hat{\rho}$ is zero, and at $\beta_{10} = 0$, $\hat{\rho} = 0.387$. The p-value is highest at $\beta_{10} = 0.5$ and less than 0.05 for all values of β_{10} less than -0.332.

```
>> n0=1000;
>> S=load('activelearning.txt');
>> trad1=S(:,1); trad2=S(:,2);
>> act1=S(:,3); act2=S(:,4);
>> trad=[trad1 trad2]; act=[act1 act2];
>> r=zeros(n0,1); p=zeros(n0,1); b=zeros(n0,1);
>> for i=1:n0
     b(i)=(i-(n0/2))/(n0/2);
     [r0 z0 p0]=spear(act1,act2-b(i)*act1);
     r(i)=r0; p(i)=p0;
     end
>> stairs(b,p,':k')
>> hold on
>> stairs(b,r,'-k')
```

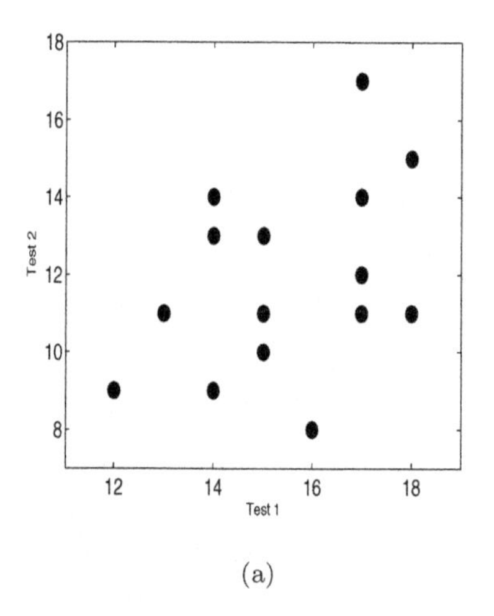

(a)

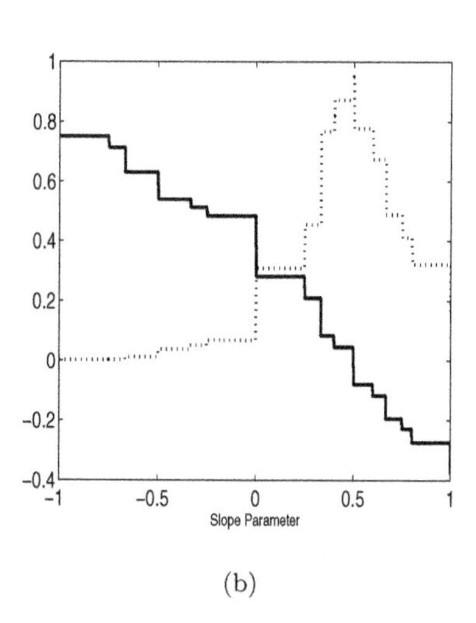

(b)

Fig. 12.1 (a) Plot of test #1 scores (during term) and test #2 scores (8 months after). (b) Plot of Spearman correlation coefficient (*solid*) and corresponding p-value (*dotted*) for nonparametric test of slope for $-1 \leq \beta_{10} \leq 1$.

12.2.1 Sen-Theil Estimator of Regression Slope

Among n bivariate observations, there are $\binom{n}{2}$ different pairs (X_i, Y_i) and (X_j, Y_j), $i \neq j$. For each pair (X_i, Y_i) and (X_j, Y_j), $1 \leq i < j \leq n$ we find the corresponding slope

$$S_{ij} = \frac{Y_j - Y_i}{X_j - X_i}.$$

Compared to ordinary least-squares regression, a more robust estimator of the slope parameter β_1 is

$$\tilde{\beta}_1 = \text{median}\{S_{ij}, 1 \leq i < j \leq n\}.$$

Corresponding to the least-squares estimate, let

$$\tilde{\beta}_0 = \text{median}\{Y\} - \tilde{\beta}_1 \text{median}\{X\}.$$

Example 12.2 If we take $n = 20$ integers $\{1, \ldots, 20\}$ as our set of predictors $X_1, \ldots, X_{20}$, let Y be $2X + \epsilon$ where ϵ is a standard normal variable. Next, we change both Y_1 and Y_{20} to be outliers with value 20 and compare the ordinary least squares regression with the more robust nonparametric method in Figure 12.2.

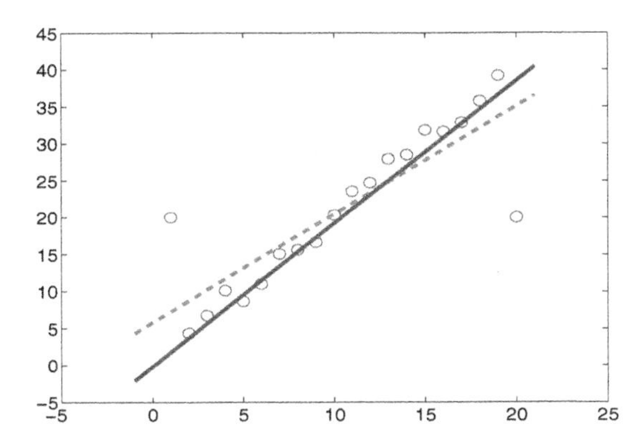

Fig. 12.2 Regression: Least squares (*dotted*) and nonparametric (*solid*).

12.3 ROBUST REGRESSION

"Robust" estimators are ones that retain desired statistical properties even when the assumptions about the data are slightly off. Robust linear regression represents a modeling alternative to regular linear regression in the case the assumptions about the error distributions are potentially invalid. In the simple linear case, we observe n independent pairs (X_i, Y_i), where the linear regression of Y on X is the conditional expectation $\mathbb{E}(Y|X) = \beta_0 + \beta_1 X$.

For rank regression, the estimator of the regression slope is considered to be robust because no single observation (or small group of observations) will have an significant influence on estimated model; the regression slope picks out the median slope out of the $\binom{n}{2}$ different pairings of data points.

One way of measuring robustness is the regression's *breakdown point*, which is the proportion of bad data needed to affect the regression adversely. For example, the sample mean has a breakdown point of 0, because a single observation can change it by an arbitrary amount. On the other hand, the sample median has a breakdown point of 50 percent. Analogous to this, ordinary least squares regression has a breakdown point of 0, while some of the robust techniques mentioned here (e.g., least-trimmed squares) have a breakdown point of 50 percent.

There is a big universe of robust estimation. We only briefly introduce some robust regression techniques here, and no formulations or derivations are given. A student who is interested in learning more should read an introductory textbook on the subject, such as *Robust Statistics* by Huber (1981).

.

12.3.1 Least Absolute Residuals Regression

By squaring the error as a measure of discrepancy, the least-squares regression is more influenced by outliers than a model based on, for example, absolute deviation errors: $\sum_i |Y_i - \hat{Y}_i|$, which is called Least Absolute Residuals Regression. By minimizing errors with a loss function that is more "forgiving" to large deviations, this method is less influenced by these outliers. In place of least-squares techniques, regression coefficients are found from linear programming.

12.3.2 Huber Estimate

The concept of robust regression is based on a more general class of estimates $(\hat{\beta}_0, \hat{\beta}_1)$ that minimize the function

$$\sum_{i=1}^{n} \frac{\psi(Y_i - \hat{Y}_i)}{\sigma},$$

where ψ is a loss function and σ is a scale factor. If $\psi(x) = x^2$, we have regular least-squares regression, and if $\psi(x) = |x|$, we have least absolute residuals regression. A general loss function introduced by Huber (1975) is

$$\psi(x) = \begin{cases} x^2 & |x| < c \\ 2c|x| - c^2 & |x| > c. \end{cases}$$

Depending on the chosen value of $c > 0$, $\psi(x)$ uses squared-error loss for small errors, but the loss function flattens out for larger errors.

12.3.3 Least Trimmed Squares Regression

Least Trimmed Squares (LTS) is another robust regression technique proposed by Rousseeuw (1985) as a robust alternative to ordinary least squares regression. Within the context of the linear model $y_i = \beta' x_i, \quad i = 1, \ldots, n$, the LTS estimator is represented by the value of β that minimizes $\sum_{i=1}^{h} r_{i:n}$. Here, x_i is a px1 vector and $r_{i:n}$ is the ith order statistic from the squared residuals $r_i = (y_i - \beta' x_i)^2$ and h is a trimming constant $(n/2 \leq h \leq n)$ chosen so that the largest $n - h$ residuals do not affect the model estimate. Rousseeuw and Leroy (1987) showed that the LTS estimator has its highest level of robustness when $h = \lfloor n/2 \rfloor + \lfloor (p+1)/2 \rfloor$. While choosing h to be low leads to a more robust estimator, there is a tradeoff of robustness for efficiency.

12.3.4 Weighted Least Squares Regression

For some data, one can improve model fit by including a scale factor (weight) in the deviation function. Weighted least squares minimizes

$$\sum_{i=1}^{n} w_i (y_i - \hat{y}_i)^2,$$

where w_i are weights that determine how much influence each response will have on the final regression. With the weights in the model, we estimate β in the linear model with

$$\hat{\beta} = (X'WX)^{-1} X'Wy,$$

where X is the design matrix made up of the vectors x_i, y is the response vector, and W is a diagonal matrix of the weights $w_1, \ldots, w_n$. This can be especially helpful if the responses seem not to have constant variances. Weights that counter the effect of heteroskedasticity, such as

$$w_i = m \left(\sum_{i=1}^{m} (y_i - \bar{y})^2 \right)^{-1},$$

work well if your data contain a lot of replicates; here m is the number of replicates at y_i. To compute this in MATLAB, the function `lscov` computes least-squares estimates with known covariance; for example, the output of

```
lscov(A,B,W)
```

returns the weighted least squares solution to the linear system $AX = B$ with diagonal weight matrix X.

12.3.5 Least Median Squares Regression

The least median of squares (LMS) regression finds the line through the data that minimizes the median (rather than the mean) of the squares of the errors. While the LMS method is proven to be robust, it cannot be easily solved like a weighted least-squares problem. The solution must be solved by searching in the space of possible estimates generated from the data, which is usually too large to do analytically. Instead, randomly chosen subsets of the data are chosen so that an approximate solution can be computed without too much trouble. The MATLAB function

```
lmsreg(y,X)
```

computes the LMS for small or medium sized data sets.

Example 12.3 Star Data. Data from Rousseeuw and Leroy (1987), p. 27, Table 3, are given in all panels of Figure 12.3 as a scatterplot of temperature versus light intensity for 47 stars. The first variable is the logarithm of the effective temperature at the surface of the star (Te) and the second one is the logarithm of its light intensity ($L/L0$). In sequence, the four panels in Figure 12.3 show plots of the bivariate data with fitted regressions based on (a) Least Squares, (b) Least Absolute Residuals, (c) Huber Loss & Least Trimmed Squares, and (d) Least Median Squares. Observations far away from most of the other observations are called *leverage points*; in this example, only the Least Median Squares approach works well because of the effect of the leverage points.

```
>> stars = load('stars.txt'); n = size(stars,1);
>> X = [ones(n,1) stars(:,2)]; y = stars(:,3);
>> bols  = X\y; [ignore,idx] = sort(stars(:,2));
>> plot(stars(:,2),stars(:,3),'o',stars(idx,2),...
    X(idx,:)*bols,'-.' )
    legend('Data','OLS')
>> %
>> %   Least Absolute Deviation
>> blad = medianregress(stars(:,2),stars(:,3));
>> plot(stars(:,2),stars(:,3),'o',stars(idx,2),...
   X(idx,:)*bols,'-.',stars(idx,2),X(idx,:)*blad,'-.')
  legend('Data','OLS','LAD');
```

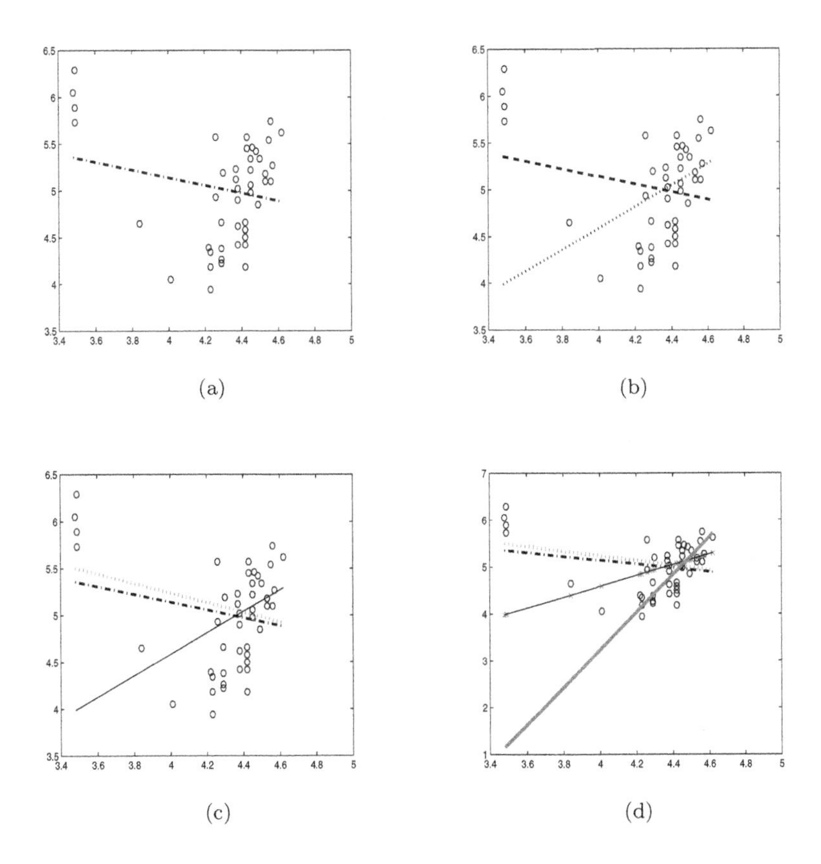

Fig. 12.3 Star data with (a) OLS Regression, (b) Least Absolute Deviation, (c) Huber Estimation and Least Trimmed Squares, (d) Least Median Squares.

```
>> %
>> % Huber Estimation
>> k = 1.345; % tuning parameters in Huber's weight function
>> wgtfun  = @(e) (k*(abs(e)>k)-abs(e).*(abs(e)>k))./abs(e)+1;
>> %  Huber's weight function
>> wgt = rand(length(y),1);   % Initial Weights
>> b0 = lscov(X,y,wgt);
>> res = y - X*b0;            % Raw Residuals
>> res = res/mad(res)/0.6745; % Standardized Residuals
>> m = 30;
>> for i=1:m
     wgt = wgtfun(res);
     % Compute the weighted estimate using these weights
     bhuber = lscov(X,y,wgt);
     if all((bhuber-b0)<.01*b0)% Stop with convergence
       return;
     else
       res = y - X*bhuber;
```

```
        res = res/mad(res)/0.6745;
      end
    end
>> plot(stars(:,2),stars(:,3),'o',stars(idx,2),X(idx,:)...
   *bols,'-.', stars(idx,2),X(idx,:)*blad,'-x',...
    stars(idx,2),X(idx,:)*bhuber,'-s')
    legend('Data','OLS','LAD','Huber');
>> %
>> % Least Trimmed Squares
>> blts = lts(stars(:,2),y);
>> plot(stars(:,2),stars(:,3),'o',stars(idx,2),X(idx,:)...
   *bols,'-.', stars(idx,2),X(idx,:)*blad,'-x',...
   stars(idx,2),X(idx,:)*bhuber,'-s', stars(idx,2),...
   X(idx,:)*blad,'-+')
   legend('Data','OLS','LAD','Huber','LTS');
>> %
>> % Least Median Squares
>> [LMSout,blms,Rsq]=LMSreg(y,stars(:,2));
>> plot(stars(:,2),stars(:,3),'o',stars(idx,2),X(idx,:)...
   *bols,'-.',stars(idx,2),X(idx,:)*blad,'-x',...
   stars(idx,2), X(idx,:)*bhuber,'-s',stars(idx,2),...
   X(idx,:)*blad,'-+', stars(idx,2),X(idx,:)*blms,'-d')
   legend('Data','OLS','LAD','Huber','LTS','LMS');
```

Example 12.4 Anscombe's Four Regressions. A celebrated example of the role of residual analysis and statistical graphics in statistical modeling was created by Anscombe (1973). He constructed four different data sets (X_i, Y_i), $i = 1, \ldots, 11$ that share the same descriptive statistics ($\bar{X}$, $\bar{Y}$, $\hat{\beta}_0$, $\hat{\beta}_1$, MSE, R^2, F) necessary to establish linear regression fit $\hat{Y} = \hat{\beta}_0 + \hat{\beta}_1 X$. The following statistics are common for the four data sets:

Sample size N	11
Mean of X ($\bar{X}$)	9
Mean of Y ($\bar{Y}$)	7.5
Intercept ($\hat{\beta}_0$)	3
Slope ($\hat{\beta}_1$)	0.5
Estimator of σ, (s)	1.2366
Correlation $r_{X,Y}$	0.816

From inspection, one can ascertain that a linear model is appropriate for Data Set 1, but the scatter plots and residual analysis suggest that the Data Sets 2–4 are not amenable to linear modeling. Plotted with the data are the lines for least-square fit (*dotted*) and rank regression (*solid line*). See Exercise 12.1 for further examination of the three regression archetypes.

						Data Set 1					
X	10	8	13	9	11	14	6	4	12	7	5
Y	8.04	6.95	7.58	8.81	8.33	9.96	7.24	4.26	10.84	4.82	5.68
						Data Set 2					
X	10	8	13	9	11	14	6	4	12	7	5
Y	9.14	8.14	8.74	8.77	9.26	8.10	6.13	3.10	9.13	7.26	4.74
						Data Set 3					
X	10	8	13	9	11	14	6	4	12	7	5
Y	7.46	6.77	12.74	7.11	7.81	8.84	6.08	5.39	8.15	6.42	5.73
						Data Set 4					
X	8	8	8	8	8	8	8	19	8	8	8
Y	6.58	5.76	7.71	8.84	8.47	7.04	5.25	12.50	5.56	7.91	6.89

12.4 ISOTONIC REGRESSION

In this section we consider bivariate data that satisfy an order or restriction in functional form. For example, if Y is known to be a decreasing function of X, a simple linear regression need only consider values of the slope parameter $\beta_1 < 0$. If we have no linear model, however, there is nothing in the empirical bivariate model to ensure such a constraint is satisfied. Isotonic regression considers a restricted class of estimators without the use of an explicit regression model.

Consider the dental study data in Table 12.16, which was used to illustrate isotonic regression by Robertson, Wright, and Dykstra (1988). The data are originally from a study of dental growth measurements of the distance (mm) from the center of the pituitary gland to the pterygomaxillary fissure (referring to the bone in the lower jaw) for 11 girls between the age of 8 and 14. It is assumed that PF increases with age, so the regression of PF on age is nondecreasing. But it is also assumed that the relationship between PF and age is not necessarily linear. The means (or medians, for that matter) are *not* strictly increasing in the PF data. Least squares regression does yield an increasing function for PF: $\hat{Y} = 0.065X + 21.89$, but the function is nearly flat and not altogether well-suited to the data.

For an isotonic regression, we impose some order of the response as a function of the regressors.

Definition 12.1 *If the regressors have a simple order $x_1 \leq \cdots \leq x_n$, a function f is isotonic with respect to x if $f(x_1) \leq \cdots \leq f(x_n)$. For our purposes, isotonic will be synonymous with monotonic. For some function g of X, we call the function g^* an isotonic regression of g with weights w if and*

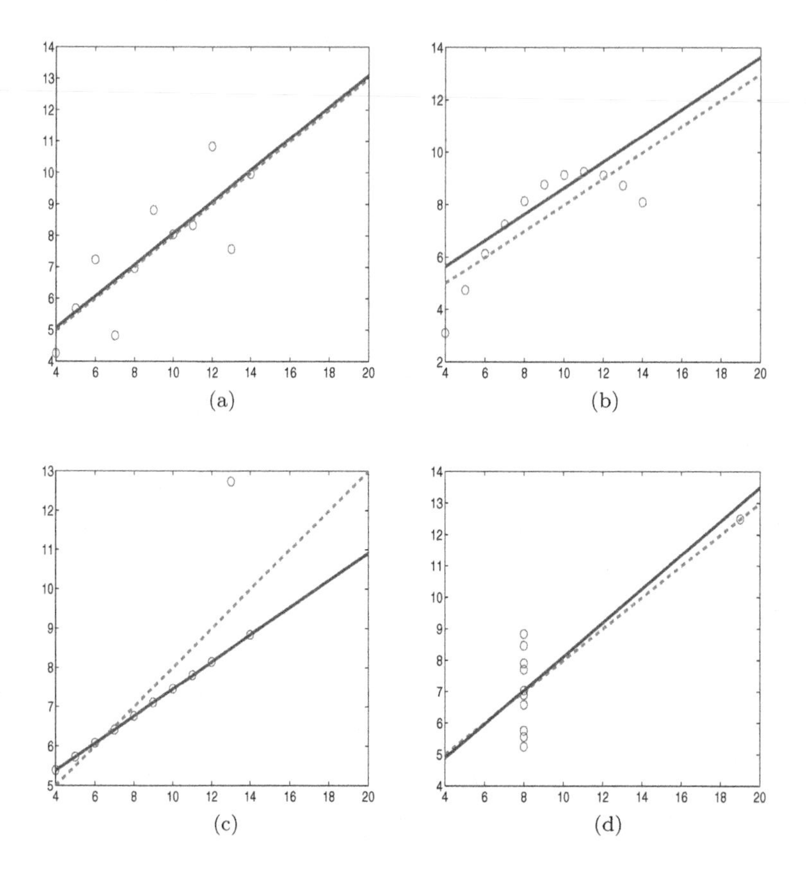

Fig. 12.4 Anscombe's regressions: LS and Robust.

only if g^* is isotonic (i.e., retains the necessary order) and minimizes

$$\sum_{i=1}^{n} w(x_i)\,(g(x_i) - f(x_i))^2 \tag{12.3}$$

in the class of all isotonic functions f.

12.4.1 Graphical Solution to Regression

We can create a simple graph to show how the isotonic regression can be solved. Let $W_k = \sum_{i=1}^{k} w(x_i)$ and $G_k = \sum_{i=1}^{k} g(x_i)w(x_i)$. In the example, the means are ordered, so $f(x_i) = \mu_i$ and $w_i = n_i$, the number of observations at each age group. We let g be the set of PF means, and the plot of W_k versus G_k, called the *cumulative sum diagram* (CSD), shows that the empirical

Table 12.16 Size of Pituitary Fissure for Subjects of Various Ages.

Age	8	10	12	14
PF	21,23.5,23	24,21,25	21.5,22,19	23.5,25
Mean	22.50	23.33	20.83	24.25
PAVA	22.22	22.22	22.22	24.25

relationship between PF and age is not isotonic.

Define G^* to be the *greatest convex minorant* (GCM) which represents the largest convex function that lies below the CSD. You can envision G^* as a taut string tied to the left most observation (W_1, G_1) and pulled up and under the CSD, ending at the last observation. The example in Figure 12.5(a) shows that the GCM for the nine observations touches only four of them in forming a tight convex bowl around the data.

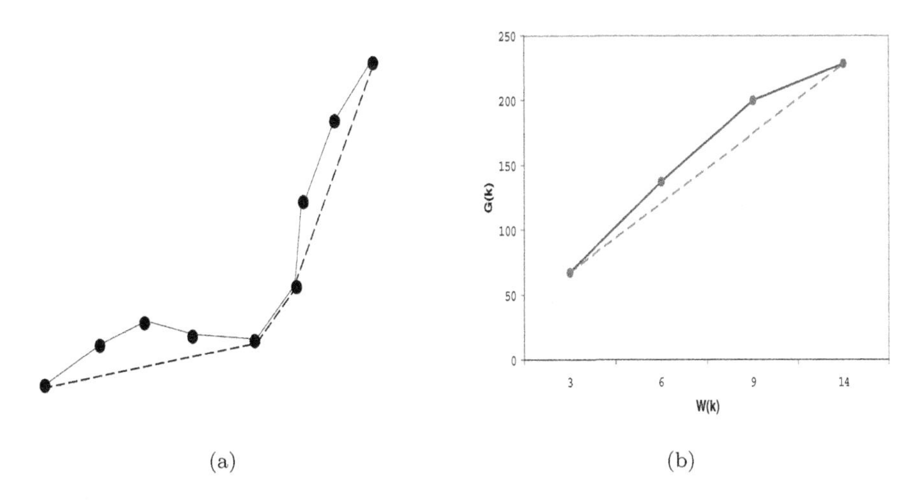

Fig. 12.5 (a) Greatest convex minorant based on nine observations. (b) Greatest convex minorant for dental data.

The GCM represents the isotonic regression. The reasoning follows below (and in the theorem that follows). Because G^* is convex, it is left differentiable at W_i. Let $g^*(x_i)$ be the left-derivative of G^* at W_i. If the graph of the GCM is under the graph of CSD at W_i, the slopes of the GCM to the left and right of W_i remain the same, i.e., if $G^*(W_i) < G_i$, then $g^*(x_{i+1}) = g^*(x_i)$. This illustrates part of the intuition of the following theorem, which is not proven here (see Chapter 1 of Robertson, Wright, and Dykstra (1988)).

Theorem 12.1 *For function f in (12.3), the left-hand derivative g^* of the greatest convex minorant is the unique isotonic regression of g on f. That is, if f is isotonic on X, then*

$$\sum_{i=1}^{n} w(x_i)\left(g(x_i) - f(x_i)\right)^2 \geq \sum_{i=1}^{n} w(x_i)\left(g(x_i) - g^*(x_i)\right)^2 + \sum_{i=1}^{n} w(x_i)\left(g^*(x_i) - f(x_i)\right)^2.$$

Obviously, this graphing technique is going to be impractical for problems of any substantial size. The following algorithm provides an iterative way of solving for the isotonic regression using the idea of the GCM.

12.4.2 Pool Adjacent Violators Algorithm

In the CSD, we see that if $g(x_{i-1}) > g(x_i)$ for some i, then g is not isotonic. To construct an isotonic g^*, take the first such pair and replace them with the weighted average

$$\bar{g}_i = \bar{g}_{i-1} = \frac{w(x_{i-1})g(x_{i-1}) + w(x_i)g(x_i)}{w(x_{i-1}) + w(x_i)}.$$

Replace the weights $w(x_i)$ and $w(x_{i-1})$ with $w(x_i) + w(x_{i-1})$. If this correction (replacing g with $\bar{g}$) makes the regression isotonic, we are finished. Otherwise, we repeat this process with until an isotonic is set. This is called the *Pool Adjacent Violators Algorithm* or PAVA.

Example 12.5 In Table 12.16, there is a decrease in PF between ages 10 and 12, which violates the assumption that pituitary fissure increases in age. Once we replace the PF averages by the average over both age groups (22.083), we still lack monotonicity because the PF average for girls of age 8 was 22.5. Consequently, these two categories, which now comprise three age groups, are averaged. The final averages are listed in the bottom row of Table 12.16

12.5 GENERALIZED LINEAR MODELS

Assume that n $(p+1)$-tuples $(y_i, x_{1i}, x_{2i}, \ldots, x_{pi})$, $i = 1, \ldots, n$ are observed. The values y_i are responses and components of vectors $x_i = (x_{1i}, x_{2i}, \ldots, x_{pi})'$ are predictors. As we discussed at the beginning of this chapter, the standard theory of linear regression considers the model

$$Y = X\beta + \epsilon, \tag{12.4}$$

where $Y = (Y_1, \ldots, Y_n)$ is the response vector, $X = (\mathbf{1}_n \; x_1 \; x_2 \; \ldots \; x_p)$ is the design matrix ($\mathbf{1}_n$ is a column vector of n 1's), and ϵ is vector of errors consisting of n i.i.d normal $\mathcal{N}(0, \sigma^2)$ random variables. The variance σ^2 is common for all Y_is and independent of predictors ir the order of observation. The parameter β is a vector of $(p+1)$ parameters in the linear relationship,

$$\mathbb{E}Y_i = x_i'\beta = \beta_0 + \beta_1 x_{1i} + \ldots \beta_p x_{pi}.$$

(a) (b)

Fig. 12.6 (a) Peter McCullagh and (b) John Nelder.

The term *generalized linear model* (GLM) refers to a large class of models, introduced by Nelder and Wedderburn (1972) and popularized by McCullagh and Nelder (1994), Figure 12.6 (a-b). In a canonical GLM, the response variable Y_i is assumed to follow an exponential family distribution with mean μ_i, which is assumed to be a function of $x_i'\beta$. This dependence can be nonlinear, but the distribution of Y_i depends on covariates only through their linear combination, $\eta_i = x_i'\beta$, called a *linear predictor.* As in the linear regression, the epithet *linear* refers to being linear in parameters, not in the explanatory variables. Thus, for example, the linear combination

$$\beta_0 + \beta_1 \; x_1 + \beta_2 \; x_2^2 + \beta_3 \; \log(x_1 + x_2) + \beta_4 \; x_1 \cdot x_2,$$

is a perfect linear predictor. What is generalized in model given in (12.4) by a GLM?

The three main generalizations concern the distributions of responses, the dependence of response on linear predictor, and variance if the error.

1. Although Y_is remain independent, their (common) distribution is generalized. Instead of normal, their distribution is selected from the exponential family of distributions (see Chapter 2). This family is quite versatile and includes normal, binomial, Poisson, negative binomial, and gamma as special cases.

2. In the linear model (12.4) the mean of Y_i, $\mu_i = \mathbb{E}Y_i$ was equal to $x_i'\beta$. The mean μ_i in GLM depends on the predictor $\eta_i = x_i'\beta$ as

$$g(\mu_i) = \eta_i \quad (= x_i'\beta). \tag{12.5}$$

The function g is called the *link* function. For the model (12.4), the link is the identity function.

3. The variance of Y_i was constant (12.4). In GLM it may not be constant and could depend on the mean μ_i.

Models and inference for categorical data, traditionally a non-parametric topic, are unified by a larger class of models which are parametric in nature and that are special cases of GLM. For example, in contingency tables, the cell counts N_{ij} could be modeled by multinomial $\mathcal{M}n(n, \{p_{ij}\})$ distribution. The standard hypothesis in contingency tables is concerning the independence of row/column factors. This is equivalent to testing $H_0 : p_{ij} = \alpha_i \beta_j$ for some unknown α_i and β_j such that $\sum_i \alpha_i = \sum_j \beta_j = 1$.

The expected cell count $\mathbb{E}N_{ij} = np_{ij}$, so that under H_0 becomes $\mathbb{E}N_{ij} = n\alpha_i\beta_j$, by taking the logarithm of both sides one obtains

$$\begin{aligned} \log \mathbb{E}N_{ij} &= \log n + \log \alpha_i + \log \beta_j \\ &= \text{const} + a_i + b_j, \end{aligned}$$

for some parameters a_i and b_j. Thus, the test of goodness of fit for this model linear and additive in parameters a and b, is equivalent to the test of the original independence hypothesis H_0 in the contingency table. More of such examples will be discussed in Chapter 18.

12.5.1 GLM Algorithm

The algorithms for fitting generalized linear models are robust and well established (see Nelder and Wedderburn (1972) and McCullagh and Nelder (1994)). The maximum likelihood estimates of β can be obtained using iterative weighted least-squares (IWLS).

(i) Given vector $\hat{\mu}^{(k)}$, the initial value of the linear predictor $\hat{\eta}^{(k)}$ is formed using the link function, and components of adjusted dependent variate (working response), $z_i^{(k)}$, can be formed as

$$z_i^{(k)} = \hat{\eta}_i^{(k)} + \left(y_i - \hat{\mu}_i^{(k)}\right) \left(\frac{d\eta}{d\mu}\right)_i^{(k)},$$

where the derivative is evaluated at the the available k^{th} value.

(ii) The quadratic (working) weights, $W_i^{(k)}$, are defined so that

$$\frac{1}{W_i^{(k)}} = \left(\frac{d\eta}{d\mu}\right)_i^2 V_i^{(k)},$$

where V is the variance function evaluated at the initial values.

(iii) The working response $z^{(k)}$ is then regressed onto the covariates x_i, with weights $W_i^{(k)}$ to produce new parameter estimates, $\hat{\beta}^{(k+1)}$. This vector is then used to form new estimates

$$\eta^{(k+1)} = X'\hat{\beta}^{(k+1)} \quad \text{and} \quad \hat{\mu}^{(k+1)} = g^{-1}(\hat{\eta}^{(k+1)}).$$

We repeat iterations until changes become sufficiently small. Starting values are obtained directly from the data, using $\hat{\mu}^{(0)} = y$, with occasional refinements in some cases (for example, to avoid evaluating $\log 0$ when fitting a log-linear model with zero counts).

By default, the scale parameter should be estimated by the *mean deviance*, $n^{-1}\sum_{i=1}^{n} D(y_i, \mu)$, from p. 44 in Chapter 3, in the case of the normal and gamma distributions.

12.5.2 Links

In the GLM the predictors for Y_i are summarized as the linear predictor $\eta_i = x_i'\beta$. The link function is a monotone differentiable function g such that $\eta_i = g(\mu_i)$, where $\mu_i = \mathbb{E}Y_i$. We already mentioned that in the normal case $\mu = \eta$ and the link is identity, $g(\mu) = \mu$.

Example 12.6 For analyzing count data (e.g., contingency tables), the Poisson model is standardly assumed. As $\mu > 0$, the identity link is inappropriate because η could be negative. However, if $\mu = e^{\eta}$, then the mean is always positive, and $\eta = \log(\mu)$ is an adequate link.

A link is called *natural* if it is connecting θ (the natural parameter in the exponential family of distributions) and μ. In the Poisson case,

$$f(y|\lambda) = \exp\left\{y \log \lambda - (\lambda + \log y!)\right\},$$

$\mu = \lambda$ and $\theta = \log \mu$. Accordingly, the log is the natural link for the Poisson distribution.

Example 12.7 For the binomial distribution,

$$f(y|\pi) = \binom{n}{y} \pi^y (1-\pi)^{n-y}$$

can be represented as

$$f(y|\pi) = \exp\left\{y \log \frac{\pi}{1-\pi} + n\log(1-\pi) + \log\binom{n}{y}\right\}.$$

The natural link $\eta = \log(\pi/(1-\pi))$ is called *logit* link. With the binomial distribution, several more links are commonly used. Examples are the *probit* link $\eta = \Phi^{-1}(\pi)$, where Φ is a standard normal CDF, and the *complementary log-log* link with $\eta = \log\{-\log(1-\pi)\}$. For these three links, the probability π of interest is expressed as $\pi = e^\eta/(1+e^\eta)$, $\pi = \Phi(\eta)$, and $\pi = 1-\exp\{-e^\eta\}$, respectively.

Distribution	$\theta(\mu)$	$b(\theta)$	ϕ
$\mathcal{N}(\mu, \sigma^2)$	μ	$\theta^2/2$	σ^2
$\mathcal{B}in(1, \pi)$	$\log(\pi/(1-\pi))$	$\log(1+\exp(\theta))$	1
$\mathcal{P}(\lambda)$	$\log\lambda$	$\exp(\theta)$	1
$\mathcal{G}amma(\mu, \nu/\mu)$	$-1/\mu$	$-\log(-\theta)$	$1/\nu$

When data y_i from the exponential family are expressed in grouped form (from which an average is considered as the group response), then the distribution for Y_i takes the form

$$f(y_i|\theta_i, \phi, \omega_i) = \exp\left\{\frac{y_i\theta_i - b(\theta_i)}{\phi}\omega_i + c(y_i, \phi, \omega_i)\right\}. \tag{12.6}$$

The weights ω_i are equal to 1 if individual responses are considered, $\omega_i = n_i$ if response y_i is an average of n_i responses, and $\omega_i = 1/n_i$ if the sum of n_i individual responses is considered.

The variance of Y_i then takes the form

$$\mathrm{Var} Y_i = \frac{b''(\theta_i)\phi}{\omega_i} = \frac{\phi V(\mu_i)}{\omega_i}.$$

12.5.3 Deviance Analysis in GLM

In GLM, the goodness of fit of a proposed model can be assessed in several ways. The customary measure is *deviance* statistics. For a data set with n observations, assume the dispersion ϕ is known and equal to 1, and consider the two extreme models, the single parameter model stating $\mathbb{E}Y_i = \hat{\mu}$ and the n parameter *saturated* model setting $\mathbb{E}Y_i = \hat{\mu}_i = Y_i$. Most likely, the interesting model is between the two extremes. Suppose $\mathcal{M}$ is the interesting model with $1 < p < n$ parameters.

If $\hat{\theta}_i^{\mathcal{M}} = \hat{\theta}_i^{\mathcal{M}}(\hat{\mu}_i)$ are predictions of the model $\mathcal{M}$ and $\hat{\theta}_i^{\mathcal{S}} = \hat{\theta}_i^{\mathcal{S}}(y_i) = y_i$ are the predictions of the saturated model, then the deviance of the model $\mathcal{M}$ is

$$D_{\mathcal{M}} = 2\sum_{i=1}^{n} \left[(y_i \hat{\theta}_i^{\mathcal{S}}(y_i) - b(\hat{\theta}_i^{\mathcal{S}})) - (y_i \hat{\theta}_i^{\mathcal{M}} - b(\hat{\theta}_i^{\mathcal{M}})) \right].$$

When the dispersion ϕ is estimated and different than 1, the *scaled deviance* of the model $\mathcal{M}$ is defined as $D^*_{\mathcal{M}} = D_{\mathcal{M}}/\phi$.

Example 12.8 For $y_i \in \{0, 1\}$ in the binomial family,

$$D = 2\sum_{1=1}^{n} \left\{ y_i \log\left(\frac{y_i}{\hat{y}_i}\right) + (n_i - y_i)\log\left(\frac{n_i - y_i}{n_i - \hat{y}_i}\right) \right\}.$$

- Deviance is minimized at saturated model $\mathcal{S}$. Equivalently, the log-likelihood $\ell^{\mathcal{S}} = \ell(y|y)$ is the maximal log-likelihood with the data y.
- The scaled deviance $D^*_{\mathcal{M}}$ is asymptotically distributed as χ^2_{n-p}. Significant deviance represents the deviation from a good model fit.
- If a model $\mathcal{K}$ with q parameters, is a subset of model $\mathcal{M}$ with p parameters $(q < p)$, then

$$\frac{D^*_{\mathcal{K}} - D^*_{\mathcal{M}}}{\phi} \sim \chi^2_{p-q}.$$

Residuals are critical for assessing the model (recall four Anscombe's regressions on p. 226). In standard normal regression models, residuals are calculated simply as $y_i - \hat{\mu}_i$, but in the context of GLMs, both predicted values and residuals are more ambiguous. For predictions, it is important to distinguish the scale: (i) predictions on the scale of $\eta = x_i'\beta$ and (ii) predictions on the scale of the observed responses y_i for which $\mathbb{E}Y_i = g^{-1}(\eta_i)$.

Regarding residuals, there are several approaches. *Response residuals* are defined as $r_i = y_i - g^{-1}(\eta_i) = y_i - \theta_i$. Also, the deviance residuals are defined as

$$r_i^D = \text{sign}(y_i - \mu_i)\sqrt{d_i},$$

where d_i are observation specific contributions to the deviance D.

Deviance residuals are ANOVA-like decompositions,

$$\sum_i (r_i^D)^2 = D,$$

thus testably assessing the contribution of each observation to the model deviance. In addition, the deviance residuals increase with $y_i - \hat{\mu}_i$ and are distributed approximately as standard normals, irrespectively of the type of GLM.

Example 12.9 For $y_i \in \{0, 1\}$ in the binomial family,

$$r_i^D = \text{sign}(y_i - \hat{y}_i)\sqrt{2\left\{y_i \log\left(\frac{y_i}{\hat{y}_i}\right) + (n_i - y_i)\log\left(\frac{n_i - y_i}{n_i - \hat{y}_i}\right)\right\}}.$$

Another popular measure of goodness of fit of GLM is Pearson statistic

$$X^2 = \sum_{i=1}^{n} \frac{(y_i - \hat{\mu}_i)^2}{V(\hat{\mu}_i)}.$$

The statistic X^2 also has a ${\chi^2}_{n-p}$ distribution.

Example 12.10 Cæsarean Birth Study. The data in this example come from Münich hospital (Fahrmeir and Tutz, 1996) and concern infection cases in births by Cæsarean section. The response of interest is occurrence of infection. Three covariates, each at two levels were considered as important for the occurrence of infection:

- `noplan` – Whether the Cæsarean section birth planned (0) or not (1);
- `riskfac` – The presence of Risk factors for the mother, such as diabetes, overweight, previous Cæsarean section birth, etc, where present = 1, not present = 0;
- `antibio` – Whether antibiotics were given (1) or not given (0) as a prophylaxis.

Table 12.17 provides the counts.

Table 12.17 Cæsarean Section Birth Data

	Planned		Not Planned	
	Infec	No Infec	Infec	No Infec
Antibiotics				
Risk Fact Yes	1	17	11	87
Risk Fact No	0	2	0	0
No Antibiotics				
Risk Fact Yes	28	30	23	3
Risk Fact No	8	32	0	9

The MATLAB function `glmfit`, described in Appendix A, is instrumental in computing the solution in the example that follows.

```
>> infection = [1 11 0 0 28 23 8 0];
>> total =     [18 98 2 0 58 26 40 9];
>> proportion = infection./total;
>> noplan =   [0 1 0 1 0 1 0 1];
>> riskfac  = [1 1 0 0 1 1 0 0];
>> antibio =  [1 1 1 1 0 0 0 0];
>> [logitCoef2,dev] = glmfit([noplan' riskfact' antibio'],...
        [infection' total'],'binomial','logit');
>> logitFit = glmval(logitCoef2,[noplan' riskfact' antibio'],'logit');
>> plot(1:8, proportion,'ks', 1:8,   logitFit,'ko');
```

The scaled deviance of this model is distributed as ${\chi^2}_3$. The number of degrees of freedom is equal to 8 (n) vector `infection` minus 5 for the five estimated parameters, $\beta_0, \beta_1, \beta_2, \beta_3, \phi$. The deviance `dev=11` is significant, yielding a p-value of `1 - chi2cdf(11,3)=0.0117`. The additive model (with no interactions) in MATLAB yields

$$\log \frac{P(\texttt{infection})}{P(\texttt{no infection})} = \beta_0 + \beta_1 \, \texttt{noplan} + \beta_2 \, \texttt{riskfac} + \beta_3 \, \texttt{antibio}.$$

The estimators of $(\beta_0, \beta_1, \beta_2, \beta_3)$ are, respectively, $(-1.89, 1.07, 2.03, -3.25)$. The interpretation of the estimators is made more clear if we look at the odds ratio

$$\frac{P(\texttt{infection})}{P(\texttt{no infection})} = e^{\beta_0} \cdot e^{\beta_1 \texttt{noplan}} \cdot e^{\beta_2 \texttt{riskfac}} \cdot e^{\beta_3 \texttt{antibio}}.$$

At the value `antibio = 1`, the antibiotics have the odds ratio of infection/no infection. This increases by the factor $\exp(-3.25) = 0.0376$, which is a decrease of more than 25 times. Figure 12.7 shows the observed proportions of infections for 16 combinations of covariates `(noplan, riskfac, antibio)` marked by squares and model-predicted probabilities for the same combinations marked by circles. We will revisit this example in Chapter 18; see Example 18.5.

12.6 EXERCISES

12.1. Using robust regression, find the intercept and slope $\tilde{\beta}_0$ and $\tilde{\beta}_1$ for each of the four data sets of Anscombe (1973) from p. 226. Plot the ordinary least squares regression along with the rank regression estimator of slope. Contrast these with one of the other robust regression techniques. For which set does $\tilde{\beta}_1$ differ the most from its LS counterpart $\hat{\beta}_1 = 0.5$? Note that in the fourth set, 10 out of 11 Xs are equal, so one should use $S_{ij} = (Y_j - Y_i)/(X_j - X_i + \epsilon)$ to avoid dividing by 0. After finding $\tilde{\beta}_0$ and $\tilde{\beta}_1$, are they different than $\hat{\beta}_0$ and $\hat{\beta}_1$? Is the hypothesis $H_0 : \beta_1 = 1/2$ rejected in a robust test against the alternative $H_1 : \beta_1 < 1/2$, for Data

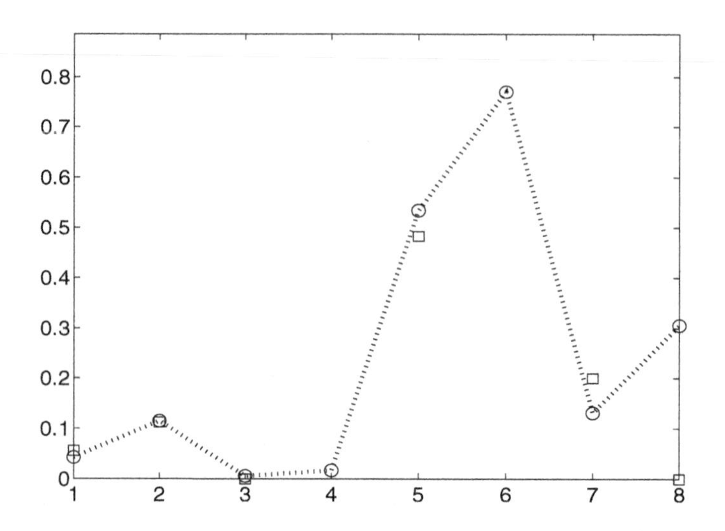

Fig. 12.7 Cæsarean Birth Infection observed proportions (*squares*) and model predictions (*circles*). The numbers 1-8 on the x-axis correspond to following combinations of covariates (**noplan, riskfac, antibio**): (0,1,1), (1,1,1), (0,0,1), (1,0,1), (0,1,0), (1,1,0), (0,0,0), and (1,0,0).

Set 3? Note, here $\beta_{10} = 1/2$.

12.2. Using the PF data in Table 12.16, compute a median squares regression and compare it to the simple linear regression curve.

12.3. Using the PF data in Table 12.16, compute a nonparametric regression and test to see if $\beta_{10} = 0$.

12.4. Consider the $\mathcal{G}amma(\alpha, \alpha/\mu)$ distribution. This parametrization was selected so that $\mathbb{E}y = \mu$. Identify θ and ϕ as functions of α and μ. Identify functions a, b and c.

Hint: The density can be represented as

$$\exp\left\{-\alpha \log \mu - \frac{\alpha y}{\mu} + \alpha \log(\alpha) + (\alpha - 1)\log y - \log(\Gamma(\alpha))\right\}$$

12.5. The zero-truncated Poisson distribution is given by

$$f(y|\lambda) = \frac{\lambda^j}{j!(e^\lambda - 1)}, \; j = 1, 2, \ldots$$

Show that f is a member of exponential family with canonical parameter $\log \lambda$.

12.6. Dalziel, Lagen and Thurston (1941) conducted an experiment to assess the effect of small electrical currents on farm animals, with the eventual goal of understanding the effects of high-voltage powerlines on livestock. The experiment was carried out with seven cows, and six shock intensities: 0, 1, 2, 3, 4, and 5 milliamps (note that shocks on the order of 15 milliamps are painful for many humans). Each cow was given 30 shocks, five at each intensity, in random order. The entire experiment was then repeated, so each cow received a total of 60 shocks. For each shock the response, mouth movement, was either present or absent. The data as quoted give the total number of responses, out of 70 trials, at each shock level. We ignore cow differences and differences between blocks (experiments).

Current (milliamps)	Number of Responses	Number of Trials	Proportion of Responses
0	0	70	0.000
1	9	70	0.129
2	21	70	0.300
3	47	70	0.671
4	60	70	0.857
5	63	70	0.900

Propose a GLM in which the probability of a response is modeled with a value of Current (in milliamps) as a covariate.

12.7. Bliss (1935) provides a table showing the number of flour beetles killed after five hours exposure to gaseous carbon disulphide at various concentrations. Propose a logistic regression model with a Dose as a covariate.

Table 12.18 Bliss Beetle Data

Dose ($\log_{10} CS_2\ mgl^{-1}$)	Number of Beetles	Number Killed
1.6907	59	6
1.7242	60	13
1.7552	62	18
1.7842	56	28
1.8113	63	52
1.8369	59	53
1.8610	62	61
1.8839	60	60

According to your model, what is the probability that a beetle will be killed if a dose of gaseous carbon disulphide is set to 1.8?

REFERENCES

Anscombe, F. (1973), "Graphs in Statistical Analysis," *American Statistician,* 27, 17-21.

Bliss, C. I. (1935), "The Calculation of the Dose-Mortality Curve," *Annals of Applied Biology,* 22, 134-167.

Dalziel, C, F. Lagen, J. B., and Thurston, J. L. (1941), "Electric Shocks," *Transactions of IEEE*, 60, 1073-1079.

Fahrmeir, L., and Tutz, G. (1994), *Multivariate Statistical Modeling Based on Generalized Linear Models,* New York: Springer Verlag

Huber, P. J. (1973), "Robust Regression: Asymptotics, Conjectures, and Monte Carlo," *Annals of Statistics*, 1, 799-821.

_______ (1981), *Robust Statistics,* New York: Wiley.

Kvam, P. H. (2000), "The Effect of Active Learning Methods on Student Retention in Engineering Statistics," *American Statistician*, 54, 2, 136-140.

Lehmann, E. L. (1998), *Nonparametrics: Statistical Methods Based on Ranks,* New Jersey: Prentice-Hall.

McCullagh, P., and Nelder, J. A. (1994), *Generalized Linear Models*, 2nd ed. London: Chapman & Hall.

Nelder, J. A., and Wedderburn, R. W. M. (1972), "Generalized Linear Models," *Journal of the Royal Statistical Society*, Ser. A, 135, 370-384.

Robertson, T., Wright, T. F., and Dykstra, R. L. (1988), *Order Restricted Statistical Inference*, New York: Wiley.

Rousseeuw, P. J. (1985), "Multivariate Estimation with High Breakdown Point," in *Mathematical Statistics and Applications B* , Eds. W. Grossmann et al., pp. 283-297, Dordrecht: Reidel Publishing Co.

Rousseeuw P. J. and Leroy A. M. (1987). *Robust Regression and Outlier Detection.* New York: Wiley.

13

Curve Fitting Techniques

"The universe is not only queerer than we imagine, it is queerer than we *can* imagine"

J.B.S. Haldane (Haldane's Law)

In this chapter, we will learn about a general class of nonparametric regression techniques that fit a response curve to input predictors without making strong assumptions about error distributions. The estimators, called *smoothing functions*, actually can be smooth or bumpy as the user sees fit. The final regression function can be made to bring out from the data what is deemed to be important to the analyst. Plots of a smooth estimator will give the user a good sense of the overall trend between the input X and the response Y. However, interesting nuances of the data might be lost to the eye. Such details will be more apparent with less smoothing, but a potentially noisy and jagged curve plotted made to catch such details might hide the overall trend of the data. Because no linear form is assumed in the model, this nonparametric regression approach is also an important component of *nonlinear regression*, which can also be parametric.

Let $(X_1, Y_1), \dots, (X_n, Y_n)$ be a set of n independent pairs of observations from the bivariate random variable (X, Y). Define the regression function $m(x)$ as $\mathbb{E}(Y|X = x)$. Let $Y_i = m(X_i) + \epsilon_i,\ i = 1, \dots, n$ when ϵ_i's are errors with zero mean and constant variance. The estimators here are *locally weighted* with the form

$$\hat{m}(x) = \sum_{i=1}^{n} a_i Y_i.$$

The local weights a_i can be assigned to Y_i in a variety of ways. The straight line in Figure 13.1 is a linear regression of Y on X that represents an extremely smooth response curve. The curvey line fit in Figure 13.1 represents an estimator that uses more local observations to fit the data at any X_i value. These two response curves represent the tradeoff we make when making a curve more or less smooth. The tradeoff is between *bias* and *variance* of the estimated curve.

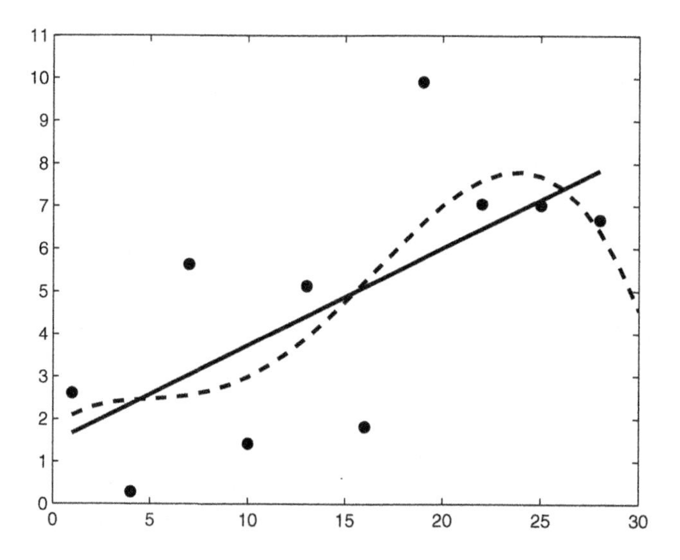

Fig. 13.1 Linear Regression and local estimator fit to data.

In the case of linear regression, the variance is estimated globally because it is assumed the unknown variance is constant over the range of the response. This makes for an optimal variance estimate. However, the linear model is often considered to be overly simplistic, so the true expected value of $\hat{m}(x)$ might be far from the estimated regression, making the estimator biased. The local (jagged) fit, on the other hand, uses only responses at the value X_i to estimate $\hat{m}(X_i)$, minimizing any potential bias. But by estimating $m(x)$ locally, one does not pool the variance estimates, so the variance estimate at X is constructed using only responses at or close to X.

This illustrates the general difference between smoothing functions; those that estimate $m(x)$ using points only at x or close to it have less bias and high variance. Estimators that use data from a large neighborhood of x will produce a good estimate of variance but risk greater bias. In the next sections, we feature two different ways of defining the local region (or neighborhood) of a design point. At an estimation point x, *kernel estimators* use fixed intervals around x such as $x \pm c_0$ for some $c_0 > 0$. *Nearest neighbor estimators* use the span produced by a fixed number of design points that are closest to x.

13.1 KERNEL ESTIMATORS

Let $K(x)$ be a real-valued function for assigning local weights to the linear estimator, that is,

$$y(x) = \sum K\left(\frac{x - x_i}{h}\right) y_i.$$

If $K(u) \propto \mathbf{1}(|u| \leq 1)$ then a fitted curve based on $K(\frac{x-x_i}{h})$ will estimate $m(x)$ using only design points within h units of x. Usually it is assumed that $\int_R K(x)dx = 1$, so any bounded probability density could serve as a kernel. Unlike kernel functions used in density estimation, now $K(x)$ also can take negative values, and in fact such unrestricted kernels are needed to achieve optimal estimators in the asymptotic sense. An example is the *beta kernel* defined as

$$K(x) = \frac{1}{B(1/2, \gamma+1)} \left(1 - x^2\right)^{\gamma} \mathbf{1}(|x| \leq 1), \ \gamma = 0, 1, 2 \ldots \tag{13.1}$$

With the added parameter γ, the beta-kernel is remarkably flexible. For $\gamma = 0$, the beta kernel becomes uniform. If $\gamma = 1$ we get the Epanechikov kernel, $\gamma = 2$ produces the biweight kernel, $\gamma = 3$ the triweight, and so on (see Figure 11.4 on p. 209). For γ large enough, the beta kernel is close the Gaussian kernel

$$K(x) = \frac{1}{\sqrt{2\pi\sigma^2}} e^{-\frac{x}{2\sigma^2}},$$

with $\sigma^2 = 1/(2\gamma + 3)$, which is the variance of densities from (13.1). For example, if $\gamma = 10$, then $\int_{-1}^{1} \left(K(x) - \sigma^{-1}\phi(x/\sigma)\right)^2 dx \approx 0.00114$, where $\sigma = 1/\sqrt{2\gamma + 3}$. Define a scaling coefficient h so that

$$K_h(x) = \frac{1}{h} K\left(\frac{x}{h}\right), \tag{13.2}$$

where h is the associated *bandwidth*. By increasing h, the kernel function spreads weight away from its center, thus giving less weight to those data points close to x and sharing the weight more equally with a larger group of design points. A family of beta kernels and the Epanechikov kernel are given in Figure 13.2.

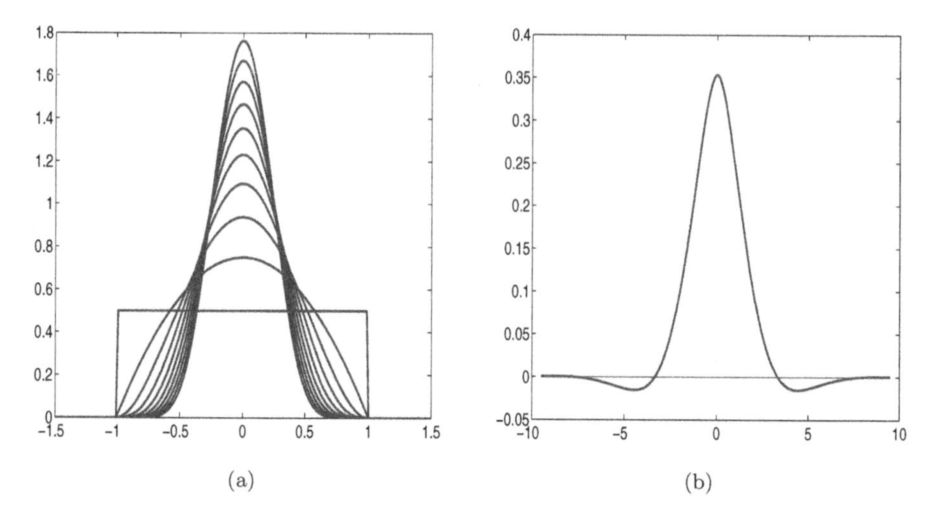

Fig. 13.2 (a) A family of symmetric beta kernels; (b) $K(x) = \frac{1}{2}\exp\{-|x|/\sqrt{2}\}\sin(|x|/\sqrt{2} - \pi/4)$.

13.1.1 Nadaraya-Watson Estimator

Nadaraya (1964) and Watson (1964) independently published the earliest results on for smoothing functions (but this is debateable), and the Nadaraya-Watson Estimator (NWE) of $m(x)$ is defined as

$$\hat{m}(x) = \frac{\sum_{i=1}^{n} K_h(X_i - x)Y_i}{\sum_{i=1}^{n} K_h(X_i - x)}. \tag{13.3}$$

For x fixed, the value $\hat{\theta}$ that minimizes

$$\sum_{i=1}^{n}(Y_i - \theta)^2 K_h(X_i - x), \tag{13.4}$$

is of the form $\sum_{i=1}^{n} a_i Y_i$. The Nadaraya-Watson estimator is the minimizer of (13.4) with $a_i = K_h(X_i - x)/\sum_{i=1}^{n} K_h(X_i - x)$.

Although several competing kernel-based estimators have been derived since, the NWE provided the basic framework for kernel estimators, including local polynomial fitting which is described later in this section. The MATLAB function

```
nada_wat(x0, X, Y, bw)
```

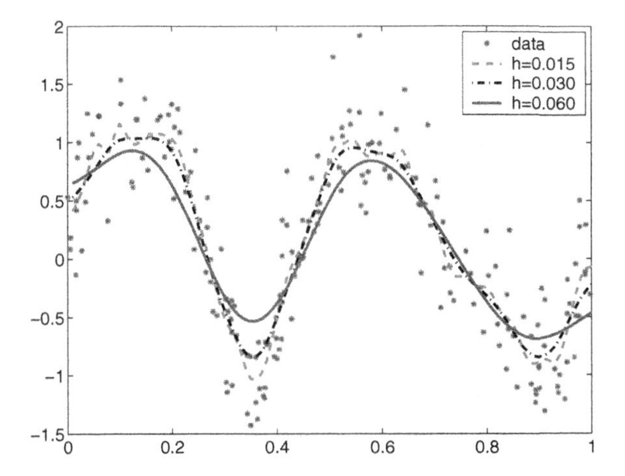

Fig. 13.3 Nadaraya-Watson Estimators for different values of bandwidth.

computes the Nadaraya-Watson kernel estimate at $x =$ `x0`. Here, (X,Y) are input data, and `bw` is the bandwidth.

Example 13.1 Noisy pairs (X_i, Y_i), $i = 1, \dots, 200$ are generated in the following way:

```
>>    x=sort(rand(1,200));
>>    y=sort(rand(1,200));
>>    y=sin(4*pi*y)+0.3*randn(1,200);
```

Three bandwidths are selected $h = 0.015, 0.030$, and 0.060. The three Nadaraya-Watson Estimators are shown in Figure 13.3. As expected, the estimators constructed with the larger bandwidths appear smoother than those with smaller bandwidths.

13.1.2 Gasser-Müller Estimator.

The Gasser-Müller estimator proposed in 1979 uses areas of the kernel for the weights. Suppose X_i are ordered, $X_1 \leq X_2 \cdots \leq X_n$. Let $X_0 = -\infty$ and $X_{n+1} = \infty$ and define midpoints $s_i = (X_i + X_{i+1})/2$. Then

$$\hat{m}(x) = \sum_{i=1}^{n} Y_i \int_{s_{i-1}}^{s_i} K_h(u - x) du. \tag{13.5}$$

The Gasser-Müller estimator is the minimizer of (13.4) with the weights $a_i = \int_{s_{i-1}}^{s_i} K_h(u - x) du$.

13.1.3 Local Polynomial Estimator

Both Nadaraya-Watson and Gasser-Müller estimators are *local constant fit* estimators, that is, they minimize weighted squared error $\sum_{i=1}^{n}(Y_i - \theta)^2 \omega_i$ for different values of weights ω_i. Assume that for z in a small neighborhood of x the function $m(z)$ can well be approximated by a polynomial of order p :

$$m(z) \approx \sum_{j=0}^{p} \beta_j (z - x)^j$$

where $\beta_j = m^{(j)}(x)/j!$. Instead of minimizing (13.4), the local polynomial (LP) estimator minimizes

$$\sum_{i=1}^{n} \left(Y_i - \sum_{j=0}^{p} \beta_j (X_i - x)^j \right)^2 K_h(X_i - x) \tag{13.6}$$

over $\beta_1, \dots, \beta_p$. Assume, for a fixed x, $\hat{\beta}_j, j = 0, \dots, p$ minimize (13.6). Then, $\hat{m}(x) = \hat{\beta}_0$, and an estimator of jth derivative of m is

$$\hat{m}^{(j)}(x) = j!\hat{\beta}_j,\ j = 0, 1, \dots, p. \tag{13.7}$$

If $p = 0$, that is, if the polynomials are constants, the local polynomial estimator is Nadaraya-Watson. It is not clear that the estimator $\hat{m}(x)$ for general p is a locally weighted average of responses, (of the form $\sum_{i=1}^{n} a_i Y_i$) as are the Nadaraya-Watson and Gasser-Müller estimators. The following representation of the LP estimator makes its calculation easy via the weighted least square problem. Consider the $n \times (p + 1)$ matrix depending on x and $X_i - x,\ i = 1, \dots, n$.

$$X = \begin{pmatrix} 1 & X_1 - x & (X_1 - x)^2 & \dots & (X_1 - x)^p \\ 1 & X_2 - x & (X_2 - x)^2 & \dots & (X_2 - x)^p \\ \dots & \dots & \dots & & \dots \\ 1 & X_n - x & (X_n - x)^2 & \dots & (X_n - x)^p \end{pmatrix}$$

Define also the diagonal weight matrix W and response vector Y:

$$W = \begin{pmatrix} K_h(X_1 - x) & & & \\ & K_h(X_2 - x) & & \\ & & \vdots & \\ & & & K_h(X_n - x) \end{pmatrix}, \quad Y = \begin{pmatrix} Y_1 \\ Y_2 \\ \vdots \\ Y_n \end{pmatrix}.$$

Then the minimization problem can be written as $(Y - X\beta)'W(Y - X\beta)$. The solution is well known: $\hat{\beta} = (X'WX)^{-1}X'W\ Y$. Thus, if $(a_1\ a_2\ \dots a_n)$ is the first row of matrix $(X'WX)^{-1}X'W$, $\hat{m}(x) = a \cdot Y = \sum_i a_i Y_i$. This representation (in matrix form) provides an efficient and elegant way to calculate the LP regression estimator. In MATLAB, use the function

```
lpfit(x, y, p, h),
```

where (x, y) is the input data, p is the order and h is the bandwidth.

For general p, the first row $(a_1\ a_2\ \ldots\ a_n)$ of $(X'WX)^{-1}X'W$ is quite complicated. Yet, for $p = 1$ (the local linear estimator), the expression for $\hat{m}(x)$ simplifies to

$$\hat{m}(x) = \frac{1}{n}\sum_{i=1}^{n} \frac{(S_2(x) - S_1(x)(X_i - x))K_h(X_i - x)}{S_2(x)S_0(x) - S_1(x)^2} Y_i,$$

where $S_j = \frac{1}{n}\sum_{i=1}^{n}(X_i - x)^j K_h(X_i - x)$, $j = 0, 1$, and 2. This estimator is implemented in MATLAB by the function

```
loc_lin.m.
```

13.2 NEAREST NEIGHBOR METHODS

As an alternative to kernel estimators, nearest neighbor estimators define points local to X_i not through a kernel bandwidth, which is a fixed strip along the x-axis, but instead on a set of points closest to X_i. For example, a neighborhood for x might be defined to be the closest k design points on either side of x, where k is a positive integer such that $k \leq n/2$. Nearest neighbor methods make sense if we have spaces with clustered design points followed by intervals with sparse design points. The nearest neighbor estimator will increase its span if the design points are spread out. There is added complexity, however, if the data includes repeated design points. for purposes of illustration, we will assume this is not the case in our examples.

Nearest neighbor and kernel estimators produce similar results, in general. In terms of bias and variance, the nearest neighbor estimator described in this section performs well if the variance decreases more than the squared bias increases (see Altman, 1992).

13.2.1 LOESS

William Cleveland (1979), Figure 13.4(a), introduced a curve fitting regression technique called LOWESS, which stands for *locally weighted regression scatter plot smoothing*. Its derivative, LOESS[1], stands more generally for a local regression, but many researchers consider LOWESS and LOESS as synonyms.

[1]Term actually defined by geologists as deposits of fine soil that are highly susceptible to wind erosion. We will stick with our less silty mathematical definition in this chapter.

(a) (b)

Fig. 13.4 (a) William S. Cleveland, Purdue University; (b) Geological Loess.

Consider a multiple linear regression set up with a set of regressors $\mathcal{X}_i = X_{i1}, \ldots, X_{ik}$ to predict Y_i, $i = 1, \ldots, n$. If $Y = f(x_1, \ldots, x_k) + \epsilon$, where $\epsilon \sim \mathcal{N}(0, \sigma^2)$. Adjacency of the regressors is defined by a distance function $d(\mathcal{X}, \mathcal{X}^*)$. For $k = 2$, if we are fitting a curve at (X_{r1}, X_{r2}) with $1 \leq r \leq n$, then for $i = 1, \ldots, n$,

$$d_i = \sqrt{(X_{i1} - X_{r1})^2 + (X_{i2} - X_{r2})^2}.$$

Each data point influences the regression at (X_{r1}, X_{r2}) according to its distance to that point. In the LOESS method, this is done with a tri-cube weight function

$$w_i = \begin{cases} \left(1 - \left(\frac{d_i}{d_q}\right)^3\right)^3 & d_i \leq d_q \\ 0 & d_i > d_q \end{cases},$$

where only q of n points closest to $\mathcal{X}_i$ are considered to be "in the neighborhood" of $\mathcal{X}_i$, and d_q is the distance of the furthest $\mathcal{X}_i$ that is in the neighborhood. Actually, many other weight functions can serve just as well as the tri-weight function; requirements for w_i are discussed in Cleveland (1979).

If q is large, the LOESS curve will be smoother but less sensitive to nuances in the data. As q decreases, the fit looks more like an interpolation of the data, and the curve is zig-zaggy. Usually, q is chosen so that $0.10 \leq q/n \leq 0.25$. Within the window of observations in the neighborhood of $\mathcal{X}$, we construct the LOESS curve $\mathrm{Y}(\mathcal{X})$ using either linear regression (called first order) or quadratic (second order).

There are great advantages to this curve estimation scheme. LOESS does not require a specific function to fit the model to the data; only a smoothing parameter ($\alpha = q/n$) and local polynomial (first or second order) are required.

Given that complex functions can be modeled with such a simple precept, the LOESS procedure is popular for constructing a regression equation with cloudy, multidimensional data.

On the other hand, LOESS requires a large data set in order for the curve-fitting to work well. Unlike least-squares regression (and, for that matter, many non-linear regression techniques), the LOESS curve does not give the user a simple math formula to relate the regressors to the response. Because of this, one of the most valuable uses of LOESS is as an exploratory tool. It allows the practitioner to visually check the relationship between a regressor and response no matter how complex or convoluted the data appear to be.

In MATLAB, use the function

```
loess(x,y,newx,a,b)
```

where `x` and `y` represent the bivariate data (vectors), `newx` is the vector of fitted points, `a` is the smoothing parameter (usually 0.10 or 0.25), and `b` is the order of polynomial (1 or 2). `loess` produces an output equal to `newx`.

Example 13.2 Consider the motorcycle accident data found in Schmidt, Matter and Schüler (1981). The first column is time, measured in milliseconds, after a simulated impact of a motorcycle. The second column is the acceleration factor of the driver's head (accel), measured in g ($9.8m/s^2$). Time versus accel is graphed in Figure 13.5. The MATLAB code below creates a LOESS curve to model acceleration as a function of time (also in the figure). Note how the smoothing parameter influences the fit of the curve.

```
>> load motorcycle.dat
>> time = motorcycle(:,1);
>> accel = motorcycle(:,2);
>> loess(time, accel, newx, 0.20, 1);
>> plot(time, accel,'o');
>> hold on
>> plot(time, newx,'-');
```

For regression with two regressors (`x,y`), use the MATLAB function:

```
loess2(x,y,z,newx,newy,a,b)
```

that contains inputs (`x,y,z`) and creates a surface fit in (`newx,newy`).

13.3 VARIANCE ESTIMATION

In constructing confidence intervals for $m(x)$, the variance estimate based on the smooth linear regression (with pooled-variance estimate) will produce the

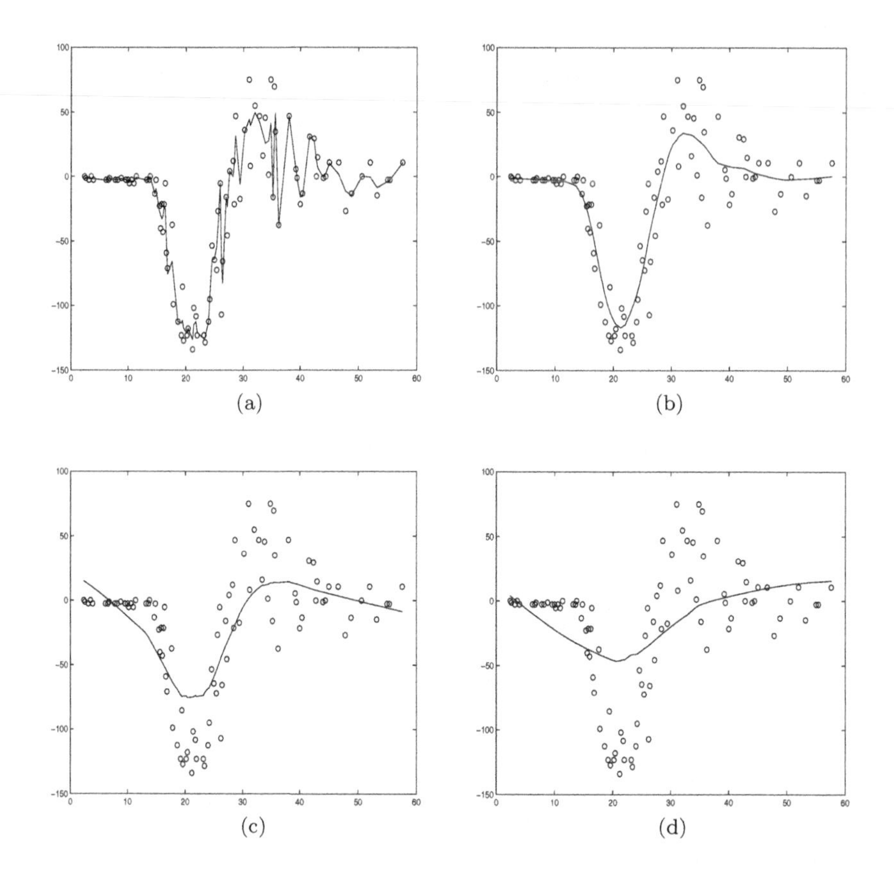

Fig. 13.5 Loess curve-fitting for Motorcycle Data using (a) $\alpha = 0.05$, (b) $\alpha = 0.20$, (c) $\alpha = 0.50$, and (d) $\alpha = 0.80$.

narrowest interval. But if the estimate is biased, the confidence interval will have poor coverage probability. An estimator of $m(x)$ based only on points near x will produce a poor estimate of variance, and as a result is apt to generate wide, uninformative intervals.

One way to avoid the worst pitfalls of these two extremes is to detrend the data locally and use the estimated variance from the detrended data. Altman and Paulson (1993) use psuedo-residuals $\tilde{e}_i = y_i - (y_{i+1} + y_{i-1})/2$ to form a variance estimator

$$\tilde{\sigma}^2 = \frac{2}{3(n-2)} \sum_{i=1}^{n-1} \tilde{e}_i^2,$$

where $\tilde{\sigma}^2/\sigma^2$ is distributed χ^2 with $(n-2)/2$ degrees of freedom. Because both the kernel and nearest neighbor estimators have linear form in y_i, a

Fig. 13.6 I. J. Schoenberg (1903–1990).

$100(1-\alpha)\%$ confidence interval for $m(t)$ can be approximated with

$$\hat{m}(t) \pm t_r(\alpha)\sqrt{\tilde{\sigma}^2 \sum a^2{}_i},$$

where $r = (n-2)/2$.

13.4 SPLINES

> **spline** (splīne) *n.* **1.** A flexible piece of wood, hard rubber, or metal used in drawing curves. **2.** A wooden or metal strip; a slat.
>
> The American Heritage Dictionary

Splines, in the mathematical sense, are concatenated piecewise polynomial functions that either interpolate or approximate the scatterplot generated by n observed pairs, $(X_1, Y_1), \dots, (X_n, Y_n)$. Isaac J. Schoenberg, the "father of splines," was born in Galatz, Romania, on April 21, 1903, and died in Madison, Wisconsin, USA, on February 21, 1990. The more than 40 papers on splines written by Schoenberg after 1960 gave much impetus to the rapid development of the field. He wrote the first several in 1963, during a year's leave in Princeton at the Institute for Advanced Study; the others are part of his prolific output as a member of the Mathematics Research Center at the University of Wisconsin-Madison, which he joined in 1965.

13.4.1 Interpolating Splines

There are many varieties of splines. Although piecewise constant, linear, and quadratic splines easy to construct, cubic splines are most commonly used because they have a desirable extremal property.

Denote the cubic spline function by $m(x)$. Assume $X_1, X_2, \dots, X_n$ are ordered and belong to a finite interval $[a, b]$. We will call $X_1, X_2, \dots, X_n$ *knots*. On each interval $[X_{i-1}, X_i]$, $i = 1, 2, \dots, n+1, X_0 = a, X_{n+1} = b$, the spline $m(x)$ is a polynomial of degree less than or equal to 3. In addition, these polynomial pieces are connected in such a way that the second derivatives are continuous. That means that at the knot points $X_i, i = 1, \dots, n$ where the two polynomials from the neighboring intervals meet, the polynomials have common tangent and curvature. We say that such functions belong to $\mathbb{C}^2[a, b]$, the space of all functions on $[a, b]$ with continuous second derivative.

The cubic spline is called *natural* if the polynomial pieces on the intervals $[a, X_1]$ and $[X_n, b]$ are of degree 1, that is, linear. The following two properties distinguish natural cubic splines from other functions in $\mathbb{C}^2[a, b]$.

Unique Interpolation. Given the n pairs, $(X_1, Y_1), \dots, (X_n, Y_n)$, with distinct knots X_i there is a *unique* natural cubic spline m that interpolates the points, that is, $m(X_i) = Y_i$.

Extremal Property. Given n pairs, $(X_1, Y_1), \dots, (X_n, Y_n)$, with distinct and ordered knots X_i, the natural cubic spline $m(x)$ that interpolates the points also minimizes the curvature on the interval $[a, b]$, where $a < X_1$ and $X_n < b$. In other words, for any other function $g \in \mathbb{C}^2[a, b]$,

$$\int_a^b (m''(t))^2 dt \leq \int_a^b (g''(t))^2 dt.$$

Example 13.3 One can "draw" the letter $\mathcal{V}$ using a simple spline. The bivariate set of points (X_i, Y_i) below lead the cubic spline to trace a shape reminiscent of the script letter $\mathcal{V}$. The result of MATLAB program is given in Figure 13.7.

```
>> x = [10 40 40 20  60 50 25 16 30 60 80 75 65 100];
>> y = [85 90 65 55 100 70 35 10 10 36 60 65 55  50];
>> t=1:length(x);
>> tt=linspace(t(1),t(end),250);
>> xx=spline(t,x,tt);
>> yy=spline(t,y,tt);
>> plot(xx,yy,'-','linewidth',2), hold on
>> plot(x,y,'o','markersize' ,6)
>> axis('equal'),axis('off')
```

Fig. 13.7 A cubic spline drawing of letter $\mathcal{V}$.

Example 13.4 In MATLAB, the function `csapi.m` computes the cubic spline interpolant, and for the following x and y,

```
>>  x = (4*pi)*[0 1 rand(1,20)]; y = sin(x);
>>  cs = csapi(x,y);
>>  fnplt(cs); hold on, plot(x,y,'o')
>>  legend('cubic spline','data'), hold off
```

the interpolation is plotted in Figure 13.8(a), along with the data. A surface interpolation by 2-d splines is demonstrated by the following MATLAB code and Figure 13.8(b).

```
>>  x = -1:.2:1; y=-1:.25:1; [xx, yy] = ndgrid(x,y);
>>  z = sin(10*(xx.^2+yy.^2)); pp = csapi({x,y},z);
>>  fnplt(pp)
```

There are important distinctions between spline regressions and regular polynomial regressions. The latter technique is applied to regression curves where the practitioner can see an interpolating quadratic or cubic equation that locally matches the relationship between the two variables being plotted. The Stone-Weierstrass theorem (Weierstrass, 1885) tells us that any continuous function in a closed interval can be approximated well by some polynomial. While a higher order polynomial will provide a closer fit at any particular point, the loss of parsimony is not the only potential problem of over fitting; unwanted oscillations can appear between data points. Spline functions avoid this pitfall.

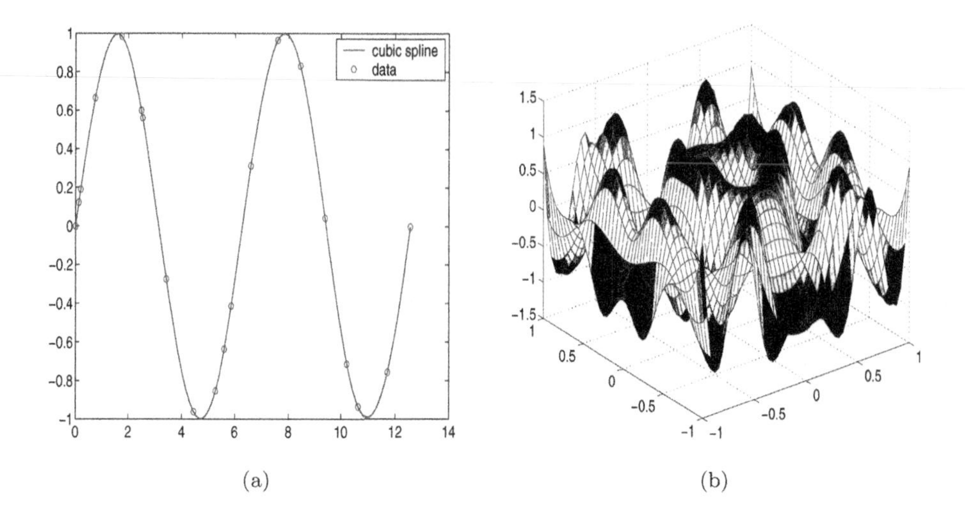

(a) (b)

Fig. 13.8 (a) Interpolating sine function; (b) Interpolating a surface.

13.4.2 Smoothing Splines

Smoothing splines, unlike interpolating splines, may not contain the points of a scatterplot, but are rather a form of nonparametric regression. Suppose we are given bivariate observations (X_i, Y_i), $i = 1, \ldots, n$. The continuously differentiable function $\hat{m}$ on $[a, b]$ that minimizes the functional

$$\sum_{i=1}^{n}(Y_i - m(X_i))^2 + \lambda \int_a^b (m''(t))^2 dt \qquad (13.8)$$

is exactly a natural cubic spline. The cost functional in (13.8) has two parts: $\sum_{i=1}^{n}(Y_i - m(X_i))^2$ is minimized by an interpolating spline, and $\int_a^b (m''(t))^2 dt$ is minimized by a straight line. The parameter λ trades off the importance of these two competing costs in (13.8). For small λ, the minimizer is close to an interpolating spline. For λ large, the minimizer is closer to a straight line.

Although natural cubic smoothing splines do not appear to be related to kernel-type estimators, they can be similar in certain cases. For a value of x that is away from the boundary, if n is large and λ small, let

$$\hat{m}(x) = \frac{1}{n} \sum_{i=1}^{n} \frac{K_{h_i}(X - i - x)}{f(X_i)} Y_i,$$

where f is the density of the X's, $h_i = [\lambda/(nf(X_i))]^{1/4}$ and the kernel K is

$$K(x) = \frac{1}{2}\exp\{-|x|/\sqrt{2}\}\ \sin(|x|/\sqrt{2} + \pi/4). \tag{13.9}$$

As an alternative to minimizing (13.8), the following version is often used:

$$p\sum_{i=1}^{n}(Y_i - m(X_i))^2 + (1-p)\int_a^b (m''(t))^2 dt \tag{13.10}$$

In this case, $\lambda = (1-p)/p$. Assume that h is an average spacing between the neighboring X's. An automatic choice for p is $6(6+h^3)$ or $\lambda = h^3/6$.

Smoothing Splines as Linear Estimators. The spline estimator is linear in the observations, $\hat{\mathbf{m}} = S(\lambda)\mathbf{Y}$, for a smoothing matrix $S(\lambda)$. The Reinsch algorithm (Reinsch, 1967) efficiently calculates S as

$$S(\lambda) = \left(I + \lambda Q R^{-1} Q'\right)^{-1}, \tag{13.11}$$

where Q and R are structured matrices of dimensions $n \times (n-2)$ and $(n-2) \times (n-2)$, respectively:

$$Q = \begin{pmatrix} q_{12} & & & \\ q_{22} & q_{23} & & \\ q_{32} & q_{33} & & \\ & q_{43} & & \\ & & \cdots & \\ & & & q_{n-2,n-1} \\ & & & q_{n-1,n-1} \\ & & & q_{n,n-1} \end{pmatrix}, \quad R = \begin{pmatrix} r_{22} & r_{23} & & \\ r_{32} & r_{33} & & \\ & r_{43} & & \\ & & \cdots & \\ & & & q_{n-2,n-1} \\ & & & q_{n-1,n-1} \end{pmatrix},$$

with entries

$$q_{ij} = \begin{cases} \frac{1}{h_{j-1}}, & i = j-1 \\ -(\frac{1}{h_{j-1}} + \frac{1}{h_j}), & i = j \\ \frac{1}{h_j}, & i = j+1 \end{cases}$$

and

$$r_{ij} = \begin{cases} \frac{1}{6}h_{j-1}, & i = j-1 \\ \frac{1}{3}(h_{j-1} + h_j), & i = j \\ \frac{1}{6}h_j, & i = j+1. \end{cases}$$

The values h_i are spacings between the X_i's, i.e., $h_i = X_{i+1} - X_i$, $i = 1, \ldots, n-1$. For details about the Reinsch Algorithm, see Green and Silverman (1994).

13.4.3 Selecting and Assessing the Regression Estimator

Let $\hat{m}_h(x)$ be the regression estimator of $m(x)$, obtained by using the set of n observations $(X_1, Y_1), \ldots, (X_n, Y_n)$, and parameter h. Note that for kernel-type estimators, h is the bandwidth, but for splines, h is λ in (13.8). Define the avarage mean-square error of the estimator $\hat{m}_h$ as

$$AMSE(h) = \frac{1}{n}\sum_{i=1}^{n} \mathbb{E}\left[\hat{m}(X_i) - m(X_i)\right]^2.$$

Let $\hat{m}_{(i)h}(x)$ be the estimator of $m(x)$, based on bandwidth parameter h, obtained by using all the observation pairs except the pair (X_i, Y_i). Define the cross-validation score $CV(h)$ depending on the bandwith/trade-off parameter h as

$$CV(h) = \frac{1}{n}\sum_{i=1}^{n} \left[Y_i - \hat{m}_{(i)h}(x)\right]^2. \qquad (13.12)$$

Because the expected $CV(h)$ score is proportional to the $AMSE(h)$ or, more precisely,

$$\mathbb{E}\left[CV(h)\right] \approx AMSE(h) + \sigma^2,$$

where σ^2 is constant variance of errors ϵ_i, the value of h that minimizes $CV(h)$ is likely, on average, to produce the best estimators.

For smoothing splines, and more generally, for linear smoothers $\hat{\mathbf{m}} = S(h)\mathbf{y}$, the computationally demanding procedure in (13.12) can be simplified by

$$CV(h) = \frac{1}{n}\sum_{i=1}^{n} \left[\frac{Y_i - \hat{m}_h(x)}{1 - S_{ii}(h)}\right]^2, \qquad (13.13)$$

where $S_{ii}(h)$ is the diagonal element in the smoother (13.11). When n is large, constructing the smoothing matrix $S(h)$ is computationally difficult. There are efficient algorithms (Hutchison and de Hoog, 1985) that calculate only needed diagonal elements $S_{ii}(h)$, for smoothing splines, with calculational cost of $O(n)$.

Another simplification in finding the best smoother is the generalized cross-validation criterion, GCV. The denominator in (13.13) $1 - S_{ii}(h)$ is replaced by overall average $1 - n^{-1}\sum_{i=1}^{n} S_{ii}(h)$, or in terms of its trace, $1 - n^{-1} tr S(h)$. Thus

$$GCV(h) = \frac{1}{n}\sum_{i=1}^{n} \left[\frac{Y_i - \hat{m}_h(x)}{1 - tr S(h)/n}\right]^2. \qquad (13.14)$$

Example 13.5 Assume that $\hat{m}$ is a spline estimator and that $\lambda_1, \ldots, \lambda_n$ are eigenvalues of matrix $QR^{-1}Q'$ from (13.11). Then, $trS(h) = \sum_{i=1}^{n}(1+h\lambda_i)^{-1}$. The GCV criterion becomes

$$GCV(h) = \frac{nRSS(h)^2}{\left[n - \sum_{i=1}^{n} \frac{1}{1+h\lambda_i}\right]^2}.$$

13.4.4 Spline Inference

Suppose that the estimator $\hat{m}$ is a linear combination of the Y_is,

$$\hat{m}(x) = \sum_{i=1}^{n} a_i(x) Y_i.$$

Then

$$\mathbb{E}(\hat{m}(x)) = \sum_{i=1}^{n} a_i(x) m(X_i), \quad \text{and} \quad \mathbb{V}\mathrm{ar}(\hat{m}(x)) = \left(\sum_{i=1}^{n} a_i(x)^2\right)\sigma^2.$$

Given $x = X_j$ we see that $\hat{m}$ is unbiased, that is, $\mathbb{E}\hat{m}(X_j) = m(X_j)$ only if all $a_i = 0,\ i \neq j$.

On the other hand, variance is minimized if all a_i are equal. This illustrates, once again, the trade off between the estimator's bias and variance. The variance of the errors is supposed to be constant. In linear regression we estimated the variance as

$$\hat{\sigma}^2 = \frac{RSS}{n-p},$$

where p is the number of free parameters in the model. Here we have an analogous estimator,

$$\hat{\sigma}^2 = \frac{RSS}{n - tr(S)},$$

where $RSS = \sum_{i=1}^{n} \left[Y_i - \hat{m}(X_i)\right]^2$.

13.5 SUMMARY

This chapter has given a brief overview of both kernel estimators and local smoothers. An example from Gasser et al. (1984) shows that choosing a smoothing method over a parametric regression model can make a crucial difference in the conclusions of a data analysis. A parametric model by Preece and Baines (1978) was constructed for predicting the future height of a human based on measuring children's heights at different stages of development. The parametric regression model they derived for was particularly complicated but provided a great improvement in estimating the human growth curve. Published six years later, the nonparametric regression by Gasser et al. (1984) brought out an important nuance of the growth data that could not be modeled with the Preece and Baines model (or any model that came before it). A subtle growth spurt which seems to occur in children around seven years in age. Altman (1992) notes that such a growth spurt was discussed in past medical papers, but had "disappeared from the literature following the development of the parametric models which did not allow for it."

13.6 EXERCISES

13.1. Describe how the LOESS curve can be equivalent to least-squares regression.

13.2. Data set `oj287.dat` is the light curve of the blazar OJ287. Blazars, also known as *BL Lac Objects* or *BL Lacertaes*, are bright, extragalactic, starlike objects that can vary rapidly in their luminosity. Rapid fluctuations of blazar brightness indicate that the energy producing region is small. Blazars emit polarized light that is featureless on a light plot. Blazars are interpreted to be active galaxy nuclei, not so different from quasars. From this interpretation it follows that blazars are in the center of an otherwise normal galaxy, and are probably powered by a supermassive black hole. Use a local-polynomial estimator to analyze the data in `oj287.dat` where column 1 is the julian time and column 2 is the brightness. How does the fit compare for the three values of p in $\{0, 1, 2\}$?

13.3. Consider the function

$$s(x) = \begin{cases} 1 - x + x^2 - x^3 & 0 < x < 1 \\ -2(x-1) - 2(x-1)^2 & 1 < x < 2 \\ -4 - 6(x-2) - 2(x-2)^2 & 2 < x < 3\ . \end{cases}$$

Does $s(x)$ define a smooth cubic spline on [0, 3] with knots 1, and 2? If so, plot the 3 polynomials on [0,3].

13.4. In MATLAB, open the data file `earthquake.dat` which contains water level records for a set of six wells in California. The measurements are made across time. Construct a LOESS smoother to examine trends in the data. Where does LOESS succeed? Where does it fail to capture the trends in the data?

13.5. Simulate a data set as follows:

```
x= r and(1,100);
x = sort(x);
y = x.^2 + 0.1 * randn(1,100);
```

Fit an interpolating spline to the simulated data as shown in Figure 13.5(a). The dotted line is $y = x^2$.

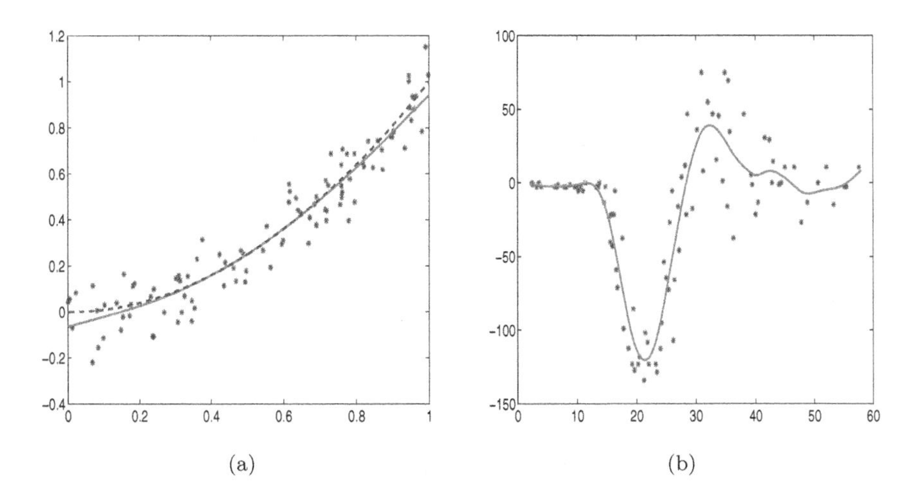

Fig. 13.9 (a) Square plus noise, (b) Motorcycle Data: Time (X_i) and Acceleration (Y_i), $i = 1, \ldots 82$.

13.6. Refer to the motorcycle data from Figure 13.5. Fit a spline to the data. Variable `time` is the time in milliseconds and `accel` is the acceleration of a head measured in (g). See Figure 13.5 (b) as an example.

13.7. Star S in the Big Dipper constellation (Ursa Major) has a regular variation in its apparent magnitude[2]:

θ	-100	-60	-20	20	60	100	140
magnitude	8.37	9.40	11.39	10.84	8.53	7.89	8.37

The magnitude is known to be periodic with period 240, so that the magnitude at $\theta = -100$ is the same as at $\theta = 140$. The m-spline `y = csape(x,y,'periodic')` constructs a cubic spline whose first and second derivatives are the same at the ends of the interval. Use it to

[2]L. Campbell and L. Jacchia, *The Story of Variable Stars*, The Blackiston Co., Philadelphia, 1941.

interpolate the data. Plot the data and the interpolating curve in the same figure. Estimate the magnitude at $\theta = 0$.

13.8. Use the smoothing splines to analyze the data in `oj287.dat` that was described in Exercise 13.2. For your reference, the data and implementation of spline smoothing are given in Figure 13.10.

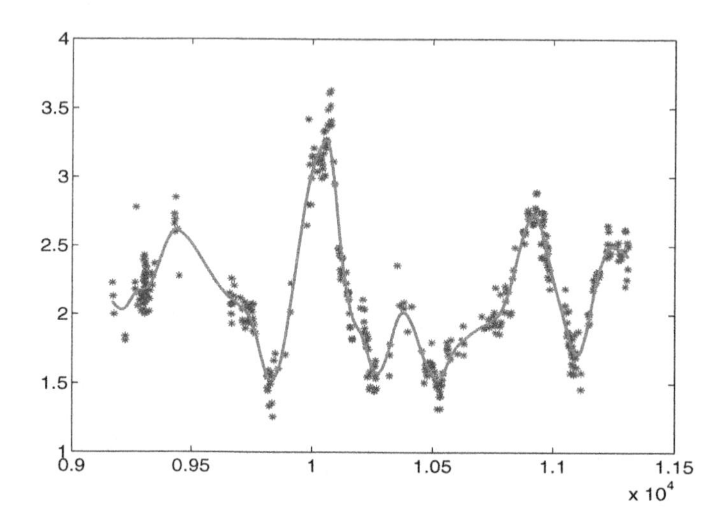

Fig. 13.10 Blazar OJ287 luminosity.

REFERENCES

Altman, N. S. (1992), "An Introduction to Kernel and Nearest Neighbor Nonparametric Regression," *American Statistician*, 46, 175-185.

Altman, N. S., and Paulson, C. P. (1993), "Some Remarks about the Gasser-Sroka-Jennen-Steinmetz Variance Estimator," *Communications in Statistics, Theory and Methods*, 22, 1045-1051.

Anscombe, F. (1973), "Graphs in Statistical Analysis," *American Statistician*, 27, 17-21.

Cleveland, W. S. (1979), "Robust Locally Weighted Regression and Smoothing Scatterplots," *Journal of the American Statistical Association*, 74, 829-836.

De Boor, C. (1978), *A Practical Guide to Splines*, New York: Springer Verlag.

Gasser, T., and Müller, H. G. (1979), "Kernel Estimation of Regression Functions," in *Smoothing Techniques for Curve Estimation*, Eds. Gasser and Rosenblatt, Heidelberg: Springer Verlag.

Gasser, T., Müller, H. G., Köhler, W., Molinari, L., and Prader, A. (1984), "Nonparametric Regression Analysis of Growth Curves," *Annals of Statistics*, 12, 210-229.

Green, P.J., and Silverman, B.W. (1994), *Nonparametric Regression and Generalized Linear Models: A Roughness Penalty Approach*, London: Chapman and Hall.

Huber, P. J. (1973), "Robust Regression: Asymptotics, Conjectures, and Monte Carlo," *Annals of Statistics*, 1, 799-821.

Hutchinson, M. F., and de Hoog, F. R. (1985), "Smoothing noisy data with spline functions," *Numerical Mathematics*, 1, 99-106.

Müller, H. G. (1987), "Weighted Local Regression and Kernel Methods for Nonparametric Curve Fitting," *Journal of the American Statistical Association*, 82, 231-238.

Nadaraya, E. A. (1964), "On Estimating Regression," *Theory of Probability and Its Applications*, 10, 186-190.

Preece, M. A., and Baines, M. J. (1978), "A New Family of Mathematical Models Describing the Human Growth Curve," *Annals of Human Biology*, 5, 1-24.

Priestley, M. B., and Chao, M. T. (1972), "Nonparametric Function Fitting," *Journal of the Royal Statistical Society*, Ser. B, 34, 385-392.

Reinsch, C. H. (1967), "Smoothing by Spline Functions," *Numerical Mathematics*, 10, 177-183.

Schmidt, G., Mattern, R., and Schüler, F. (1981), "Biomechanical Investigation to Determine Physical and Traumatological Differentiation Criteria for the Maximum Load Capacity of Head and Vertebral Column with and without Helmet under Effects of Impact," *EEC Research Program on Biomechanics of Impacts. Final Report Phase III, 65*, Heidelberg, Germany: Institut fur Rechtsmedizin.

Silverman, B. W. (1985), "Some Aspects of the Spline Smoothing Approach to Non-parametric Curve Fitting," *Journal of the Royal Statistical Society*, Ser. B, 47, 152.

Tufte, E. R. (1983), *The Visual Display of Quantitative Information*, Cheshire, CT: Graphic Press.

Watson, G. S. (1964), "Smooth Regression Analysis," *Sankhya, Series A*, 26, 359-372.

Weierstrass, K. (1885), "Über die analytische Darstellbarkeit sogenannter willkürlicher Functionen einer reellen Vernderlichen." Sitzungsberichte der Königlich Preußischen Akademie der Wissenschaften zu Berlin, 1885 (II). Erste Mitteilung (part 1) 633639; Zweite Mitteilung (part 2) 789-805.

14

Wavelets

It is error only, and not truth, that shrinks from inquiry.

Thomas Paine (1737–1809)

14.1 INTRODUCTION TO WAVELETS

Wavelet-based procedures are now indispensable in many areas of modern statistics, for example in regression, density and function estimation, factor analysis, modeling and forecasting of time series, functional data analysis, data mining and classification, with ranges of application areas in science and engineering. Wavelets owe their initial popularity in statistics to *shrinkage*, a simple and yet powerful procedure efficient for many nonparametric statistical models.

Wavelets are functions that satisfy certain requirements. The name *wavelet* comes from the requirement that they integrate to zero, "waving" above and below the x-axis. The diminutive in *wavelet* suggest its good localization. Other requirements are technical and needed mostly to ensure quick and easy calculation of the direct and inverse wavelet transform.

There are many kinds of wavelets. One can choose between smooth wavelets, compactly supported wavelets, wavelets with simple mathematical expressions, wavelets with short associated filters, etc. The simplest is the *Haar wavelet,* and we discuss it as an introductory example in the next section.

Examples of some wavelets (from Daubechies' family) are given in Figure 14.1. Note that scaling and wavelet functions in panels (a, b) in Figure 14.1 (Daubechies 4) are supported on a short interval (of length 3) but are not smooth; the other family member, Daubechies 16 (panels (e, f) in Figure 14.1) is smooth, but its support is much larger.

Like sines and cosines in Fourier analysis, wavelets are used as atoms in representing other functions. Once the wavelet (sometimes informally called *the mother wavelet*) $\psi(x)$ is fixed, one can generate a family by its translations and dilations, $\{\psi(\frac{x-b}{a}), (a, b) \in \mathbb{R}^+ \times \mathbb{R}\}$. It is convenient to take special values for a and b in defining the wavelet basis: $a = 2^{-j}$ and $b = k \cdot 2^{-j}$, where k and j are integers. This choice of a and b is called *critical sampling* and generates a sparse basis. In addition, this choice naturally connects multiresolution analysis in discrete signal processing with the mathematics of wavelets.

Wavelets, as building blocks in modeling, are localized well in both time and scale (frequency). Functions with rapid local changes (functions with discontinuities, cusps, sharp spikes, etc.) can be well represented with a minimal number of wavelet coefficients. This parsimony does not, in general, hold for other standard orthonormal bases which may require many "compensating" coefficients to describe discontinuity artifacts or local bursts.

Heisenberg's principle states that time-frequency models cannot be precise in the time and frequency domains simultaneously. Wavelets, of course, are subject to Heisenberg's limitation, but can adaptively distribute the time-frequency precision depending on the nature of function they are approximating. The economy of wavelet transforms can be attributed to this ability.

The above already hints at how the wavelets can be used in statistics. Large and noisy data sets can be easily and quickly transformed by a discrete wavelet transform (the counterpart of discrete Fourier transform). The data are coded by their wavelet coefficients. In addition, the descriptor "fast" in Fast Fourier transforms can, in most cases, be replaced by "faster" for the wavelets. It is well known that the computational complexity of the fast Fourier transformation is $O(n \cdot \log_2(n))$. For the fast wavelet transform the computational complexity goes down to $O(n)$. This means that the complexity of algorithm (in terms either of number of operations, time, or memory) is proportional to the input size, n.

Various data-processing procedures can now be done by processing the corresponding wavelet coefficients. For instance, one can do function smoothing by shrinking the corresponding wavelet coefficients and then back-transforming the shrunken coefficients to the original domain (Figure 14.2). A simple shrinkage method, thresholding, and some thresholding policies are discussed later.

An important feature of wavelet transforms is their *whitening* property. There is ample theoretical and empirical evidence that wavelet transforms reduce the dependence in the original signal. For example, it is possible, for any given stationary dependence in the input signal, to construct a biorthogonal

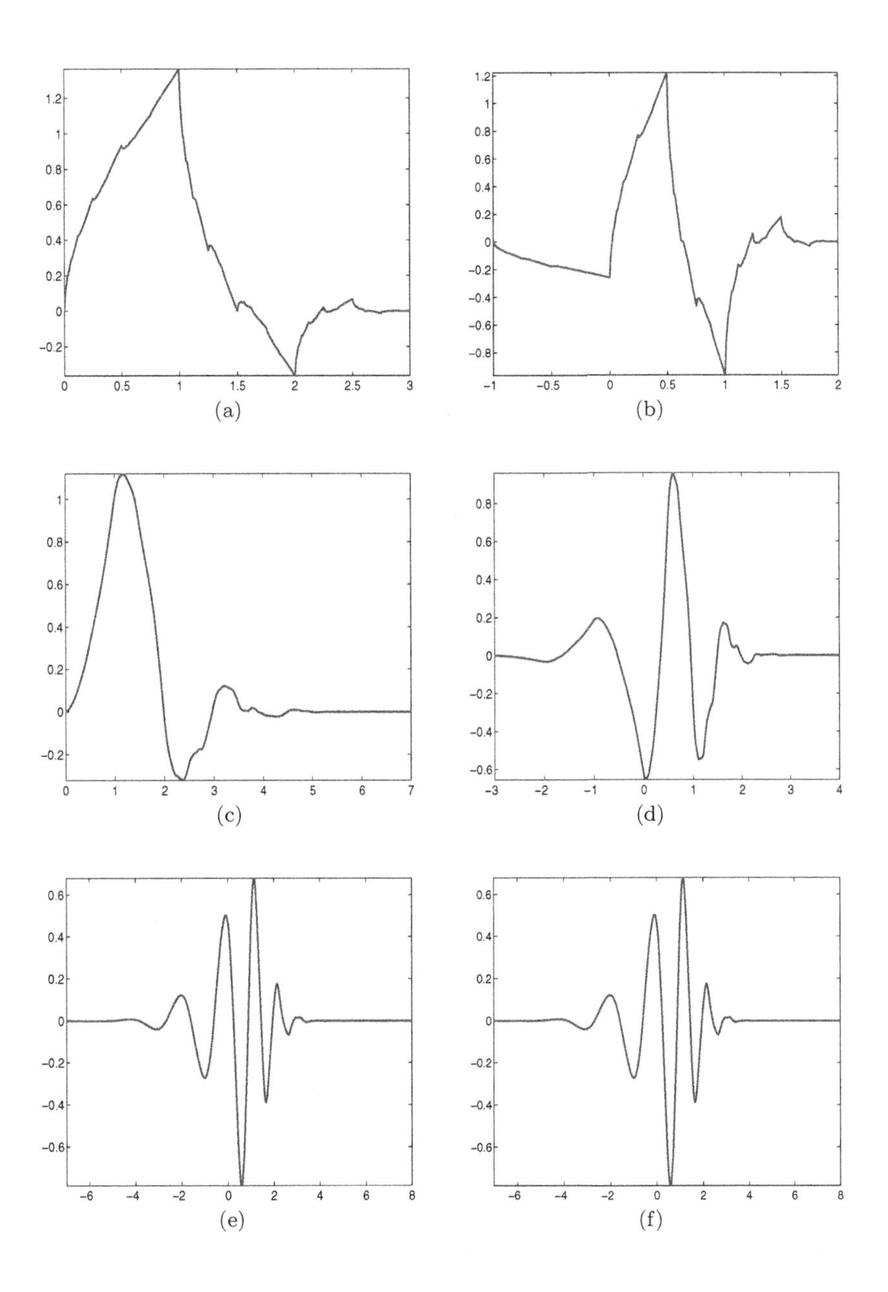

Fig. 14.1 Wavelets from the Daubechies family. Depicted are scaling functions (*left*) and wavelets (*right*) corresponding to (a, b) 4, (c, d) 8, and (e, f) 16 tap filters.

Fig. 14.2 Wavelet-based data processing.

wavelet basis such that the corresponding in the transform are uncorrelated (a wavelet counterpart of the so called Karhunen-Loève transform). For a discussion and examples, see Walter and Shen (2001).

We conclude this incomplete inventory of wavelet transform features by pointing out their sensitivity to self-similarity in data. The scaling regularities are distinctive features of self-similar data. Such regularities are clearly visible in the wavelet domain in the wavelet spectra, a wavelet counterpart of the Fourier spectra. More arguments can be provided: computational speed of the wavelet transform, easy incorporation of prior information about some features of the signal (smoothness, distribution of energy across scales), etc.

Basics on wavelets can be found in many texts, monographs, and papers at many different levels of exposition. Student interested in the exposition that is beyond this chapter coverage should consult monographs by Daubechies (1992), Ogden (1997), and Vidakovic (1999), and Walter and Shen (2001), among others.

14.2 HOW DO THE WAVELETS WORK?

14.2.1 The Haar Wavelet

To explain how wavelets work, we start with an example. We choose the simplest and the oldest of all wavelets (we are tempted to say: grandmother of all wavelets!), the Haar wavelet, $\psi(x)$. It is a step function taking values 1 and -1, on intervals $[0, \frac{1}{2})$ and $[\frac{1}{2}, 1)$, respectively. The graphs of the Haar wavelet and some of its dilations/translations are given in Figure 14.4.

The Haar wavelet has been known for almost 100 years and is used in various mathematical fields. Any continuous function can be approximated uniformly by Haar functions, even though the "decomposing atom" is discontinuous.

Dilations and translations of the function ψ,

$$\psi_{jk}(x) = \text{const} \cdot \psi(2^j x - k), \; j, k \in \mathbb{Z},$$

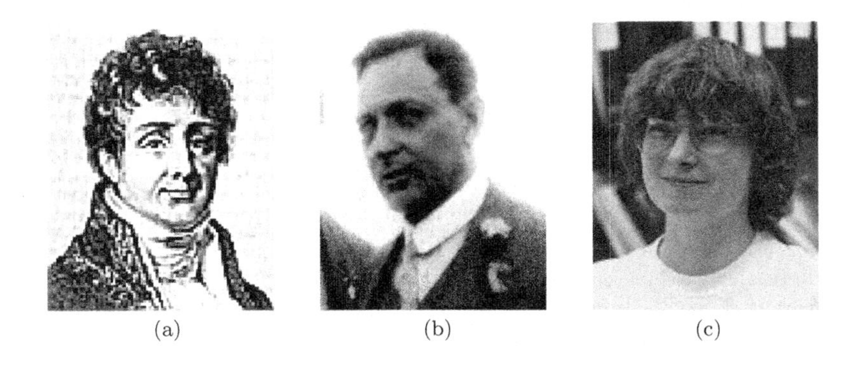

Fig. 14.3 (a) Jean Baptiste Joseph Fourier 1768–1830, Alfred Haar 1885–1933, and (c) Ingrid Daubechies, Professor at Princeton

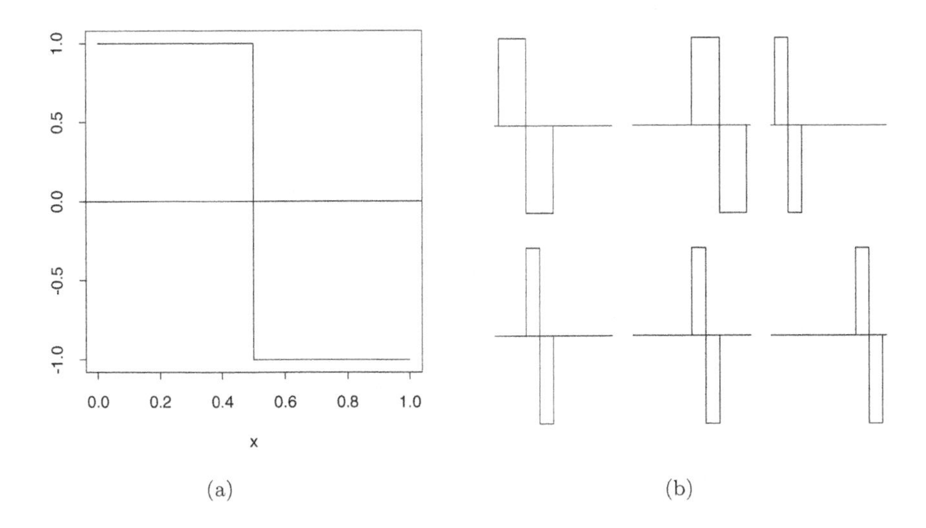

Fig. 14.4 (a) Haar wavelet $\psi(x) = \mathbf{1}(0 \le x < \frac{1}{2}) - \mathbf{1}(\frac{1}{2} < x \le 1)$; (b) Some dilations and translations of Haar wavelet on [0,1].

where $\mathbb{Z} = \{\dots, -2, -1, 0, 1, 2, \dots\}$ is set of all integers, define an orthogonal basis of $L^2(\mathbb{R})$ (the space of all square integrable functions). This means that any function from $L^2(\mathbb{R})$ may be represented as a (possibly infinite) linear combination of these basis functions.

The orthogonality of ψ_{jk}'s is easy to check. It is apparent that

$$\int \psi_{jk} \cdot \psi_{j'k'} = 0, \tag{14.1}$$

whenever $j = j'$ and $k = k'$ are not satisfied simultaneously. If $j \neq j'$ (say $j' < j$), then nonzero values of the wavelet $\psi_{j'k'}$ are contained in the set where the wavelet ψ_{jk} is constant. That makes integral in (14.1) equal to zero: If $j = j'$, but $k \neq k'$, then at least one factor in the product $\psi_{j'k'} \cdot \psi_{jk}$ is zero. Thus the functions ψ_{ij} are mutually orthogonal. The constant that makes this orthogonal system orthonormal is $2^{j/2}$. The functions $\psi_{10}, \psi_{11}, \psi_{20}, \psi_{21}, \psi_{22}, \psi_{23}$ are depicted in Figure 14.4(b).

The family $\{\psi_{jk}, j \in \mathbb{Z}, k \in \mathbb{Z}\}$ defines an orthonormal basis for $\mathbb{L}^2$. Alternatively we will consider orthonormal bases of the form $\{\phi_{L,k}, \psi_{jk}, j \geq L, k \in \mathbb{Z}\}$, where ϕ is called the *scaling function* associated with the wavelet basis ψ_{jk}, and $\phi_{jk}(x) = 2^{j/2}\phi(2^j x - k)$. The set of functions $\{\phi_{L,k}, k \in \mathbb{Z}\}$ spans the same subspace as $\{\psi_{jk}, j < L, k \in \mathbb{Z}\}$. For the Haar wavelet basis the scaling function is simple. It is an indicator of the interval [0,1), that is,

$$\phi(x) = \mathbf{1}(0 \leq x < 1).$$

The data analyst is mainly interested in wavelet representations of functions generated by data sets. Discrete wavelet transforms map the data from the time domain (the original or input data, signal vector) to the wavelet domain. The result is a vector of the same size. Wavelet transforms are linear and they can be defined by matrices of dimension $n \times n$ when they are applied to inputs of size n. Depending on a boundary condition, such matrices can be either orthogonal or "close" to orthogonal. A wavelet matrix W is close to orthogonal when the orthogonality is violated by non-periodic handling of boundaries resulting in a small, but non-zero value of the norm $||WW' - I||$, where I is the identity matrix. When the matrix is orthogonal, the corresponding transform can be thought is a rotation in $\mathbb{R}^n$ space where the data vectors represent coordinates of points. For a fixed point, the coordinates in the new, rotated space comprise the discrete wavelet transformation of the original coordinates.

Example 14.1 Let $y = (1, 0, -3, 2, 1, 0, 1, 2)$. The associated function f is given in Fig. 14.5. The values $f(k) = y_k$, $k = 0, 1, \dots, 7$ are interpolated by a piecewise constant function. The following matrix equation gives the connection between y and the wavelet coefficients d, $y = W'd$,

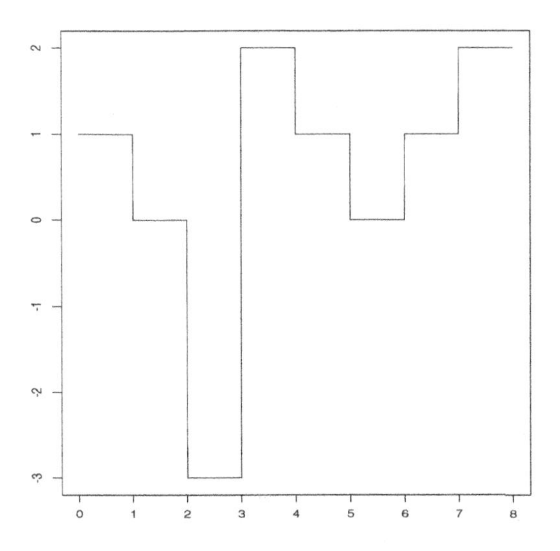

Fig. 14.5 A function interpolating y on [0,8).

$$
\begin{bmatrix} 1 \\ 0 \\ -3 \\ 2 \\ 1 \\ 0 \\ 1 \\ 2 \end{bmatrix} = \begin{bmatrix}
\frac{1}{2\sqrt{2}} & \frac{1}{2\sqrt{2}} & \frac{1}{2} & 0 & \frac{1}{\sqrt{2}} & 0 & 0 & 0 \\
\frac{1}{2\sqrt{2}} & \frac{1}{2\sqrt{2}} & \frac{1}{2} & 0 & -\frac{1}{\sqrt{2}} & 0 & 0 & 0 \\
\frac{1}{2\sqrt{2}} & \frac{1}{2\sqrt{2}} & -\frac{1}{2} & 0 & 0 & \frac{1}{\sqrt{2}} & 0 & 0 \\
\frac{1}{2\sqrt{2}} & \frac{1}{2\sqrt{2}} & -\frac{1}{2} & 0 & 0 & -\frac{1}{\sqrt{2}} & 0 & 0 \\
\frac{1}{2\sqrt{2}} & -\frac{1}{2\sqrt{2}} & 0 & \frac{1}{2} & 0 & 0 & \frac{1}{\sqrt{2}} & 0 \\
\frac{1}{2\sqrt{2}} & -\frac{1}{2\sqrt{2}} & 0 & \frac{1}{2} & 0 & 0 & -\frac{1}{\sqrt{2}} & 0 \\
\frac{1}{2\sqrt{2}} & -\frac{1}{2\sqrt{2}} & 0 & -\frac{1}{2} & 0 & 0 & 0 & \frac{1}{\sqrt{2}} \\
\frac{1}{2\sqrt{2}} & -\frac{1}{2\sqrt{2}} & 0 & -\frac{1}{2} & 0 & 0 & 0 & -\frac{1}{\sqrt{2}}
\end{bmatrix} \cdot \begin{bmatrix} c_{00} \\ d_{00} \\ d_{10} \\ d_{11} \\ d_{20} \\ d_{21} \\ d_{22} \\ d_{23} \end{bmatrix} \tag{14.2}
$$

The solution is $d = Wy$,

$$
\begin{bmatrix} c_{00} \\ d_{00} \\ d_{10} \\ d_{11} \\ d_{20} \\ d_{21} \\ d_{22} \\ d_{23} \end{bmatrix} = \begin{bmatrix} \sqrt{2} \\ -\sqrt{2} \\ 1 \\ -1 \\ \frac{1}{\sqrt{2}} \\ -\frac{5}{\sqrt{2}} \\ \frac{1}{\sqrt{2}} \\ -\frac{1}{\sqrt{2}} \end{bmatrix}.
$$

Accordingly

$$\begin{aligned} f(x) &= \sqrt{2}\phi_{0,0}(x) - \sqrt{2}\psi_{0,0}(x) + \psi_{1,0}(x) - \psi_{1,1}(x) \\ &\quad + \frac{1}{\sqrt{2}}\psi_{2,0}(x) - \frac{5}{\sqrt{2}}\psi_{2,1}(x) + \frac{1}{\sqrt{2}}\psi_{2,2}(x) - \frac{1}{\sqrt{2}}\psi_{2,3}(x). \end{aligned} \tag{14.3}$$

The solution is easy to verify. For example, when $x \in [0, 1)$,

$$f(x) = \sqrt{2} \cdot \frac{1}{2\sqrt{2}} - \sqrt{2} \cdot \frac{1}{2\sqrt{2}} + 1 \cdot \frac{1}{2} + \frac{1}{\sqrt{2}} \cdot \frac{1}{\sqrt{2}} = 1/2 + 1/2 = 1 \ (= y_0).$$

The MATLAB m-file `WavMat.m` forms the wavelet matrix W, for a given wavelet base and dimension which is a power of 2. For example, `W = WavMat(h, n, k0, shift)` will calculate $n \times n$ wavelet matrix, corresponding to the filter h (connections between wavelets and filtering will be discussed in the following section), and `k0` and `shift` are given parameters. We will see that Haar wavelet corresponds to a filter $h = \{\sqrt{2}/2, \sqrt{2}/2\}$. Here is the above example in MATLAB:

```
>> W = WavMat([sqrt(2)/2 sqrt(2)/2],2^3,3,2);
>> W'
ans =
    0.3536    0.3536    0.5000         0    0.7071         0         0         0
    0.3536    0.3536    0.5000         0   -0.7071         0         0         0
    0.3536    0.3536   -0.5000         0         0    0.7071         0         0
    0.3536    0.3536   -0.5000         0         0   -0.7071         0         0
    0.3536   -0.3536         0    0.5000         0         0    0.7071         0
    0.3536   -0.3536         0    0.5000         0         0   -0.7071         0
    0.3536   -0.3536         0   -0.5000         0         0         0    0.7071
    0.3536   -0.3536         0   -0.5000         0         0         0   -0.7071

>> dat=[1 0 -3 2 1 0 1 2];
>> wt = W * dat'; wt'

ans =
    1.4142   -1.4142    1.0000   -1.0000    0.7071   -3.5355    0.7071   -0.7071
>> data = W' * wt;  data'
ans =
    1.0000    0.0000   -3.0000    2.0000    1.0000    0.0000    1.0000    2.0000
```

Performing wavelet transformations via the product of wavelet matrix W and input vector y is conceptually straightforward, but of limited practical value. Storing and manipulating wavelet matrices for inputs exceeding tens of thousands in length is not feasible.

14.2.2 Wavelets in the Language of Signal Processing

Fast discrete wavelet transforms become feasible by implementing the so called *cascade algorithm* introduced by Mallat (1989). Let $\{h(k), k \in Z\}$ and $\{g(k), k \in Z\}$ be the *quadrature mirror filters* in the terminology of signal processing. Two filters h and g form a quadrature mirror pair when:

$$g(n) = (-1)^n h(1-n).$$

The filter $h(k)$ is a *low pass* or *smoothing* filter while $g(k)$ is the *high pass* or *detail* filter. The following properties of $h(n), g(n)$ can be derived by using so called scaling relationship, Fourier transforms and orthogonality: $\Sigma_k h(k) = \sqrt{2}$, $\Sigma_k g(k) = 0$, $\Sigma_k h(k)^2 = 1$, and $\Sigma_k h(k) k(k-2m) = \mathbf{1}(m=0)$.

The most compact way to describe the cascade algorithm, as well to give efficient recipe for determining discrete wavelet coefficients is by using *operator representation of filters*. For a sequence $a = \{a_n\}$ the operators H and G are defined by the following coordinate-wise relations:

$$\begin{aligned}(Ha)_n &= \Sigma_k h(k-2n) a_k \\ (Ga)_n &= \Sigma_k g(k-2n) a_k.\end{aligned}$$

The operators H and G perform filtering and down-sampling (omitting every second entry in the output of filtering), and correspond to a single step in the wavelet decomposition. The wavelet decomposition thus consists of subsequent application of operators H and G in the particular order on the input data.

Denote the original signal y by $c^{(J)}$. If the signal is of length $n = 2^J$, then $c^{(J)}$ can be understood as the vector of coefficients in a series $f(x) = \Sigma_{k=0}^{2^J-1} c_k^{(J)} \phi_{nk}$, for some scaling function ϕ. At each step of the wavelet transform we move to a coarser approximation $c^{(j-1)}$ with $c^{(j-1)} = Hc^{(j)}$ and $d^{(j-1)} = Gc^{(j)}$. Here, $d^{(j-1)}$ represent the "details" lost by degrading $c^{(j)}$ to $c^{(j-1)}$. The filters H and G are decimating, thus the length of $c^{(j-1)}$ or $d^{(j-1)}$ is half the length of $c^{(j)}$. The discrete wavelet transform of a sequence $y = c^{(J)}$ of length $n = 2^J$ can then be represented as another sequence of length 2^J (notice that the sequence $c^{(j-1)}$ has half the length of $c^{(j)}$):

$$(c^{(0)}, d^{(0)}, d^{(1)}, \ldots, d^{(J-2)}, d^{(J-1)}). \tag{14.4}$$

In fact, this decomposition may not be carried until the singletons $c^{(0)}$ and $d^{(0)}$ are obtained, but could be curtailed at $(J-L)^{th}$ step,

$$(c^{(L)}, d^{(L)}, d^{(L+1)}, \ldots, d^{(J-2)}, d^{(J-1)}), \tag{14.5}$$

for any $0 \le L \le J-1$. The resulting vector is still a valid wavelet transform. See Exercise 14.4 for Haar wavelet transform "by hand."

```
function dwtr = dwtr(data, L, filterh)
%  function dwtr = dwt(data, L, filterh);
%  Calculates the DWT of periodic data set
%  with scaling filter  filterh  and  L  detail levels.
%
%   Example of Use:
```

```
%   data = [1 0 -3 2 1 0 1 2]; filter = [sqrt(2)/2 sqrt(2)/2];
%   wt = DWTR(data, 3, filter)
%-----------------------------------------------------------------------
n = length(filterh);                    %Length of wavelet filter
C = data(:)';                           %Data (row vector) live in V_j
dwtr = [];                              %At the beginning dwtr empty
H  =  fliplr(filterh);                  %Flip because of convolution
G  =  filterh;                          %Make quadrature mirror
G(1:2:n) = -G(1:2:n);                   %     counterpart
for j = 1:L                             %Start cascade
   nn = length(C);                      %Length needed to
   C = [C(mod((-(n-1):-1),nn)+1)  C]; % make periodic
   D = conv(C,G);                       %Convolve,
   D = D([n:2:(n+nn-2)]+1);             %     keep periodic and decimate
   C = conv(C,H);                       %Convolve,
   C = C([n:2:(n+nn-2)]+1);             %     keep periodic and decimate
   dwtr = [D,dwtr];                     %Add detail level to dwtr
end;                                    %Back to cascade or end
dwtr = [C, dwtr];                       %Add the last ``smooth'' part
```

As a result, the discrete wavelet transformation can be summarized as:

$$\boldsymbol{y \longrightarrow (H^{J-L}y, GH^{J-1-L}y, GH^{J-2-L}y, \ldots, GHy, Gy),\ 0 \le L \le J-1.}$$

The MATLAB program `dwtr.m` performs discrete wavelet transform:

```
> data = [1 0 -3 2 1 0 1 2]; filter = [sqrt(2)/2 sqrt(2)/2];
>  wt = dwtr(data, 3, filter)

wt =
1.4142 -1.4142 1.0000 -1.0000 0.7071 -3.5355 0.7071 -0.7071
```

The reconstruction formula is also simple in terms of H and G; we first define adjoint operators $H^\star$ and $G^\star$ as follows:

$$\begin{aligned}(H^\star a)_k &= \Sigma_n h(k-2n)a_n \\ (G^\star a)_k &= \Sigma_n g(k-2n)a_n.\end{aligned}$$

Recursive application leads to:

$$\boldsymbol{(c^{(L)}, d^{(L)}, d^{(L+1)}, \ldots, d^{(J-2)}, d^{(J-1)}) \longrightarrow y = (H^\star)^J c^{(L)} + \Sigma_{j=L}^{J-1}(H^\star)^j G^\star d^{(j)},}$$

for some $0 \le L \le J-1$.

```
function  data = idwtr(wtr, L, filterh)
% function data = idwt(wtr, L, filterh); Calculates the IDWT of wavelet
% transformation wtr using wavelet filter  "filterh"  and  L  scales.
% Use
%>> max(abs(data - IDWTR(DWTR(data,3,filter), 3,filter)))
%
```

```
%ans = 4.4409e-016

nn = length(wtr);   n = length(filterh);        % Lengths
if nargin==2, L = round(log2(nn)); end;        % Depth of transformation
H = filterh;                                    % Wavelet H filter
G = fliplr(H); G(2:2:n) = -G(2:2:n);            % Wavelet G filter
LL = nn/(2^L);                                  % Number of scaling coeffs
C =  wtr(1:LL);                                 % Scaling coeffs
for j = 1:L                                     % Cascade algorithm
   w  = mod(0:n/2-1,LL)+1;                      % Make periodic
   D  = wtr(LL+1:2*LL);                         % Wavelet coeffs
   Cu(1:2:2*LL+n) = [C C(1,w)];                 % Upsample & keep periodic
   Du(1:2:2*LL+n) = [D D(1,w)];                 % Upsample & keep periodic
   C  = conv(Cu,H) + conv(Du,G);                % Convolve & add
   C  = C([n:n+2*LL-1]-1);                      % Periodic part
   LL = 2*LL;                                   % Double the size of level
end;
data = C;                                       % The inverse DWT
```

Because wavelet filters uniquely correspond to selection of the wavelet orthonormal basis, we give a table a few common (and short) filters commonly used. See Table 14.19 for filters from the Daubechies, Coiflet and Symmlet families [1]. See Exercise 14.5 for some common properties of wavelet filters.

The careful reader might have already noticed that when the length of the filter is larger than two, boundary problems occur (there are no boundary problems with the Haar wavelet). There are several ways to handle the boundaries, two main are: *symmetric* and *periodic*, that is, extending the original function or data set in a symmetric or periodic manner to accommodate filtering that goes outside of domain of function/data.

14.3 WAVELET SHRINKAGE

Wavelet shrinkage provides a simple tool for nonparametric function estimation. It is an active research area where the methodology is based on optimal shrinkage estimators for the location parameters. Some references are Donoho and Johnstone (1994, 1995), Vidakovic (1999), Antoniadis, and Bigot and Sapatinas (2001). In this section we focus on the simplest, yet most important shrinkage strategy – wavelet thresholding.

In discrete wavelet transform the filter H is an "averaging" filter while its mirror counterpart G produces details. The wavelet coefficients correspond to details. When detail coefficients are small in magnitude, they may be

[1] Filters are indexed by the number of taps and rounded at seven decimal places.

Table 14.19 Some Common Wavelet Filters from the Daubechies, Coiflet and Symmlet Families.

Name	h_0	h_1	h_2	h_3	h_4	h_5
Haar	$1/\sqrt{2}$	$1/\sqrt{2}$				
Daub 4	0.4829629	0.8365163	0.2241439	-0.1294095		
Daub 6	0.3326706	0.8068915	0.4598775	-0.1350110	-0.0854413	0.0352263
Coif 6	0.0385808	-0.1269691	-0.0771616	0.6074916	0.7456876	0.2265843
Daub 8	0.2303778	0.7148466	0.6308808	-0.0279838	-0.1870348	0.0308414
Symm 8	-0.0757657	-0.0296355	0.4976187	0.8037388	0.2978578	-0.0992195
Daub 10	0.1601024	0.6038293	0.7243085	0.1384281	-0.2422949	-0.0322449
Symm 10	0.0273331	0.0295195	-0.0391342	0.1993975	0.7234077	0.6339789
Daub 12	0.1115407	0.4946239	0.7511339	0.3152504	-0.2262647	-0.1297669
Symm 12	0.0154041	0.0034907	-0.1179901	-0.0483117	0.4910559	0.7876411

Name	h_6	h_7	h_8	h_9	h_{10}	h_{11}
Daub 8	0.0328830	-0.0105974				
Symm 8	-0.0126034	0.0322231				
Daub 10	0.0775715	-0.0062415	-0.0125808	0.0033357		
Symm 10	0.0166021	-0.1753281	-0.0211018	0.0195389		
Daub 12	0.0975016	0.0275229	-0.0315820	0.0005538	0.0047773	-0.0010773
Symm 12	0.3379294	-0.0726375	-0.0210603	0.0447249	0.0017677	-0.0078007

omitted without substantially affecting the general picture. Thus the idea of thresholding wavelet coefficients is a way of cleaning out unimportant details that correspond to noise.

An important feature of wavelets is that they provide unconditional bases[2] for functions that are more regular, smooth have fast decay of their wavelet coefficients. As a consequence, wavelet shrinkage acts as a smoothing operator. The same can not be said about Fourier methods. Shrinkage of Fourier coefficients in a Fourier expansion of a function affects the result globally due to the non-local nature of sines and cosines. However, trigonometric bases can be localized by properly selected window functions, so that they provide local, wavelet-like decompositions.

Why does wavelet thresholding work? Wavelet transforms disbalanced data. Informally, the "energy" in data set (sum of squares of the data) is preserved (equal to sum of squares of wavelet coefficients) but this energy is packed in a few wavelet coefficients. This *disbalancing property* ensures that the function of interest can be well described by a relatively small number of wavelet coefficients. The normal i.i.d. noise, on the other hand, is invariant with respect to orthogonal transforms (e.g., wavelet transforms) and passes to the wavelet domain structurally unaffected. Small wavelet coefficients likely

[2]Informally, a family $\{\psi_i\}$ is an unconditional basis for a space of functions S if one can determine if the function $f = \Sigma_i a_i \psi_i$ belongs to S by inspecting only the magnitudes of coefficients, $|a_i|$s.

correspond to a noise because the signal part gets transformed to a few big-magnitude coefficients.

The process of thresholding wavelet coefficients can be divided into two steps. The first step is the policy choice, which is the choice of the threshold function T. Two standard choices are: *hard* and *soft* thresholding with corresponding transformations given by:

$$T^{hard}(d, \lambda) = d\ \mathbf{1}(|d| > \lambda),$$
$$T^{soft}(d, \lambda) = (d - sign(d)\lambda)\ \mathbf{1}(|d| > \lambda), \tag{14.6}$$

where λ denotes the threshold, and d generically denotes a wavelet coefficient. Figure 14.6 shows graphs of (a) hard- and (b) soft-thresholding rules when the input is wavelet coefficient d.

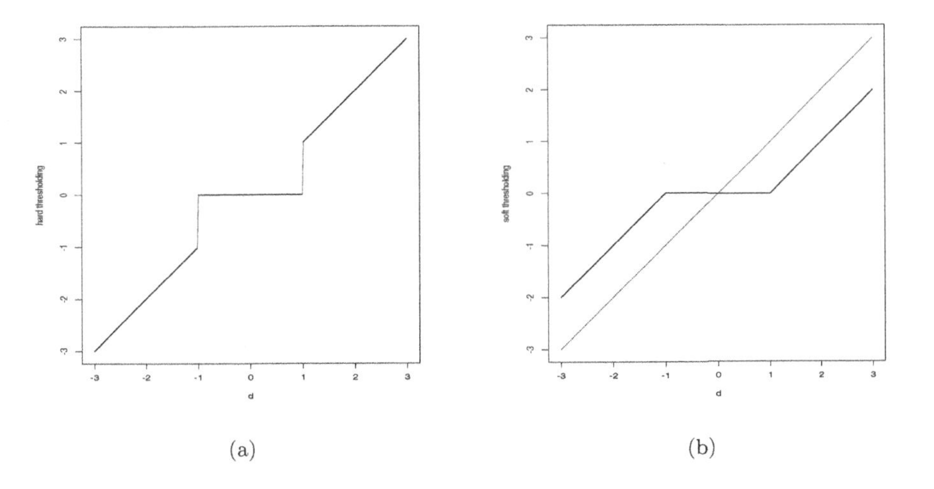

(a) (b)

Fig. 14.6 (a) Hard and (b) soft thresholding with $\lambda = 1$.

Another class of useful functions are general shrinkage functions. A function S from that class exhibits the following properties:

$$S(d) \approx 0, \text{ for } d \text{ small}; \quad S(d) \approx d, \text{ for } d \text{ large}.$$

Many state-of-the-art shrinkage strategies are in fact of type $S(d)$.

The second step is the choice of a threshold if the shrinkage rule is thresholding or appropriate parameters if the rule has S-functional form. In the following subsection we briefly discuss some of the standard methods of selecting a threshold.

14.3.1 Universal Threshold

In the early 1990s, Donoho and Johnstone proposed a threshold λ (Donoho and Johnstone, 1993; 1994) based on the result in theory of extrema of normal random variables.

Theorem 14.1 *Let $Z_1, \dots, Z_n$ be a sequence of i.i.d. standard normal random variables. Define*

$$A_n = \{\max_{i=1,\dots,n} |Z_i| \le \sqrt{2\log n}\}.$$

Then

$$\pi_n = P(A_n) \to 0, n \to \infty.$$

In addition, if

$$B_n(t) = \{\max_{i=1,\dots,n} |Z_i| > t + \sqrt{2\log n}\},$$

then $P(B_n(t)) < e^{-\frac{t^2}{2}}$.

Informally, the theorem states that the Z_is are "almost bounded" by $\pm\sqrt{2\log n}$. Anything among the n values larger in magnitude than $\sqrt{2\log n}$ does not look like the i.i.d. normal noise. This motivates the following threshold:

$$\lambda^U = \sqrt{2\log n}\ \hat{\sigma}, \tag{14.7}$$

which Donoho and Johnstone call *universal.* This threshold is one of the first proposed and provides an easy and automatic thresholding.

In the real-life problems the level of noise σ is not known, however wavelet domains are suitable for its assessment. Almost all methods for estimating the variance of noise involve the wavelet coefficients at the scale of finest detail. The signal-to-noise ratio is smallest at this level for almost all reasonably behaved signals, and the level coefficients correspond mainly to the noise.

Some standard estimators of σ are:

$$(i)\ \hat{\sigma} = \sqrt{\frac{1}{N/2-1}\Sigma_{k=1}^{N/2}(d_{n-1,k} - \bar{d})^2}, \text{ with } \bar{d} = \frac{1}{N/2}; \Sigma d_{n-1,k} \tag{14.8}$$

or a more robust MAD estimator,

$$(ii) \qquad \hat{\sigma} = 1/0.6745 \quad median_k|d_{n-1,k} - median_m(d_{n-1,m})|, \tag{14.9}$$

where $d_{n-1,k}$ are coefficients in the level of finest detail. In some situations, for instance when data sets are large or when σ is over-estimated, the universal thresholding oversmooths.

Example 14.2 The following MATLAB script demonstrates how the wavelets smooth the functions. A Doppler signal of size 1024 is generated and random normal noise of size $\sigma = 0.1$ is added. By using the Symmlet wavelet 8-tap filter the noisy signal is transformed. After thresholding in the wavelet domain the signal is back-transformed to the original domain.

```
% Demo of wavelet-based function estimation
clear all
close all
% (i) Make "Doppler" signal on [0,1]
t=linspace(0,1,1024);
sig = sqrt(t.*(1-t)).*sin((2*pi*1.05) ./(t+.05));
% and plot it
figure(1); plot(t, sig)


% (ii) Add noise of size 0.1. We are fixing
% the seed of random number generator for repeatability
% of example. We add the random noise to the signal
% and make a plot.
randn('seed',1)
sign = sig + 0.1 * randn(size(sig));
figure(2); plot(t, sign)

% (iii) Take the filter H, in this case this is SYMMLET 8

filt = [ -0.07576571478934  -0.02963552764595  ...
          0.49761866763246   0.80373875180522  ...
          0.29785779560554  -0.09921954357694  ...
         -0.01260396726226   0.03222310060407];

% (iv) Transform the noisy signal in the wavelet domain.
% Choose L=8, eight detail levels in the  decomposition.

sw = dwtr(sign, 8, filt);

% At this point you may view the sw. Is it disbalanced?
% Is it decorrelated?
%(v) Let's now threshold the small coefficients.
% The universal threshold is determined as
% lambda = sqrt(2 * log(1024)) * 0.1 = 0.3723
%
% Here we assumed $sigma=0.1$ is known. In real life
% this is not the case and we estimate sigma.
% A robust estimator is 'MAD'  from the finest level of detail
% believed to be mostly transformed noise.

finest = sw(513:1024);
sigma_est = 1/0.6745 * median(abs( finest - median(finest)));
lambda = sqrt(2 * log(1024)) * sigma_est;

% hard threshold in the wavelet domain
swt=sw .* (abs(sw) > lambda );
figure(3);  plot([1:1024], swt, '-')
```

```
% (vi) Back-transform the thresholded object to the time
% domain. Of course, retain the same filter and value L.

sig_est = idwtr(swt, 8, filt);
figure(4);  plot(t, sig_est, '-'); hold on; plot(t, sig, ':');
```

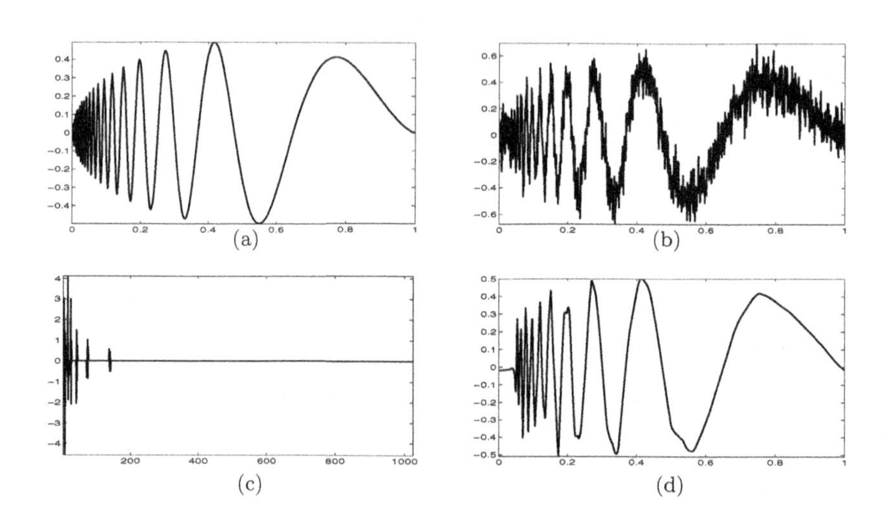

Fig. 14.7 Demo output (a) Original **doppler** signal, (b) Noisy **doppler**, (c) Wavelet coefficients that "survived" thresholding, (d) Inverse-transformed thresholded coefficients.

Example 14.3 A researcher was interested in predicting earthquakes by the level of water in nearby wells. She had a large ($8192 = 2^{13}$ measurements) data set of water levels taken every hour in a period of time of about one year in a California well. Here is the description of the problem:

> The ability of water wells to act as strain meters has been observed for centuries. Lab studies indicate that a seismic slip occurs along a fault prior to rupture. Recent work has attempted to quantify this response, in an effort to use water wells as sensitive indicators of volumetric strain. If this is possible, water wells could aid in earthquake prediction by sensing precursory earthquake strain.
>
> We obtained water level records from a well in southern California, collected over a year time span. Several moderate size earthquakes (magnitude 4.0 - 6.0) occurred in close proximity to the well during this time interval. There is a a significant amount of noise in the water level record which must first be filtered out. Environmental factors

such as earth tides and atmospheric pressure create noise with frequencies ranging from seasonal to semidiurnal. The amount of rainfall also affects the water level, as do surface loading, pumping, recharge (such as an increase in water level due to irrigation), and sonic booms, to name a few. Once the noise is subtracted from the signal, the record can be analyzed for changes in water level, either an increase or a decrease depending upon whether the aquifer is experiencing a tensile or compressional volume strain, just prior to an earthquake.

This data set is given in `earthquake.dat`. A plot of the raw data for hourly measurements over one year ($8192 = 2^{13}$ observations) is given in Figure 14.8(a). The detail showing the oscillation at the earthquake time is presented in Figure 14.8(b).

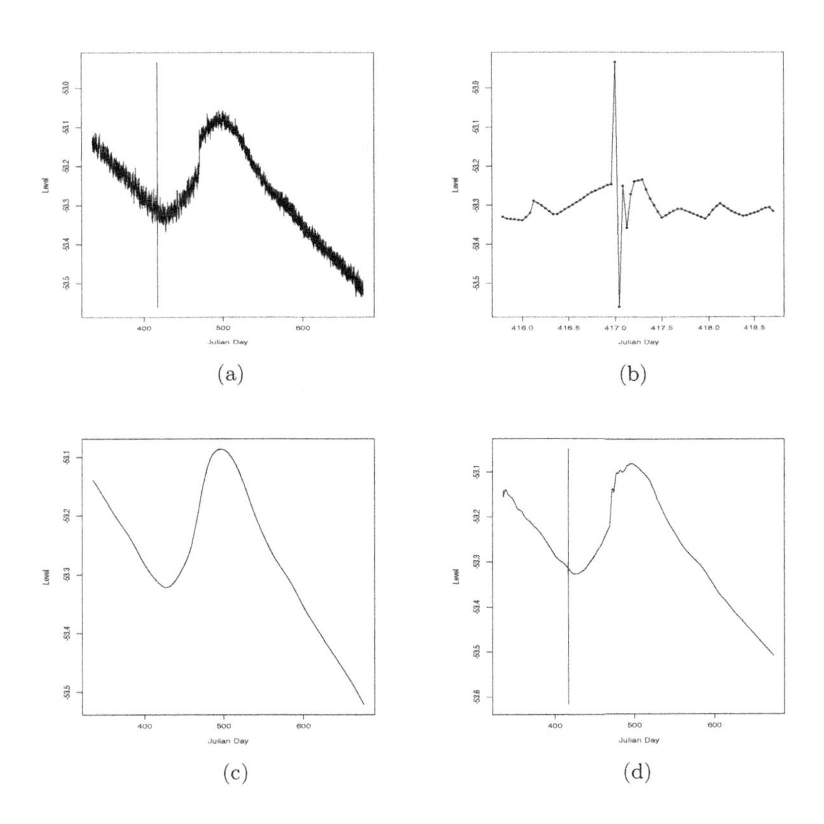

Fig. 14.8 Panel (a) shows $n = 8192$ hourly measurements of the water level for a well in an earthquake zone. Notice the wide range of water levels at the time of an earthquake around $t = 417$. Panel (b) focusses on the data around the earthquake time. Panel (c) shows the result of LOESS, and (d) gives a wavelet based reconstruction.

Application of LOESS smoother captured trend but the oscillation artifact is smoothed out as evident from Figure 14.8(c). After applying the Daubechies 8 wavelet transform and universal thresholding we got a fairly smooth baseline function with preserved jump at the earthquake time. The processed data are presented in Figure 14.8(d). This feature of wavelet methods demonstrated data adaptivity and locality.

How this can be explained? The wavelet coefficients corresponding to the earthquake feature (big oscillation) are large in magnitude and are located at all even the finest detail level. These few coefficients "survived" the thresholding, and the oscillation feature shows in the inverse transformation. See Exercise 14.6 for the suggested follow-up.

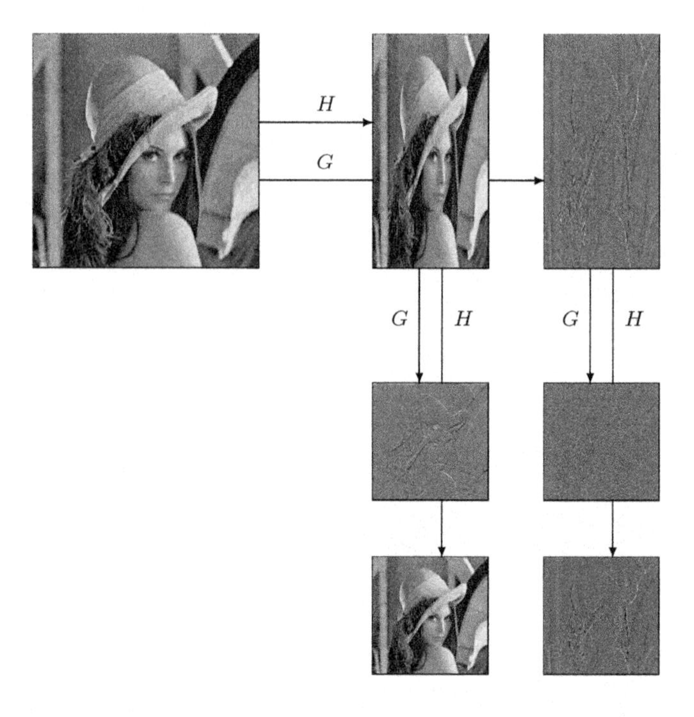

Fig. 14.9 One step in wavelet transformation of 2-D data exemplified on celebrated Lenna image.

Example 14.4 The most important application of 2-D wavelets is in image processing. Any gray-scale image can be represented by a matrix A in which the entries a_{ij} correspond to color intensities of the pixel at location (i, j). We assume as standardly done that A is a square matrix of dimension $2^n \times 2^n$, n integer.

The process of wavelet decomposition proceeds as follows. On the rows of the matrix A the filters H and G are applied. Two resulting matrices $H_r A$ and $G_r A$ are obtained, both of dimension $2^n \times 2^{n-1}$ (Subscript r suggest that the filters are applied on rows of the matrix A, 2^{n-1} is obtained in the dimension of $H_r A$ and $G_r A$ because wavelet filtering decimate). Now, the filters H and G are applied on the columns of $H_r A$ and $G_r A$ and matrices $H_c H_r A, G_c H_r A, H_c G_r A$ and $G_c G_r A$ of dimension $2^{n-1} \times 2^{n-1}$ are obtained. The matrix $H_c H_r A$ is the average, while the matrices $G_c H_r A, H_c G_r A$ and $G_c G_r A$ are details (see Figure 14.9).[3]

The process could be continued in the same fashion with the *smoothed* matrix $H_c H_r A$ as an input, and can be carried out until a single number is obtained as an overall "smooth" or can be stopped at any step. Notice that in decomposition exemplified in Figure 14.9, the matrix is decomposed to one smooth and three detail submatrices.

A powerful generalization of wavelet bases is the concept of wavelet packets. Wavelet packets result from applications of operators H and G, discussed on p. 271, in *any* order. This corresponds to an overcomplete system of functions from which the best basis for a particular data set can be selected.

14.4 EXERCISES

14.1. Show that the matrix W' in (14.2) is orthogonal.

14.2. In (14.1) we argued that ψ_{jk} and $\psi_{j'k'}$ are orthogonal functions whenever $j = j'$ and $k = k'$ is not satisfied simultaneously. Argue that ϕ_{jk} and $\psi_{j'k'}$ are orthogonal whenever $j' \geq j$. Find an example in which ϕ_{jk} and $\psi_{j'k'}$ are not orthogonal if $j' < j$.

14.3. In Example 14.1 it was verified that in (14.3) $f(x) = 1$ whenever $x \in [0, 1)$. Show that $f(x) = 0$ whenever $x \in [1, 2)$.

14.4. Verify that $(\sqrt{2}, -\sqrt{2}, 1, -1, \frac{1}{\sqrt{2}}, -\frac{5}{\sqrt{2}}, \frac{1}{\sqrt{2}}, -\frac{1}{\sqrt{2}})$ is *a* Haar wavelet transform of data set $y = (1, 0, -3, 2, 1, 0, 1, 2)$ by using operators H and G from (14.4).

[3]This image of Lenna (Sjooblom) Soderberg, a Playboy centerfold from 1972, has become one of the most widely used standard test images in signal processing.

Hint. For the Haar wavelet, low- and high-pass filters are $h = (1/\sqrt{2}\ \ 1/\sqrt{2})$ and $g = (1/\sqrt{2}\ \ -1/\sqrt{2})$, so

$$\begin{aligned} Hy &= H((1,0,-3,2,1,0,1,2)) \\ &= (1\cdot 1/\sqrt{2}+0\cdot 1/\sqrt{2},\quad -3\cdot 1/\sqrt{2}+2\cdot 1/\sqrt{2}, \\ &\qquad 1\cdot 1/\sqrt{2}+0\cdot 1/\sqrt{2},\ 1\cdot 1/\sqrt{2}+2\cdot 1/\sqrt{2}) \\ &= \left(\frac{1}{\sqrt{2}},-\frac{1}{\sqrt{2}},\frac{1}{\sqrt{2}},\frac{3}{\sqrt{2}}\right), \quad \text{and} \\ Gy &= G((1,0,-3,2,1,0,1,2)) = \left(\frac{1}{\sqrt{2}},-\frac{5}{\sqrt{2}},\frac{1}{\sqrt{2}},-\frac{1}{\sqrt{2}}\right). \end{aligned}$$

Repeat the G operator on Hy and $H(Hy)$. The final filtering is $H(H(Hy))$. Organize result as

$$(H(H(Hy)), G(H(Hy)), G(Hy), Gy)\,.$$

14.5. Demonstrate that all filters in Table 14.19 satisfy the following properties (up to rounding error):

$$\Sigma_i h_i = \sqrt{2},\quad \Sigma_i h_i^2 = 1,\ \text{ and } \Sigma_i h_i h_{i+2} = 0.$$

14.6. Refer to Example 14.3 in which wavelet-based smoother exhibited notable difference from the standard smoother LOESS. Read the data `earthquake.dat` into MATLAB, select the wavelet filter, and apply the wavelet transform to the data.

(a) Estimate the size of the noise by estimating σ using MAD from page 276 and find the universal threshold λ_U.

(b) Show that finest level of detail contains coefficients exceeding the universal threshold.

(c) Threshold the wavelet coefficients using hard thresholding rule with λ_U that you have obtained in (b), and apply inverse wavelet transform. Comment. How do you explain oscillations at boundaries?

REFERENCES

Antoniadis, A., Bigot, J., and Sapatinas, T. (2001), "Wavelet Estimators in Nonparametric Regression: A Comparative Simulation Study," *Journal of Statistical Software*, 6, 1-83.

Daubechies, I. (1992), *Ten Lectures on Wavelets.* Philadelphia: S.I.A.M.

Donoho, D., and Johnstone, I. (1994), "Ideal Spatial Adaptation by Wavelet Shrinkage," *Biometrika,* 81, 425-455.

Donoho, D., and Johnstone, I. (1995), Adapting to Unknown Smoothness via Wavelet Shrinkage," *Journal of the American Statistical Association,* 90, 1200-1224.

Donoho, D., Johnstone, I., Kerkyacharian, G., and Pickard, D. (1996), "Density Estimation by Wavelet Thresholding," *Annals of Statistics*, 24, 508-539.

Mallat, S. (1989), "A Theory for Multiresolution Signal Decomposition: The Wavelet Representation," *IEEE Transactions on Pattern Analysis and Machine Intelligence,* 11, 674-693.

Ogden, T. (1997), *Essential Wavelets for Statistical Applications and Data Analysis.* Boston: Birkhäuser.

Vidakovic, B. (1999), *Statistical Modeling by Wavelets,* New York: Wiley.

Walter, G.G., and Shen X. (2001), *Wavelets and Others Orthogonal Systems,* 2nd ed. Boca Raton, FL: Chapman & Hall/CRC.

15

Bootstrap

Confine! I'll confine myself no finer than I am:
these clothes are good enough to drink in; and so be these boots too:
an they be not, let them hang themselves in their own straps.

William Shakespeare (*Twelfth Night*, Act 1, Scene III)

15.1 BOOTSTRAP SAMPLING

Bootstrap resampling is one of several controversial techniques in statistics and according to some, the most controversial. By resampling, we mean to take a random sample *from the sample*, as if your sampled data $X_1, \ldots, X_n$ represented a finite population of size n. This new sample (typically of the same size n) is taken by "sampling with replacement", so some of the n items from the original sample can appear more than once. This new collection is called a *bootstrap sample*, and can be used to assess statistical properties such as an estimator's variability and bias, predictive performance of a rule, significance of a test, and so forth, when the exact analytic methods are impossible or intractable.

By simulating directly from the data, the bootstrap avoids making unnecessary assumptions about parameters and models – we are figuratively pulling ourselves up by our bootstraps rather than relying on the outside help of parametric assumptions. In that sense, the bootstrap is a nonparametric procedure. In fact, this resampling technique includes both parametric and

(a) (b)

Fig. 15.1 (a) Bradley Efron, Stanford University; (b) Prasanta Chandra Mahalanobis (1893–1972)

nonparametric forms, but it is essentially empirical.

The term *bootstrap* was coined by Bradley Efron (Figure 15.1(a)) at his 1977 Stanford University Reitz Lecture to describe a resampling method that can help us to understand characteristics of an estimator (e.g., uncertainty, bias) without the aid of additional probability modeling. The bootstrap described by Efron (1979) is not the first resampling method to help out this way (e.g., permutation methods of Fisher (1935) and Pitman (1937), spatial sampling methods of Mahalanobis (1946), or jack-knife methods of Quenouille (1949)), but it's the most popular resampling tool used in statistics today.

So what good is a bootstrap sample? For any direct inference on the underlying distribution, it is obviously inferior to the original sample. If we estimate a parameter $\theta = \theta(F)$ from a distribution F, we obviously prefer to use $\theta_n = \theta(F_n)$. What the bootstrap sample *can* tell us, is how θ_n might change from sample to sample. While we can only compute θ_n once (because we have just the one sample of n), we can resample (and form a bootstrap sample) an infinite amount of times, in theory. So a meta-estimator built from a bootstrap sample (say $\tilde{\theta}$) tells us not about θ, but about θ_n. If we generate repeated bootstrap samples $\tilde{\theta}_1, \ldots, \tilde{\theta}_B$, we can form an indirect picture of how θ_n is distributed, and from this we generate confidence statements for θ. B is not really limited – it's as large as you want as long as you have the patience for generating repeated bootstrap samples.

For example, $\bar{x} \pm z_{\alpha/2}\sigma_{\bar{x}}$ constitutes an exact (1-α)100% confidence interval for μ if we know $X_1, \ldots, X_n \sim \mathcal{N}(\mu, \sigma^2)$ and $\sigma_{\bar{x}} = \sigma/\sqrt{n}$. We are essentially finding the appropriate quantiles from the sampling distribution of point estimate $\bar{x}$. Unlike this simple example, characteristics of the sample estimator often are much more difficult to ascertain, and even an interval based on a normal approximation seems out of reach or provide poor coverage probability. This is where resampling comes in most useful.

Fig. 15.2 Baron Von Munchausen: the first bootstrapper.

The idea of bootstrapping was met with initial trepidation. After all, it might seem to be promising something for nothing. The stories of Baron Von Munchausen (Raspe, 1785), based mostly on folk tales, include astounding feats such as riding cannonballs, travelling to the Moon and being swallowed by a whale before escaping unharmed. In one adventure, the baron escapes from a swamp by pulling himself up by his own hair. In later versions he was using his own bootstraps to pull himself out of the sea, which gave rise to the term *bootstrapping*.

15.2 NONPARAMETRIC BOOTSTRAP

The *percentile bootstrap* procedure provides a 1-α nonparametric confidence interval for θ directly. We examine the EDF from the bootstrap sample for $\tilde{\theta}_1 - \theta_n, \ldots, \tilde{\theta}_B - \theta_n$. If θ_n is a good estimate of θ, then we *know* $\tilde{\theta} - \theta_n$ is a good estimate of $\theta_n - \theta$. We don't know the distribution of $\theta_n - \theta$ because we don't know θ, so we cannot use the quantiles from $\theta_n - \theta$ to form a confidence interval. But we do know the distribution of $\tilde{\theta} - \theta_n$, and the quantiles serve the same purpose. Order the outcomes of the bootstrap sample $(\tilde{\theta}_1 - \theta_n, \ldots, \tilde{\theta}_B - \theta_n)$. Choose the $\alpha/2$ and $1 - \alpha/2$ sample quantiles from the bootstrap sample: $[\tilde{\theta}(1-\alpha/2)-\theta_n, \tilde{\theta}(\alpha/2)-\theta_n]$. Then

$$P\left(\tilde{\theta}(1-\alpha/2)-\theta_n < \theta-\theta_n < \tilde{\theta}(\alpha/2)-\theta_n\right)$$
$$= P\left(\tilde{\theta}(1-\alpha/2) < \theta < \tilde{\theta}(\alpha/2)\right) \approx 1-\alpha.$$

The quantiles of the bootstrap samples form an approximate confidence interval for θ that is computationally simple to construct.

Parametric Case. If the actual data are assumed to be generated from a distribution $F(x;\theta)$ (with unknown θ), we can improve over the nonparametric bootstrap. Instead of resampling from the data, we can generate a more efficient bootstrap sample by simulating data from $F(x;\theta_n)$.

Example 15.1 Hubble Telescope and Hubble Correlation. The Hubble constant (H) is one of the most important numbers in cosmology because it is instrumental in estimating the size and age of the universe. This long-sought number indicates the rate at which the universe is expanding, from the primordial "Big Bang." The Hubble constant can be used to determine the intrinsic brightness and masses of stars in nearby galaxies, examine those same properties in more distant galaxies and galaxy clusters, deduce the amount of dark matter present in the universe, obtain the scale size of faraway galaxy clusters, and serve as a test for theoretical cosmological models.

In 1929, Edwin Hubble (Figure 15.3(a)) investigated the relationship between the distance of a galaxy from the earth and the velocity with which it appears to be receding. Galaxies appear to be moving away from us no matter which direction we look. This is thought to be the result of the Big Bang. Hubble hoped to provide some knowledge about how the universe was formed and what might happen in the future. The data collected include distances (megaparsecs[1]) to $n = 24$ galaxies and their recessional velocities (km/sec). The scatter plot of the pairs is given in Figure 15.3(b). Hubble's law claims that Recessional Velocity is directly proportional to the Distance and the coefficient of proportionality is Hubble's constant, H. By working backward in time, the galaxies appear to meet in the same place. Thus $1/H$ can be used to estimate the time since the Big Bang – a measure of the age of the universe. Thus, because of this simple linear model, it is important to estimate correlation between distances and velocities and see if the no-intercept linear regression model is appropriate.

[1] 1 parsec = 3.26 light years.

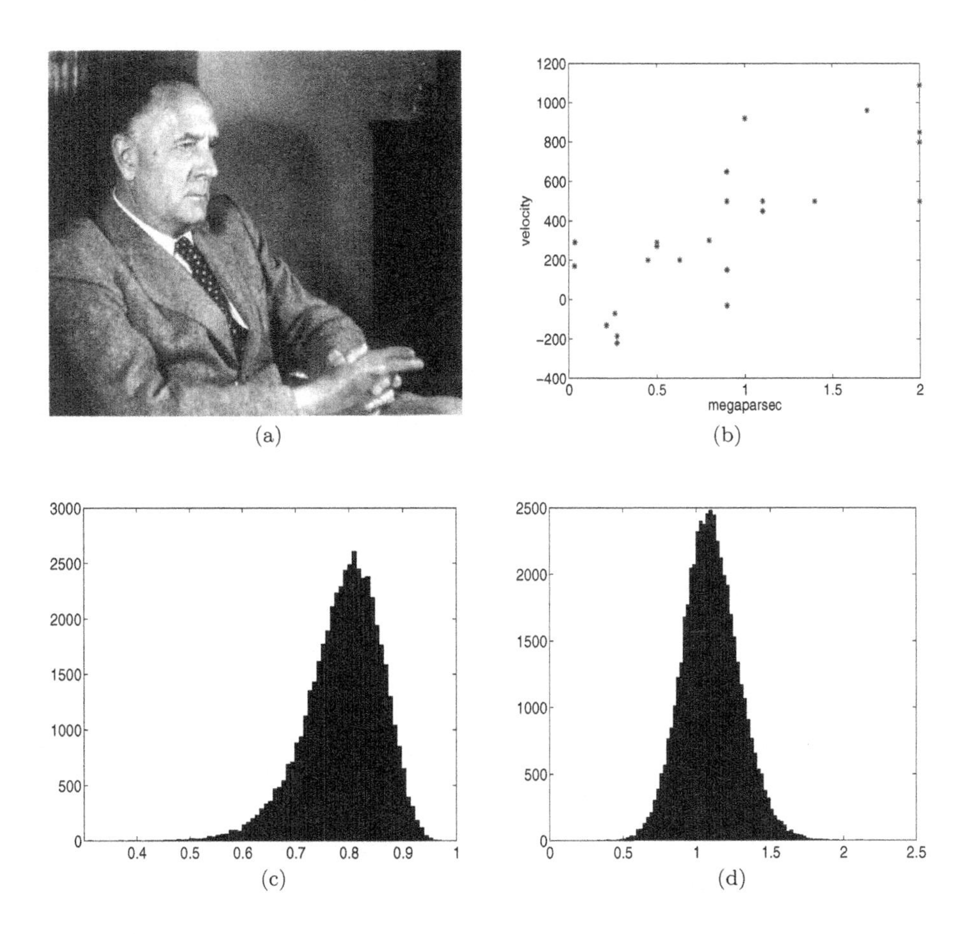

Fig. 15.3 (a) Edwin Powell Hubble (1889–1953), American astronomer who is considered the founder of extragalactic astronomy and who provided the first evidence of the expansion of the universe; (b) Scatter plot of 24 distance-velocity pairs. Distance is measured in parsecs and velocity in km/h; (c) Histogram of correlations from 50000 bootstrap samples; (d) Histogram of correlations of Fisher's z transformations of the bootstrap correlations.

Distance in megaparsecs ([Mpc])	.032	.034	.214	.263	.275	.275
	.45	.5	.5	.63	.8	.9
	.9	.9	.9	1.0	1.1	1.1
	1.4	1.7	2.0	2.0	2.0	2.0
The recessional velocity ([km/sec])	170	290	−130	−70	−185	−220
	200	290	270	200	300	−30
	650	150	500	920	450	500
	500	960	500	850	800	1090

The correlation coefficient between *mpc* and v, based on $n = 24$ pairs is 0.7896. How confident are we about this estimate? To answer this question we resample data and obtain $B = 50000$ subrogate samples, each consisting of 24 randomly selected (with repeating) pairs from the original set of 24 pairs. The histogram of all correlations $r_i^*, i = 1, \ldots 50000$ among bootstrap samples is shown in Figure 15.3(c). From the bootstrap samples we find that the standard deviation of r can be estimated by 0.0707. From the empirical density for r, we can generate various bootstrap summaries about r.

Figure 15.3(d) shows the Fisher z-transform of the r^*s, $z_i^* = 0.5 \log[(1 + r_i^*)/(1 - r_i^*)]$ which are bootstrap replicates of $z = 0.5 \log[(1 + r)/(1 - r)]$. Theoretically, when normality is assumed, the standard deviation of z is $(n - 3)^{-1/2}$. Here, we estimate standard deviation of z using bootstrap samples as 0.1906 which is close to $(24 - 3)^{-1/2} = 0.2182$. The core of the MATLAB program calculating bootstrap estimators is

```
>> bsam=[];
>> B=50000;
>> for b = 1:B
     bs =  bootsample(pairs);
     ccbs = corrcoef(bs);
     bsam = [bsam ccbs(1,2)];
>>    end
```

where the function

`bootsample(x)`

is a simple m-file resampling the `vecin` that is $n \times p$ data matrix with n equal to number of observations and p equal to dimension of a single observation.

```
function  vecout = bootsample(vecin)
[n, p] = size(vecin);
selected_indices = floor(1+n.*(rand(1,n)));
vecout = vecin(selected_indices,:);
```

Example 15.2 Trimmed Mean. For robust estimation of the population mean, outliers can be trimmed off the sample, ensuring the estimator will be less influenced by tails of the distribution. If we trim off almost all of the data, we will end up using the sample median. Suppose we trim off 50% of the data by excluding the smallest and largest 25% of the sample. Obviously, the standard error of this estimator is not easily tractable, so no exact confidence interval can be constructed. This is where the bootstrap technique can help out. In this example, we will focus on constructing a two-sided 95% confidence interval for μ, where

$$\mu = \frac{\int_{x_{1/4}}^{x_{3/4}} t dF(t)}{F(x_{3/4}) - F(x_{1/4})} = 2\int_{x_{1/4}}^{x_{3/4}} t dF(t)$$

is an alternative measure of central tendency, the same as the population mean if the distribution is symmetric.

If we compute the trimmed mean from the sample as μ_n, it is easy to generate bootstrap samples and do the same. In this case, limiting B to 1000 or 2000 will make computing easier, because each repeated sample must be ranked and trimmed before $\tilde{\mu}$ can be computed. Let $\tilde{\mu}(.025)$ and $\tilde{\mu}(.975)$ be the lower and upper quantiles from the bootstrap sample $\tilde{\mu}_1, \dots, \tilde{\mu}_B$.

The MATLAB m-file `trimmean(x,P)` trims P% (so $0 < P < 100$) of the data, or $P/2$% of the biggest and smallest observations. The MATLAB m-file

```
ciboot(x,'trimmean',5,.90,1000,10)
```

acquires 1000 bootstrap samples from x, performs the *trimmean* function (its additional argument, P=10, is left on the end) and a 90% (2-sided) confidence interval is generated. The middle value is the point estimate. Below, the vector x represents a skewed sample of test scores, and a 90% confidence interval for the trimmed mean is (57.6171, 89.9474). The third argument in the `ciboot` function can take on integer values between one and six, and this input dictates the type of bootstrap to construct. The input options are

1. Normal approximation (std is bootstrap).
2. Simple bootstrap principle (bad, don't use).
3. Studentized, std is computed via jackknife.
4. Studentized, std is 30 samples' bootstrap.
5. Efron's pctl method.
6. Efron's pctl method with bias correction(default)

```
>> x = [11,13,14,32,55,58,61,67,69,73,73,89,90,93,94,94,95,96,99,99];
>> m = trimmean(x,10)
m =
   71.7895
>> m2 = mean(x)
m2 =
   68.7500
>> ciboot(x,'trimmean',5,.90,1000,10)
ans =
   57.6171   71.7895   82.9474
```

Estimating Standard Error. The most common application of a simple bootstrap is to estimate the standard error of the estimator θ_n. The algorithm is similar to the general nonparametric bootstrap:

- Generate B bootstrap samples of size n.
- Evaluate the bootstrap estimators $\tilde{\theta}_1, \ldots, \tilde{\theta}_B$.
- Estimate standard error of θ_n as

$$\hat{\sigma}_{\theta_n} = \sqrt{\frac{\Sigma_{i=1}^{B}\left(\tilde{\theta}_i - \tilde{\theta}^*\right)^2}{B-1}},$$

where $\tilde{\theta}^* = B^{-1}\Sigma\tilde{\theta}_i$.

15.3 BIAS CORRECTION FOR NONPARAMETRIC INTERVALS

The percentile method described in the last section is simple, easy to use, and has good large sample properties. However, the coverage probability is not accurate for many small sample problems. The *Acceleration and Bias-Correction* (or BC_a) method improves on the percentile method by adjusting the percentiles (e.g., $\tilde{\theta}(1-\alpha/2, \tilde{\theta}(\alpha/2))$ chosen from the bootstrap sample. A detailed discussion is provided in Efron and Tibshirani (1993).

The BC_a interval is determined by the proportion of the bootstrap estimates $\tilde{\theta}$ less than θ_n, i.e., $p_0 = B^{-1}\Sigma I(\tilde{\theta}_i < \theta_n)$ define the *bias factor* as

$$z_0 = \Phi^{-1}(p_0)$$

express this bias, where Φ is the standard normal CDF, so that values of z_0

away from zero indicate a problem. Let

$$a_0 = \frac{\Sigma_{i=1}^{B}\left(\tilde{\theta}^* - \tilde{\theta}_i\right)^3}{6\left(\Sigma_{i=1}^{B}(\tilde{\theta}^* - \tilde{\theta}_i)^2\right)^{3/2}}$$

be the *acceleration factor*, where $\tilde{\theta}^*$ is the average of the bootstrap estimates $\tilde{\theta}_1, \ldots, \tilde{\theta}_B$. It gets this name because it measures the rate of change in σ_{θ_n} as a function of θ.

Finally, the $100(1-\alpha)\%$ $\mathrm{BC_a}$ interval is computed as

$$[\ \tilde{\theta}(q_1),\ \tilde{\theta}(q_2)\ ],$$

where

$$q_1 = \Phi\left(z_0 + \frac{z_0 + z_{\alpha/2}}{1 - a_0(z_0 + z_{\alpha/2})}\right),$$

$$q_2 = \Phi\left(z_0 + \frac{z_0 + z_{1-\alpha/2}}{1 - a_0(z_0 + z_{1-\alpha/2})}\right).$$

Note that if $z_0 = 0$ (no measured bias) and $a_0 = 0$, then (15.1) is the same as the percentile bootstrap interval. In the MATLAB m-file `ciboot`, the $\mathrm{BC_a}$ is an option (6) for the nonparametric interval. For the trimmed mean example, the bias corrected interval is shifted upward:

```
>> ciboot(x,'trimmean',6,.90,1000,10)
ans =
   60.0412   71.7895   84.4211
```

Example 15.3 Recall the data from Crowder et al. (1991) which was discussed in Example 10.2. The data contain strength measurements (in coded units) for 48 pieces of weathered cord. Seven of the pieces of cord were damaged and yielded strength measurements that are considered right censored. The following MATLAB code uses a bias-corrected bootstrap to calculate a 95% confidence interval for the probability that the strength measure is equal to or less than 50, that is, $F(50)$.

```
>> data = [36.3, 41.7, 43.9, 49.9, 50.1, 50.8, 51.9, 52.1, 52.3, 52.3,...
        52.4, 52.6, 52.7, 53.1, 53.6, 53.6, 53.9, 53.9, 54.1, 54.6,...
        54.8, 54.8, 55.1, 55.4, 55.9, 56.0, 56.1, 56.5, 56.9, 57.1,...
        57.1, 57.3, 57.7, 57.8, 58.1, 58.9, 59.0, 59.1, 59.6, 60.4,...
        60.7, 26.8, 29.6, 33.4, 35.0, 40.0, 41.9, 42.5];
>> censor=[ones(1,41), zeros(1,7)];
>> [kmest, sortdat, sortcen] = KMcdfSM(data', censor', 0);
>> prob = kmest( sum( 50.0 >=data), 1)
   prob =
```

```
      0.0949
>> function fkmt = kme_at_50(dt)
    % this function performs Kaplan-Meier
    % estimation with given parameter
    % and produces estimated F(50.0)
   [kmest sortdat] = KMcdfSM(dt(:,1), dt(:,2), 0);
   fkmt = kmest(sum(50.0 >= sortdat), 1);
```

Using kme_at_50.m and ciboot functions we obtain a confidence interval for $F(50)$ based on 1000 bootstrap replicates:

```
 >> ciboot([data' censor'], 'kme_at_50', 5, .95, 1000)
ans =
   0.0227    0.0949    0.1918
 >> % a 95% CI for F(50) is (0.0227, 0.1918)
 >> function fkmt = kme_all_x(dt)
    % this function performs Kaplan-Meier estimation with given parameter
    % and gives estimated F() for all data points
    [kmest sortdat] = KMcdfSM(dt(:,1), dt(:,2), 0);
    data = [36.3, 41.7, //...deleted...//, 41.9, 42.5];
    temp_val = [];
    %calculate each CDF F() value for all data points
    for i=1:length(data)
      if sum(data(i) >= sortdat) > 0
       temp_val = [temp_val kmest(sum(data(i) >= sortdat), 1)];
      else   % when there is no observation, CDF  is simply 0
       temp_val = [temp_val 0];
      end
    end
    fkmt = temp_val;
```

The MATLAB functions ciboot and kme_all_x are used to produce Figure 15.4:

```
 >> ci = ciboot([data' censor'], 'kme_all_x', 5, .95, 1000);
 >> figure;
 >> plot(data', ci(:,2)', '.');
 >> hold on;
 >> plot(data', ci(:,1)', '+');
 >> plot(data', ci(:,3)', '*');
```

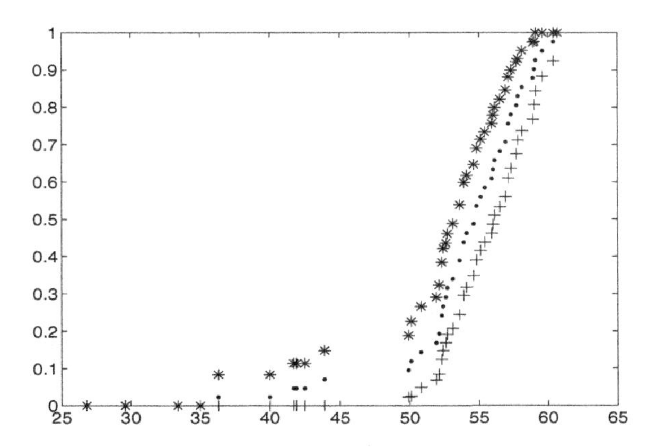

Fig. 15.4 95% confidence band the CDF of Crowder's data using 1000 bootstrap samples. Lower boundary of the confidence band is plotted with marker '+', while the upper boundary is plotted with marker '*'.

15.4 THE JACKKNIFE

The *jackknife* procedure, introduced by Quenouille (1949), is a resampling method for estimating bias and variance in θ_n. It predates the bootstrap and actually serves as a special case. The resample is based on the "leave one out" method, which was computationally easier when computing resources were limited.

The ith jackknife sample is $(x_1, \ldots, x_{i-1}, x_{i+1}, ..., x_n)$. Let $\hat{\theta}_{(i)}$ be the estimator of θ based only on the ith jackknife sample. The jackknife estimate of the *bias* is defined as

$$\hat{b}_J = (n-1)\left(\hat{\theta}_n - \hat{\theta}^*\right),$$

where $\hat{\theta}^* = n^{-1}\Sigma\hat{\theta}_{(i)}$. The jackknife estimator for the variance of θ_n is

$$\sigma_J{}^2 = \frac{n-1}{n}\Sigma_{i=1}^n\left(\hat{\theta}_{(i)} - \hat{\theta}^*\right)^2.$$

The jackknife serves as a poor man's version of the bootstrap. That is, it estimates bias and variance the same, but with a limited resampling mechanism. In MATLAB, the m-file

```
jackknife(x,function,p1,...)
```

produces the jackknife estimate for the input function. The function `jackrsp(x,k)` produces a matrix of jackknife samples (taking k elements out, with default of $k = 1$).

```
>> [b,v,f]=jackknife('trimmean', x', 10)
                    %note: row vector input
b =
   -0.1074              % Jackknife estimate of bias
v =
   65.3476              % Jackknife estimate of variance
f =
   71.8968              % Jackknife corrected estimate
```

The jackknife performs well in most situations, but poorly in some. In case θ_n can change significantly with slight changes to the data, the jackknife can be temperamental. This is true with $\theta =$ median, for example. In such cases, it is recommended to augment the resampling by using a *delete-d jackknife*, which leaves out d observations for each jackknife sample. See Chapter 11 of Efron and Tibshirani (1993) for details.

15.5 BAYESIAN BOOTSTRAP

The Bayesian bootstrap (BB), a Bayesian analogue to the bootstrap, was introduced by Rubin (1981). In Efron's standard bootstrap, each observation X_i from the sample $X_1, \dots, X_n$ has a probability of $1/n$ to be selected and after the selection process the relative frequency f_i of X_i in the bootstrap sample belongs to the set $\{0, 1/n, 2/n, \dots, (n-1)/n, 1\}$. Of course, $\Sigma_i f_i = 1$. Then, for example, if the statistic to be evaluated is the sample mean, its bootstrap replicate is $\bar{X}^* = \Sigma_i f_i X_i$.

In Bayesian bootstrapping, at each replication a discrete probability distribution $\mathbf{g} = \{g_1, \dots, g_n\}$ on $\{1, 2, \dots, n\}$ is generated and used to produce bootstrap statistics. Specifically, the distribution $\mathbf{g}$ is generated by generating $n-1$ uniform random variables $U_i \sim \mathcal{U}(0,1)$, $i = 1, \dots, n-1$, and ordering them according to $\tilde{U}_j = U_{j:n-1}$ with $\tilde{U}_0 \equiv 0$ and $\tilde{U}_n \equiv 1$. Then the probability of X_i is defined as

$$g_i = \tilde{U}_i - \tilde{U}_{i-1}, \quad i = 1, \dots, n.$$

If the sample mean is the statistic of interest, its Bayesian bootstrap replicate is a weighted average if the sample, $\bar{X}^* = \Sigma_i g_i X_i$. The following example explains why this resampling technique is Bayesian.

Example 15.4 Suppose that $X_1, \dots, X_n$ are i.i.d. $\mathcal{B}er(p)$, and we seek a BB estimator of p. Let n_1 be the number of ones in the sample and $n - n_1$ the number of zeros. If the BB distribution $\mathbf{g}$ is generated then let $P_1 = \Sigma g_i \mathbf{1}(X_i = 1)$ be the probability of 1 in the sample. The distribution for P_1 is simple, because the gaps in the $U_1, \dots, U_{n-1}$ follow the $(n-1)$-variate Dirichlet distribution, $\mathcal{D}ir(1, 1, \dots, 1)$. Consequently, P_1 is the sum of n_1 gaps and is distrubted $\mathcal{B}e(n_1, n - n_1)$. Note that $\mathcal{B}e(n_1, n - n_1)$ is, in fact, the posterior for P_1 if the prior is $\propto [P_1(1-P_1)]^{-1}$. That is, for $x \in \{0,1\}$,

$$P(X = x|P_1) = P_1^x(1-P_1)^{1-x}, \quad P_1 \propto [P_1(1-P_1)]^{-1},$$

then the posterior is

$$[P_1|X_1, \dots, X_n] \sim \mathcal{B}e(n_1, n-n_1).$$

For general case when X_i take $d \leq n$ different values the Bayesian interpretation is still valid; see Rubin's (1981) article.

Example 15.5 We revisit Hubble's data and give a BB estimate of variability of observed coefficient of correlation r. For each BB distribution $\mathbf{g}$ calculate

$$r^* = \frac{\Sigma_{i=1}^n g_i X_i Y_i - (\Sigma_{i-1}^n g_i X_i)(\Sigma_{i-1}^n g_i X_i)}{[\Sigma_{i=1}^n g_i X_i^2 - (\Sigma_{i=1}^n g_i X_i)^2]^{1/2}\,[\Sigma_{i=1}^n g_i Y_i^2 - (\Sigma_{i=1}^n g_i Y_i)^2]^{1/2}},$$

where $(X_i, Y_i), i = 1, \dots 24$ are observed pairs of distances and velocities. The MATLAB program below performs the BB resampling.

```
>> x = [0.032 0.034 0.214 0.263 0.275 0.275 0.45 0.5 0.5 0.63
     0.8 0.9 0.9 0.9 0.9 1.0 1.1 1.1 1.4 1.7 2.0 2.0 2.0 2.0 ]; %Mpc
>> y = [170 290 -130 -70 -185 -220 200 290 270 200 300 -30 ...
     650 150 500 920 450 500 500 960 500 850 800 1090];    %velocity
>> n=24; corr(x', y');
>> B=50000; %number of BB replicates
>> bbcorr = []; %store BB correlation replicates
>> for i = 1:B
      sampl = (rand(1,n-1));
      osamp = sort(sampl);
      all = [0 osamp 1];
      gis = diff(all, 1);
         % gis is BB distribution,  corrbb is correlation
         % with gis as weights
      ssx  = sum(gis .* x);      ssy = sum( gis .* y);
      ssx2 = sum(gis .* x.^2); ssy2 = sum(gis .* y.^2);
      ssxy = sum(gis .* x .* y);
      corrbb = (ssxy - ssx * ssy)/...
       sqrt((ssx2 - ssx^2)*(ssy2 - ssy^2)); %correlation replicate
   bbcorr=[bbcorr corrbb];  %add replicate to the storage sequence
>> end
>> figure(1)
>> hist(bbcorr,80)
>> std(bbcorr)
>> zs = 1/2 * log((1+bbcorr)./(1-bbcorr)); %Fisher's z
>> figure(2)
>> hist(zs,80)
>> std(zs)
```

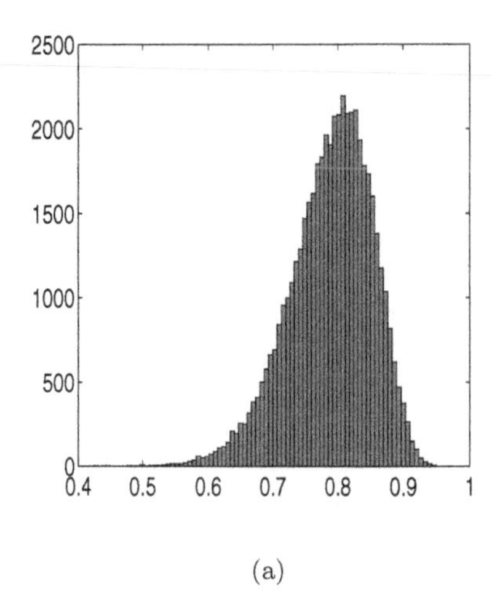

(a)

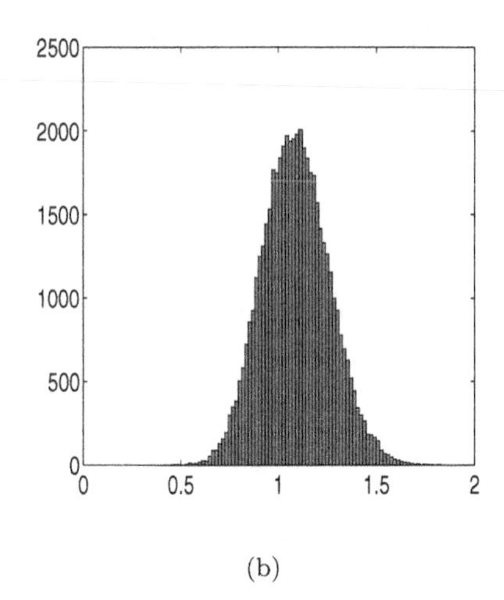

(b)

Fig. 15.5 The histogram of 50,000 BB resamples for the correlation between the distance and velocity in the Hubble data; (b) Fisher z-transform of the BB correlations.

The histograms of correlation bootstrap replicates and their z-transforms in Figure 15.5 (a-b) look similar to the those in Figure 15.3 (c-d). Numerically, $B = 50,000$ replicates gave standard deviation of observed r as 0.0635 and standard deviation of $z = 1/2 \log((1+r)/(1-r))$ as 0.1704 slightly smaller than theoretical $24 - 3^{-1/2} = 0.2182$.

15.6 PERMUTATION TESTS

Suppose that in a statistical experiment the sample or samples are taken and a statistic S is constructed for testing a particular hypothesis H_0. The values of S that seem extreme from the viewpoint of H_0 are critical for this hypothesis. The decision if the observed value of statistics S is extreme is made by looking at the distribution of S when H_0 is true. But what if such distribution is unknown or too complex to find? What if the distribution for S is known only under stringent assumptions that we are not willing to make?

Resampling methods consisting of permuting the original data can be used to approximate the null distribution of S. Given the sample, one forms the permutations that are *consistent with experimental design and* H_0, and then calculates the value of S. The values of S are used to estimate its density (often as a histogram) and using this empirical density we find an approximate p-value, often called a *permutation p-value.*

What permutations are consistent with H_0? Suppose that in a two-sample problem we want to compare the means of two populations based on two independent samples $X_1, \ldots, X_m$ and $Y_1, \ldots, Y_n$. The null hypothesis H_0 is $\mu_X = \mu_Y$. The permutations consistent with H_0 would be all permutations of a combined (concatenated) sample $X_1, \ldots, X_m, Y_1, \ldots, Y_n$. Or suppose we a repeated measures design in which observations are triplets corresponding to three treatments, i.e., $(X_{11}, X_{12}, X_{13}), \ldots, (X_{n1}, X_{n2}, X_{n3})$, and that H_0 states that the three treatment means are the same, $\mu_1 = \mu_2 = \mu_3$. Then permutations consistent with this experimental design are random permutations among the triplets (X_{i1}, X_{i2}, X_{i3}), $i = 1, \ldots, n$ and a possible permutation might be

$$(X_{13}, X_{11}, X_{12}), (X_{21}, X_{23}, X_{22}), (X_{32}, X_{33}, X_{31}), \ldots, (X_{n2}, X_{n1}, X_{n3}).$$

Thus, depending on the design and H_0, consistent permutations can be quite different.

Example 15.6 Byzantine Coins. To illustrate the spirit of permutation tests we use data from a paper by Hendy and Charles (1970) (see also Hand et al, 1994) that represent the silver content (%Ag) of a number of Byzantine coins discovered in Cyprus. The coins (Figure 15.6) are from the first and fourth coinage in the reign of King Manuel I, Comnenus (1143–1180).

1st coinage	5.9	6.8	6.4	7.0	6.6	7.7	7.2	6.9	6.2
4th coinage	5.3	5.6	5.5	5.1	6.2	5.8	5.8		

The question of interest is whether or not there is statistical evidence to suggest that the silver content of the coins was significantly different in the later coinage.

Fig. 15.6 A coin of Manuel I Comnenus (1143–1180).

Of course, the two-sample t-test or one of its nonparametric counterparts is possible to apply here, but we will use the permutation test for purposes of illustration. The following MATLAB commands perform the test:

```
>> coins=[5.9  6.8  6.4  7.0  6.6  7.7  7.2  6.9  6.2 ...
>>           5.3  5.6  5.5  5.1  6.2  5.8  5.8];
>> coins1=coins(1:9); coins2=coins(10:16);
>> S = (mean(coins1)-mean(coins2))/sqrt(var(coins1)+var(coins2))
>> Sps =[]; asl=0; %Sps is permutation S,
>>                 %asl is achieved significance level
>> N=10000;
>> for i = 1:N
      coinsp=coins(randperm(16));
      coinsp1=coinsp(1:9); coinsp2=coinsp(10:16);
      Sp = (mean(coinsp1)-mean(coinsp2))/  ...
                sqrt(var(coinsp1)+var(coinsp2));
      Sps = [Sps Sp ];
      asl = asl + (abs(Sp) > S );
   end
>> asl = asl/N
```

The value for S is 1.7301, and the permutation p-value or the achieved significance level is `asl` = 0.0004. Panel (a) in Figure 15.7 shows the permutation null distribution of statistics S and the observed value of S is indicated by the dotted vertical line. Note that there is nothing special about selecting

$$S = \frac{\bar{X}_1 - \bar{X}_2}{\sqrt{s_1^2/n_1 + s_2^2/n_2}},$$

and that any other statistics that sensibly measures deviation from $H_0 : \mu_1 = \mu_2$ could be used. For example, one could use $S = \text{median}(X_1)/s_1 - \text{median}(X_2)/s_2$, or simply $S = \bar{X}_1 - \bar{X}_2$.

To demonstrate how the choice what to permute depends on statistical design, we consider again the two sample problem but with paired observations. In this case, the permutations are done within the pairs, independently from pair to pair.

Example 15.7 Left-handed Grippers. Measurements of the left- and right-hand gripping strengths of 10 left-handed writers are recorded.

Person	1	2	3	4	5	6	7	8	9	10
Left hand (X)	140	90	125	130	95	121	85	97	131	110
Right hand (Y)	138	87	110	132	96	120	86	90	129	100

Do the data provide strong evidence that people who write with their left hand have greater gripping strength in the left hand than they do in the right hand?

In the MATLAB solution provided below, dataL and dataR are paired measurements and pdataL and pdataR are random permutations, either $\{1, 2\}$ or $\{2, 1\}$ of the 10 original pairs. The statistics S is the difference of the sample means. The permutation null distribution is shown as non-normalized histogram in Figure 15.7(b). The position of S with respect to the histogram is marked by dotted line.

```
>> dataL =[ 140 , 90 , 125 , 130 , 95 , 121 , 85 , 97 , 131 , 110 ];
>> dataR =[ 138 , 87 , 110 , 132 , 96 , 120 , 86 , 90 , 129 , 100 ];
>> S=mean(dataL - dataR)
>> data =[dataL; dataR];
>> means=[]; asl =0; N=10000;
>> for i = 1:N
       pdata=[];
       for j=1:10
           pairs = data(randperm(2),j);
           pdata = [pdata pairs];
       end
       pdataL = pdata(1,:);
       pdataR = pdata(2,:);
       pmean=mean(pdataL - pdataR);
       means= [means pmean];
       asl = asl + (abs(pmean) > S);
   end
```

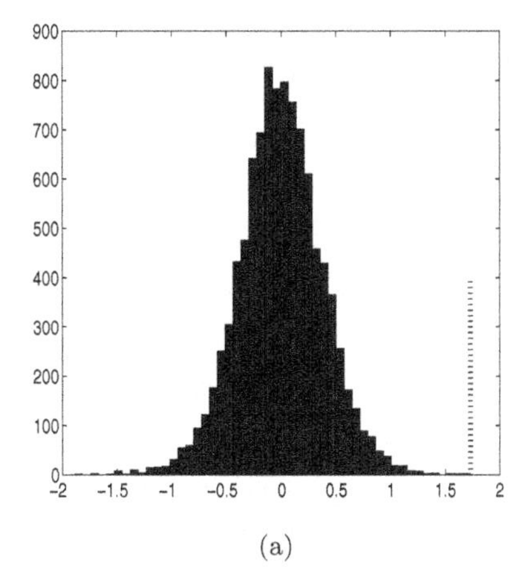

(a)

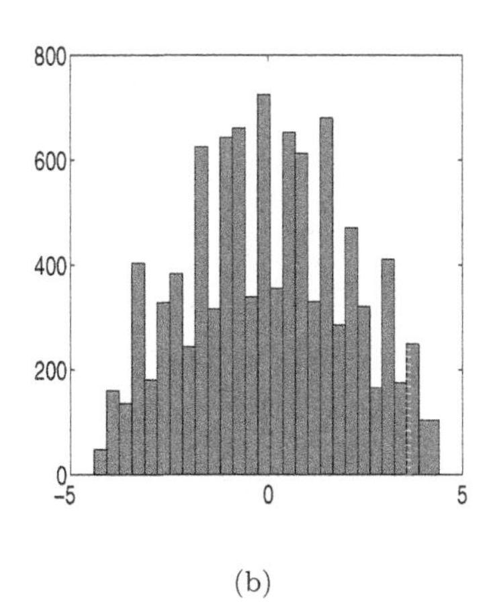

(b)

Fig. 15.7 Panels (a) and (b) show permutation null distribution of statistics S and the observed value of S (marked by *dotted* line) for the cases of (a) Bizantine coins, and (b) Left-handed grippers.

15.7 MORE ON THE BOOTSTRAP

There are several excellent resources for learning more about bootstrap techniques, and there are many different kinds of bootstraps that work on various problems. Besides Efron and Tibshirani (1993), books by Chernick (1999) and Davison and Hinkley (1997) provide excellent overviews with numerous helpful examples. In the case of dependent data various bootstrapping strategies are proposed such as block bootstrap, stationary bootstrap, wavelet-based bootstrap (wavestrap), and so on. A monograph by Good (2000) gives a comprehensive coverage of permutation tests.

Bootstrapping is not infallible. Data sets that might lead to poor performance include those with missing values and excessive censoring. Choice of statistics is also critical; see Exercise 15.6. If there are few observations in the tail of the distribution, bootstrap statistics based on the EDF perform poorly because they are deduced using only a few of those extreme observations.

15.8 EXERCISES

15.1. Generate a sample of 20 from the gamma distribution with $\lambda = 0.1$ and r=3. Compute a 90% confidence interval for the mean using (a) the standard normal approximation, (b) the percentile method and (c) the bias-corrected method. Repeat this 1000 times and report the actual *coverage probability* of the three intervals you constructed.

15.2. For the case of estimating the sample mean with $\bar{X}$, derive the expected value of the jackknife estimate of bias and variance.

15.3. Refer to insect waiting times for the *female* Western White Clematis in Table 10.15. Use the percentile method to find a 90% confidence interval for $F(30)$, the probability that the waiting time is less than or equal to 30 minutes.

15.4. In a data set of size n generated from a continuous F, how many *distinct* bootstrap samples are possible?

15.5. Refer to the dominance-submissiveness data in Exercise 7.3. Construct a 95% confidence interval for the correlation using the percentile bootstrap and the jackknife. Compare your results with the normal approximation described in Section 2 of Chapter 7.

15.6. Suppose we have three observations from $\mathcal{U}(0, \theta)$. If we are interested in estimating θ, the MLE for it is $\hat{\theta} = X_{3:3}$, the largest observation. If we obtain a bootstrap sampling procedure to estimate the variance of the MLE, what is the distribution of the bootstrap estimator for θ?

15.7. Seven patients each underwent three different methods of kidney dialysis. The following values were obtained for weight change in kilograms between dialysis sessions:

Patient	Treatment 1	Treatment 2	Treatment 3
1	2.90	2.97	2.67
2	2.56	2.45	2.62
3	2.88	2.76	1.84
4	2.73	2.20	2.33
5	2.50	2.16	1.27
6	3.18	2.89	2.39
7	2.83	2.87	2.39

Test the null hypothesis that there is no difference in mean weight change among treatments. Use properly designed permutation test.

15.8. In a controlled clinical trial *Physician's Health Study I* which began in 1982 and ended in 1987, more that 22,000 physicians participated. The participants were randomly assigned to two groups: (i) *Aspirin* and (ii) *Placebo*, where the aspirin group have been taking 325 mg aspirin every second day. At the end of trial, the number of participants who suffered from Myocardial Infarction was assessed. The counts are given in the following table:

	MyoInf	No MyoInf	Total
Aspirin	104	10933	11037
Placebo	189	10845	11034

The popular measure in assessing results in clinical trials is Risk Ratio (RR) which is the ratio of proportions of cases (risks) in the two groups/treatments. From the table,

$$RR = R_a/R_p = \frac{104/11037}{189/11034} = 0.55.$$

Interpretation of RR is that the risk of Myocardial Infarction for the Placebo group is approximately $1/0.55 = 1.82$ times higher than that for the Aspirin group. With MATLAB, construct a bootstrap estimate for the variability of RR. *Hint:*

```
aspi = [zeros(10933,1); ones(104,1)];
plac = [zeros(10845,1); ones(189,1)];
RR = (sum(aspi)/length(aspi))/(sum(plac)/length(plac));
```

```
BRR = [ ];  B=10000;
for b = 1:B
    baspi =  bootsample(aspi);
    bplac =  bootsample(plac);
BRR = [BRR (sum(baspi)/length(baspi))/(sum(bplac)/length(bplac))];
end
```

(ii) Find the variability of the difference of the risks $R_a - R_p$, and of logarithm of the odds ratio, $\log(R_a/(1-R_a)) - \log(R_p/(1-R_p))$.

(iii) Using the Bayesian bootstrap, estimate the variability of RR, $R_a - R_p$, and $\log(R_a/(1-R_a)) - \log(R_p/(1-R_p))$.

15.9. Let f_i and g_i be frequency/probability of the observation X_i in an ordinary/Bayesian bootstrap resample from $X_1, \ldots, X_n$. Prove that $\mathbb{E}f_i = \mathbb{E}g_i = 1/n$, i.e., the expected probability distribution is discrete uniform, $\mathbb{V}\text{ar}f_i = (n+1)/n$, $\mathbb{V}\text{ar}g_i = (n-1)/n^2$, and for $i \neq j$, $\mathbb{C}\text{orr}(f_i, f_j) = \mathbb{C}\text{orr}(g_i, g_j) = -1/(n-1)$.

REFERENCES

Davison, A. C., and Hinkley, D. V. (1997), *Bootstrap Methods and Their Applications,* Boston: Cambridge University Press.

Chernick, M. R., (1999), *Bootstrap Methods – A Practitioner's Guide*, New York: Wiley.

Efron, B., and Tibshirani, R. J. (1993), *An Introduction to the Bootstrap,* Boca Raton, FL: CRC Press.

Efron, B. (1979), "Bootstrap Methods: Another Look at the Jackknife," *Annals of Statistics*, 7, 1-26

Fisher, R.A. (1935), *The Design of Experiments,* New York: Hafner.

Good, P. I. (2000), *Permutation Tests: A Practical Guide to Resampling Methods for Testing Hypotheses*, 2nd ed., New York: Springer Verlag.

Hand, D.J., Daly, F., Lunn, A.D., McConway, K.J., and Ostrowski, E. (1994), *A Handbook of Small Datasets*, New York: Chapman & Hall.

Hendy, M. F., and Charles, J. A. (1970), "The Production Techniques, Silver Content, and Circulation History of the Twelfth-Century Byzantine Trachy," *Archaeometry*, 12, 13-21.

Mahalanobis, P. C. (1946), "On Large-Scale Sample Surveys," *Philosophical Transactions of the Royal Society of London,* Ser. B, 231, 329–451.

Pitman, E. J. G., (1937), "Significance Tests Which May Be Applied to Samples from Any Population," *Royal Statistical Society Supplement,* 4, 119–130 and 225–232 (parts I and II).

Quenouille, M. H. (1949), "Approximate Tests of Correlation in Time Series," *Journal of the Royal Statistical Society*, Ser. B, 11, 18–84.

Raspe, R. E. (1785). *The Travels and Surprising Adventures of Baron Munchausen*, London: Trubner, 1859 [1st Ed. 1785].

Rubin, D. (1981), "The Bayesian Bootstrap," *Annals of Statistics*, 9, 130-134.

16

EM Algorithm

Insanity is doing the same thing over and over again and expecting different results.

Albert Einstein

The Expectation-Maximization (EM) algorithm is broadly applicable statistical technique for maximizing complex likelihoods while handling problems with incomplete data. Within each iteration of the algorithm, two steps are performed: (i) the E-Step consisting of projecting an appropriate functional containing the augmented data on the space of the original, incomplete data, and (ii) the M-Step consisting of maximizing the functional.

The name EM algorithm was coined by Dempster, Laird, and Rubin (1979) in their fundamental paper, referred to here as the DLR paper. But as is usually the case, if one comes to a smart idea, one may be sure that other smart guys in the history had already thought about it. Long before, McKendrick (1926) and Healy and Westmacott (1956) proposed iterative methods that are examples of the EM algorithm. In fact, before the DLR paper appeared in 1997, dozens of papers proposing various iterative solvers were essentially applying the EM Algorithm in some form.

However, the DLR paper was the first to formally recognize these separate algorithms as having the same fundamental underpinnings, so perhaps their 1977 paper prevented further reinventions of the same basic math tool. While the algorithm is not guaranteed to converge in every type of problem (as mistakenly claimed by DLR), Wu (1983) showed convergence is guaranteed if the densities making up the full data belong to the exponential family.

This does not prevent the EM method from being helpful in nonparametric problems; Tsai and Crowley (1985) first applied it to a general nonparametric setting and numerous applications have appeared since.

16.0.1 Definition

Let Y be a random vector corresponding to the observed data y and having a postulated PDF $f(y, \psi)$, where $\psi = (\psi_1, \ldots, \psi_d)$ is a vector of unknown parameters. Let x be a vector of augmented (so called complete) data, and let z be the missing data that completes x, so that $x = [y, z]$.

Denote by $g_c(x, \psi)$ the PDF of the random vector corresponding to the complete data set x. The log-likelihood for ψ, if x were fully observed, would be

$$\log L_c(\psi) = \log g_c(x, \psi).$$

The incomplete data vector y comes from the "incomplete" sample space $\mathcal{Y}$. There is an one-to-one correspondence between the complete sample space $\mathcal{X}$ and the incomplete sample space $\mathcal{Y}$. Thus, for $x \in \mathcal{X}$, one can uniquely find the "incomplete" $y = y(x) \in \mathcal{Y}$. Also, the incomplete pdf can be found by properly integrating out the complete pdf,

$$g(y, \psi) = \int_{\mathcal{X}(y)} g_c(x, \psi) dx,$$

where $\mathcal{X}(y)$ is the subset of $\mathcal{X}$ constrained by the relation $y = y(x)$.

Let $\psi^{(0)}$ be some initial value for ψ. At the k-th step the EM algorithm one performs the following two steps:

E-Step. Calculate

$$Q(\psi, \psi^{(k)}) = \mathbb{E}_{\psi^{(k)}}\{\log L_c(\psi)|y\}.$$

M-Step. Choose any value $\psi^{(k+1)}$ that maximizes $Q(\psi, \psi^{(k)})$, that is,

$$(\forall\psi) Q(\psi^{(k+1)}, \psi^{(k)}) \geq Q(\psi, \psi^{(k)}).$$

The E and M steps are alternated until the difference

$$L(\psi^{(k+1)}) - L(\psi^{(k)})$$

becomes small in absolute value.

Next we illustrate the EM algorithm with a famous example first considered by Fisher and Balmukand (1928). It is also discussed in Rao (1973), and later by Mclachlan and Krishnan (1997) and Slatkin and Excoffier (1996).

16.1 FISHER'S EXAMPLE

The following genetics example was recognized by as an application of the EM algorithm by Dempster et al. (1979). The description provided here essentially follows a lecture by Terry Speed of UC at Berkeley. In basic genetics terminology, suppose there are two linked bi-allelic loci, A and B, with alleles A and a, and B and b, respectively, where A is dominant over a and B is dominant over b. A double heterozygote $AaBb$ will produce gametes of four types: AB, Ab, aB and ab. As the loci are linked, the types AB and ab will appear with a frequency different from that of Ab and aB, say $1-r$ and r, respectively, in males, and $1-r'$ and r' respectively in females.

Here we suppose that the parental origin of these heterozygotes is from the mating $AABB \times aabb$, so that r and r' are the male and female recombination rates between the two loci. The problem is to estimate r and r', if possible, from the offspring of selfed double heterozygotes. Because gametes AB, Ab,aB and ab are produced in proportions $(1-r)/2, r/2, r/2$ and $(1-r)/2$, respectively, by the male parent, and $(1-r')/2, r'/2, r'/2$ and $(1-r')/2$, respectively, by the female parent, zygotes with genotypes $AABB, AaBB, \ldots$ etc, are produced with frequencies $(1-r)(1-r')/4, (1-r)r'/4$, etc.

The problem here is this: although there are 16 distinct offspring genotypes, taking parental origin into account, the dominance relations imply that we only observe 4 distinct phenotypes, which we denote by A^*B^*, A^*b^*, a^*B^* and a^*b^*. Here A^*(respectively B^*) denotes the dominant, while a^* (respectively b^*) denotes the recessive phenotype determined by the alleles at A (respectively B).

Thus individuals with genotypes $AABB, AaBB, AABb$ or $AaBb$, (which account for 9/16 of the gametic combinations) exhibit the phenotype A^*B^*, i.e. the dominant alternative in both characters, while those with genotypes $AAbb$ or $Aabb$ (3/16) exhibit the phenotype A^*b^*, those with genotypes $aaBB$ and $aaBb$ (3/16) exhibit the phenotype a^*B^*, and finally the double recessive $aabb$ (1/16) exhibits the phenotype a^*b^*. It is a slightly surprising fact that the probabilities of the four phenotypic classes are definable in terms of the parameter $\psi = (1-r)(1-r')$, as follows: a^*b^* has probability $\psi/4$ (easy to see), a^*B^* and A^*b^* both have probabilities $(1-\psi)/4$, while A^*B^* has rest of the probability, which is $(2+\psi)/4$. Now suppose we have a random sample of n offspring from the selfing of our double heterozygote. The 4 phenotypic classes will be represented roughly in proportion to their theoretical probabilities, their joint distribution being multinomial

$$\mathcal{M}n\left(n; \frac{2+\psi}{4}, \frac{1-\psi}{4}, \frac{1-\psi}{4}, \frac{\psi}{4}\right). \tag{16.1}$$

Note that here neither r nor r' will be separately estimable from these data, but only the product $(1-r)(1-r')$. Because we know that $r \le 1/2$ and $r' \le 1/2$, it follows that $\psi \ge 1/4$.

How do we estimate ψ? Fisher and Balmukand listed a variety of methods that were in the literature at the time, and compare them with maximum likelihood, which is the method of choice in problems like this. We describe a variant on their approach to illustrate the EM algorithm.

Let $y = (125, 18, 20, 34)$ be a realization of vector $y = (y_1, y_2, y_3, y_4)$ believed to be coming from the multinomial distribution given in (16.1). The probability mass function, given the data, is

$$g(y, \psi) = \frac{n!}{y_1!y_2!y_3!y_4!}(1/2 + \psi/4)^{y_1}\ (1/4 - \psi/4)^{y_2+y_3}\ (\psi/4)^{y_4}.$$

The log likelihood, after omitting an additive term not containing ψ is

$$\log L(\psi) = y_1 \log(2 + \psi) + (y_2 + y_3)\log(1 - \psi) + y_4 \log(\psi).$$

By differentiating with respect to ψ one gets

$$\partial \log L(\psi)/\partial\psi = \frac{y_1}{2+\psi} - \frac{y_2 + y_3}{1-\psi} + \frac{y_4}{\psi}.$$

The equation $\partial \log L(\psi)/\partial\psi = 0$ can be solved and solution is $\psi = (15 + \sqrt{53809})/394 \approx 0.626821$.

Now assume that instead of original value y_1 the counts y_{11} and y_{12}, such that $y_{11} + y_{12} = y_1$, could be observed, and that their probabilities are $1/2$ and $\psi/4$, respectively. The complete data can be defined as $x = (y_{11}, y_{12}, y_2, y_3, y_4)$. The probability mass function of incomplete data y is $g(y, \psi) = \Sigma g_c(x, \psi)$, where

$$g_c(x, \psi) = c(x)(1/2)^{y_{11}}(\psi/4)^{y_{12}}(1/4 - \psi/4)^{y_2+y_3}(\psi/4)^{y_4},$$

$c(x)$ is free of ψ, and the summation is taken over all values of x for which $y_{11} + y_{12} = y_1$.

The complete log likelihood is

$$\log L_c(\psi) = (y_{12} + y_4)\log(\psi) + (y_2 + y_3)\log(1 - \psi). \tag{16.2}$$

Our goal is to find the conditional expectation of $\log L_c(\psi)$ given y, using the starting point for $\psi^{(0)}$,

$$Q(\psi, \psi^{(0)}) = \mathbb{E}_{\psi^{(0)}}\{\log L_c(\psi)|y\}.$$

As $\log L_c$ is linear function in y_{11} and y_{12}, the *E-Step* is done by simply by replacing y_{11} and y_{12} by their conditional expectations, given y. If Y_{11} is the random variable corresponding to y_{11}, it is easy to see that

$$Y_{11} \sim \mathcal{B}in\left(y_1, \frac{1/2}{1/2 + \psi^{(0)}/4}\right)$$

so that the conditional expectation of Y_{11} given y_1 is

$$\mathbb{E}_{\psi^{(0)}}(Y_{11}|y_1) = \frac{\frac{y_1}{2}}{\frac{1}{2} + \frac{\psi^{(0)}}{4}} = y_{11}^{(0)}.$$

Of course, $y_{12}^{(0)} = y_1 - y_{11}^{(0)}$. This completes the *E-Step* part.

In the *M-Step* one chooses $\psi^{(1)}$ so that $Q(\psi, \psi^{(0)})$ is maximized. After replacing y_{11} and y_{12} by their conditional expectations $y_{11}^{(0)}$ and $y_{12}^{(0)}$ in the Q-function, the maximum is obtained at

$$\psi^{(1)} = \frac{y_{12}^{(0)} + y_4}{y_{12}^{(0)} + y_2 + y_3 + y_4} = \frac{y_{12}^{(0)} + y_4}{n - y_{11}^{(0)}}.$$

The EM-Algorithm is composed of alternating these two steps. At the iteration k we have

$$\psi^{(k+1)} = \frac{y_{12}^{(k)} + y_4}{n - y_{11}^{(k)}},$$

where $y_{11}^{(k)} = \frac{1}{2}y_1/(1/2 + \psi^{(k)}/4)$ and $y_{12}^{(k)} = y_1 - y_{11}^{(k)}$. To see how the EM algorithm computes the MLE for this problem, see the MATLAB function `emexample.m`.

16.2 MIXTURES

Recall from Chapter 2 that mixtures are compound distributions of the form $F(x) = \int F(x|t)dG(t)$. The CDF $G(t)$ serves as a mixing distribution on kernel distribution $F(x|t)$. Recognizing and estimating mixtures of distributions is an important task in data analysis. Pattern recognition, data mining and other modern statistical tasks often call for mixture estimation.

For example, suppose an industrial process that produces machine parts with lifetime distribution F_1, but a small proportion of the parts (say, ω) are defective and have CDF $F_2 \gg F_1$. If we cannot sort out the good ones from the defective ones, the lifetime of a randomly chosen part is

$$F(x) = (1 - \omega)F_1(x) + \omega F_2(x).$$

This is a simple two-point mixture where the mixing distribution has two discrete points of positive mass. With (finite) discrete mixtures like this, the probability points of G serve as weights for the kernel distribution. In the nonparametric likelihood, we see immediately how difficult it is to solve for the MLE in the presence of the weight ω, especially if ω is unknown.

Suppose we want to estimate the weights of a fixed number k of fully known distributions. We illustrate EM approach which introduces unobserved indicators with the goal of simplifying the likelihood. The weights are estimated by maximum likelihood. Assume that a sample $X_1, X_2, \ldots, X_n$ comes from the mixture

$$f(x, \omega) = \Sigma_{j=1}^{k} \omega_j f_j(x),$$

where $f_1, \ldots, f_k$ are continuous and the weights $0 \le \omega_j \le 1$ are unknown and constitute $(k-1)$-dimensional vector $\omega = (\omega_1, \ldots, \omega_{k-1})$ and $\omega_k = 1 - \omega_1 - \cdots - \omega_{k-1}$. The class-densities $f_j(x)$ are fully specified.

Even in this simplest case when $f_1, \ldots, f_k$ are given and the only parameters are the weights ω, the log-likelihood assumes a complicated form,

$$\Sigma_{i=1}^{n} \log f(x_i, \omega) = \Sigma_{i=1}^{n} \log \left(\Sigma_{j=1}^{k} \omega_j f_j(x_i) \right).$$

The derivatives with respect to ω_j lead to the system of equations, not solvable in a closed form.

Here is a situation where the EM Algorithm can be applied with a little creative foresight. Augment the data $x = (x_1, \ldots, x_n)$ by an unobservable matrix $z = (z_{ij}, i = 1, \ldots, n; j = 1, \ldots, k)$. The values z_{ij} are indicators, defined as

$$z_{ij} = \begin{cases} 1, & x_i \text{ from } f_j \\ 0, & \text{otherwise} \end{cases}$$

The unobservable matrix z (our "missing value") tells us (in an oracular fashion) where the ith observation x_i comes from. Note that each row of z contains a single 1 and $k-1$ 0's. With augmented data, $x = (y, z)$ the (complete) likelihood takes quite a simple form,

$$\prod_{i=1}^{n} \prod_{j=1}^{k} \left(\omega_j f_j(x_i) \right)^{z_{ij}}.$$

The complete log-likelihood is simply

$$\log L_c(\omega) = \Sigma_{i=1}^{n} \Sigma_{j=1}^{k} z_{ij} \log \omega_j + C,$$

where $C = \Sigma_i \Sigma_j z_{ij} \log f_j(x_i)$ is free of ω. This is easily solved.

Assume that m^{th} iteration of the weight estimate $\omega^{(m)}$ is already obtained. The mth *E-Step* is

$$\mathbb{E}_{\omega^{(m)}}(z_{ij}|x) = \mathbb{P}_{\omega^{(m)}}(z_{ij} = 1|x) = z_{ij}^{(k)},$$

where $z_{ij}^{(m)}$ is the posterior probability of ith observation coming from the jth

mixture-component, f_j, in the iterative step m.

$$z_{ij}^{(m)} = \frac{\omega_j^{(m)} f_j(x_i)}{f(x_i, \omega^{(m)})}.$$

Because $\log L_c(\omega)$ is linear in z_{ij}, $Q(\omega, \omega^{(m)})$ is simply $\Sigma_{i=1}^n \Sigma_{j=1}^k z_{ij}^{(m)} \log \omega_j + C$. The subsequent *M-Step* is simple: $Q(\omega, \omega^{(m)})$ is maximized by

$$\omega_j^{(m+1)} = \frac{\Sigma_{i=1}^n z_{ij}^{(m)}}{n}.$$

The MATLAB script (`mixture_cla.m`) illustrates the algorithm above. A sample of size 150 is generated from the mixture $f(x) = 0.5\mathcal{N}(-5, 2^2) + 0.3\mathcal{N}(0, 0.5^2) + 0.2\mathcal{N}(2, 1)$. The mixing weights are estimated by the EM algorithm. $M = 20$ iterations of EM algorithm yielded $\hat{\omega} = (0.4977, 0.2732, 0.2290)$. Figure 16.1 gives histogram of data, theoretical mixture and EM estimate.

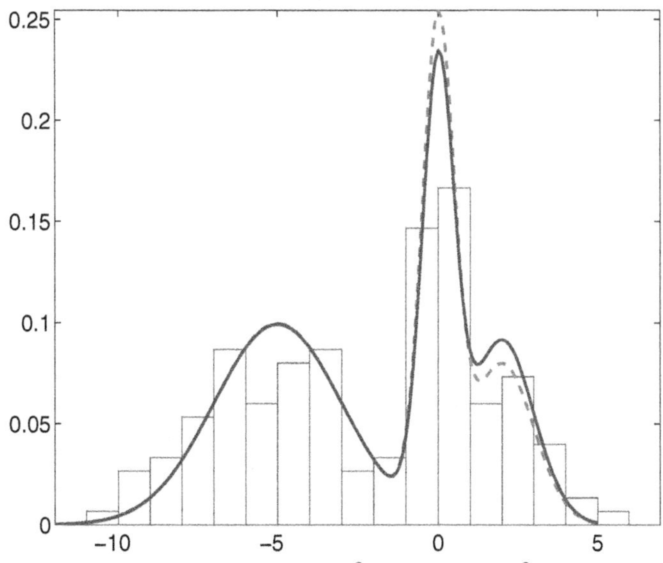

Fig. 16.1 Observations from the $0.5\mathcal{N}(-5, 2^2) + 0.3\mathcal{N}(0, 0.5^2) + 0.2\mathcal{N}(2, 1)$ mixture (histogram), the mixture (*dotted line*) and EM estimated mixture (*solid line*).

Example 16.1 As an example of a specific mixture of distributions we consider application of EM algorithm in the so called Zero Inflated Poisson (ZIP) model. In ZIP models the observations come from two populations, one in which all values are identically equal to 0 and the other Poisson $\mathcal{P}(\lambda)$. The "zero" population is selected with probability ξ, and the Poisson population with complementary probability of $1-\xi$. Given the data, both λ and ξ are to be estimated. To illustrate EM algorithm in fitting ZIP models, we consider data set (Thisted, 1988) on distribution of number of children in a sample of $n = 4075$ widows, given in Table 16.20.

Table 16.20 Frequency Distribution of the Number of Children Among 4075 Widows

Number of Children (`number`)	0	1	2	3	4	5	6
Number of Widows (`freq`)	3062	587	284	103	33	4	2

At first glance the Poisson model for this data seems to be appropriate, however, the sample mean and variance are quite different (theoretically, in Poisson models they are the same).

```
>> number = 0:6;              %number of children
>> freqs =[3062 587 284 103 33 4 2];
>> n = sum(freqs)
>> sum(freqs .* number)/n  %sample mean
   ans =
       0.3995
>> sum(freqs .* (number-0.3995).^2)/(n-1)
                              %sample variance

   ans =
       0.6626
```

This indicates presence of *over-dispersion* and the ZIP model can account for the apparent excess of zeros. The ZIP model can be formalized as

$$
\begin{aligned}
P(X=0) &= \xi + (1-\xi)\frac{\lambda^0}{0!}e^{-\lambda} = \xi + (1-\xi)e^{-\lambda} \\
P(X=i) &= (1-\xi)\frac{\lambda^i}{i!}e^{-\lambda}, \; i = 1, 2, \ldots,
\end{aligned}
$$

and the estimation of ξ and λ is of interest. To apply the EM algorithm, we treat this problem as an *incomplete data* problem. The complete data would involve knowledge of frequencies of zeros from both populations, n_{00} and n_{01}, such that the observed frequency of zeros n_0 is split as $n_{00} + n_{01}$. Here n_{00} is number of cases coming from the the point mass at 0-part and n_{01} is number of cases coming from the Poisson part of the mixture. If values of n_{00} and n_{01} are available, the estimation of ξ and λ is straightforward. For example, the MLEs are

$$
\hat{\xi} = \frac{n_{00}}{n} \quad \text{and} \quad \hat{\lambda} = \frac{\Sigma_i i\, n_i}{n - n_{00}},
$$

where n_i is the observed frequency of i children. This will be a basis for M-step in the EM implementation, because the estimator of ξ comes from the fact that $n_{00} \sim \mathcal{B}in(n, \xi)$, while the estimator of λ is the sample mean of the Poisson part. The E-step involves finding $\mathbb{E}n_{00}$ if ξ and λ are known. With $n_{00} \sim \mathcal{B}in(n_0, p_{00}/(p_{00}+p_{01}))$, where $p_{00} = \xi$ and $p_{01} = (1-\xi)e^{-\lambda}$, the expectation of n_{00} is

$$\mathbb{E}(n_{00}|\text{ observed frequencies}, \xi, \lambda) = n_0 \times \frac{\xi}{\xi + (1-\xi)e^{-\lambda}}.$$

From this expectation, the iterative procedure can be set with

$$\begin{aligned} n_{00}^{(t)} &= n_0 \times \frac{\xi^{(t)}}{\xi^{(t)} + (1-\xi^{(t)})\exp\{-\lambda^{(t)}\}} \\ \xi^{(t+1)} &= n_{00}^{(t)}/n, \text{ and} \\ \lambda^{(t+1)} &= \frac{1}{n - n_{00}^{(t)}} \Sigma_i i\, n_i, \end{aligned}$$

where t is the iteration step. The following MATLAB code performs 20 iterations of the algorithm and collects the calculated values of n_{00}, ξ and λ in three sequences `newn00s`, `newxis`, and `newlambdas`. The initial values are given for ξ and λ as $\xi_0 = 3/4$ and $\lambda_0 = 1/4$.

```
>> newxi =3/4;  newlambda = 1/4; %initial values
>> newn00s=[]; newxis=[]; newlambdas=[];
>> for i = 1:20
      newn00 = freqs(1) * newxi/(newxi + ...
               (1-newxi)*exp(-newlambda) );
      newxi = newn00/n;
      newlambda = sum((1:6).* freqs(2:7))/(n-newn00);
      %collect the values in three sequences
      newn00s=[newn00s newn00]; newxis=[newxis newxi];
      newlambdas=[newlambdas newlambda];
end
```

Table 16.21 gives the partial output of the MATLAB program. The values for `newxi`, `newlambda`, and `newn00` will stabilize after several iteration steps.

16.3 EM AND ORDER STATISTICS

When applying nonparametric maximum likelihood to data that contain (independent) order statistics, the EM Algorithm can be applied by assuming that with the observed order statistic $X_{i:k}$ (the ith smallest observation from an i.i.d. sample of k), there are associated with it $k-1$ missing values: $i-1$ values smaller than $X_{i:k}$ and $k-i$ values that are larger. Kvam and Samaniego (1994) exploited this opportunity to use the EM for finding the nonparametric

Table 16.21 Some of the Twenty Steps in the EM Implementation of ZIP Modeling on Widow Data

Step	`newxi`	`newlambda`	`newn00`
0	1/4	3/4	2430.9
1	0.5965	0.9902	2447.2
2	0.6005	1.0001	2460.1
3	0.6037	1.0081	2470.2
⋮			
18	0.6149	1.0372	2505.6
19	0.6149	1.0373	2505.8
20	0.6149	1.0374	2505.9

MLE for i.i.d. component lifetimes based on observing only k-out-of-n system lifetimes. Recall a k-out-of-n system needs k or more working components to operate, and fails after $n-k+1$ components fail, hence the system lifetime is equivalent to $X_{n-k+1:n}$.

Suppose we observe independent order statistics $X_{r_i:k_i}$, $i = 1, \ldots, n$ where the unordered values are independently generated from F. When F is absolutely continuous, the density for $X_{r_i:k_i}$ is expressed as

$$f_{r_i:k_i}(x) = r_i \binom{k_i}{r_i} F^{r_i - 1}(x)\, (1 - F(x))^{k_i - r_i}\, f(x).$$

For simplicity, let $k_i = k$. In this application, we assign the complete data to be $X_i = \{X_{i1}, \ldots, X_{ik}, Z_i\}$, $i = 1, \ldots, n$ where Z_i is defined as the rank of the value observed from X_i. The observed data can be written as $Y_i = \{W_i, Z_i\}$, where W_i is the ${Z_i}^{th}$ smallest observation from X_i.

With the complete data, the MLE for $F(x)$ is the EDF, which we will write as $N(x)/(nk)$ where $N(x) = \Sigma_i \Sigma_j \mathbf{1}(X_{ij} \le x)$. This makes the *M-step* simple, but for the *E-step*, N is estimated through the log-likelihood. For example, if $Z_i = z$, we observe W_i distributed as $X_{z:k}$. If $W_i \le x$, out of the subgroup of size k from which W_i was measured,

$$z + (k - z) \frac{F(t) - F(W_i)}{1 - F(W_i)}$$

are expected to be less than or equal to x. On the other hand, if $W_i > x$, we know $k - z + 1$ elements from X_i are larger than x, and

$$(z-1) \frac{F(W_i)}{F(x)}$$

are expected in $(-\infty, x]$.

The *E-Step* is completed by summing all of these expected counts out of the complete sample of nk based on the most recent estimator of F from the *M-Step*. Then, if $F^{(j)}$ represents our estimate of F after j iterations of the EM Algorithm, it is updated as

$$\begin{aligned} F^{(j+1)}(x) &= \frac{1}{nk}\Sigma_{i=1}^{n}\left[Z_i + (k - Z_i)\frac{F^{(j)}(x) - F^{(j)}(W_i)}{1 - F^{(j)}(x)}\mathbf{1}(W_i \le x)\right. \\ &\qquad \left. + (Z_i - 1)\frac{F^{(j)}(x)}{F^{(j)}(W_i)}\mathbf{1}(W_i > x)\right]. \end{aligned} \tag{16.3}$$

Equation (16.3) essentially joins the two steps of the EM Algorithm together. All that is needed is a initial estimate $F^{(0)}$ to start it off. The observed sample EDF suffices. Because the full likelihood is essentially a multinomial distribution, convergence of $F^{(j)}$ is guaranteed. In general, the speed of convergence is dependent upon the amount of information. Compared to the mixtures application, there is a great amount of missing data here, and convergence is expected to be relatively slow.

16.4 MAP VIA EM

The EM algorithm can be readily adapted to Bayesian context to maximize the posterior distribution. A maximum of the posterior distribution is the so called MAP (maximum a posteriori) estimator, used widely in Bayesian inference. The benefit of MAP estimators over some other posterior parameters was pointed out on p. 53 of Chapter 4 in the context of Bayesian estimators. The maximum of the posterior $\pi(\psi|y)$, if it exists, coincides with the maximum of the product of the likelihood and prior $f(y|\psi)\pi(\psi)$. In terms of logarithms, finding the MAP estimator amounts to maximizing

$$\log \pi(\psi|y) = \log L(\psi) + \log \pi(\psi).$$

The EM algorithm can be readily implemented as follows:

E-Step. At $(k+1)^{st}$ iteration calculate

$$\mathbb{E}_{\psi^{(k)}}\{\log \pi(\psi|x)|y\} = Q(\psi, \psi^{(k)}) + \log \pi(\psi).$$

The *E-Step* coincides with the traditional EM algorithm, that is, $Q(\psi, \psi^{(k)})$ has to be calculated.

M-Step. Choose $\psi^{(k+1)}$ to maximize $Q(\psi, \psi^{(k)}) + \log \pi(\psi)$. The *M-Step* here differs from that in the EM, because the objective function to be maximized withe respect to ψ's contains additional term, logarithm of the prior. However, the presence of this additional term contributes to the concavity of the objective function thus improving the speed of convergence.

Example 16.2 MAP Solution to Fisher's Genomic Example. Assume that we elicit a $\mathcal{B}e(\nu_1, \nu_2)$ prior on ψ,

$$\pi(\psi) = \frac{1}{B(\nu_1, \nu_2)} \psi^{\nu_1 - 1}(1 - \psi)^{\nu_2 - 1}.$$

The beta distribution is a natural conjugate for the missing data distribution, because $y_{12} \sim \mathcal{B}in(y_1, (\psi/4)/(1/2 + \psi/4))$. Thus the log-posterior (additive constants ignored) is

$$\begin{aligned} \log \pi(\psi|x) &= \log L(\psi) + \log \pi(\psi) \\ &= (y_{12} + y_4 + \nu_1 - 1)\log \psi + (y_2 + y_3 + \nu_2 - 1)\log(1 - \psi). \end{aligned}$$

The *E-step* is completed by replacing y_{12} by its conditional expectation $y_1 \times (\psi^{(k)}/4)/(1/2 + \psi^{(k)}/4)$. This step is the same as in the standard EM algorithm.

The *M-Step*, at $(k+1)$st iteration, is

$$\psi^{(k+1)} = \frac{y_{12}^{(k)} + y_4 + \nu_1 - 1}{y_{12}^{(k)} + y_2 + y_3 + y_4 + \nu_1 + \nu_2 - 2}.$$

When the beta prior coincides with uniform distribution (that is, when $\nu_1 = \nu_2 = 1$), the MAP and MLE solutions coincide.

16.5 INFECTION PATTERN ESTIMATION

Reilly and Lawlor (1999) applied the EM Algorithm to identify contaminated lots in blood samples. Here the observed data contain the disease exposure history of a person over k points in time. For the ith individual, let

$$X_i = \mathbf{1}(i^{th} \text{ person infected by end of trial}),$$

where $P_i = P(X_i = 1)$ is the probability that the ith person was infected at least once during k exposures to the disease. The exposure history is defined as a vector $y_i = \{y_{i1}, \dots, y_{ik}\}$, where

$$y_{ij} = \mathbf{1}(i^{th} \text{ person exposed to disease at } j^{th} \text{ time point } k).$$

Let λ_j be the rate of infection at time point j. The probability of not being infected in time point j is $1 - y_{ij}\lambda_j$, so we have $P_i = 1 - \prod(1 - y_{ij}\lambda_j)$. The corresponding likelihood for $\lambda = \{\lambda_1, \dots, \lambda_k\}$ from observing n patients is a bit daunting:

$$\begin{aligned} L(\lambda) &= \prod_{i=1}^{n} p_i^{x_i} (1-p_i)^{1-x_i} \\ &= \prod_{i=1}^{n} \left(1 - \prod_{j=1}^{k}(1-y_{ij}\lambda_j)\right)^{x_i} \left(\prod_{j=1}^{k}(1-y_{ij}\lambda_j)\right)^{1-x_i}. \end{aligned}$$

The EM Algorithm helps if we assign the unobservable

$$Z_{ij} = \mathbf{1}(\text{person } i \text{ infected at time point } j),$$

where $P(Z_{ij}=1) = \lambda_j$ if y_{ij}=1 and $P(Z_{ij}=1) = 0$ if y_{ij}=0. Averaging over y_{ij}, $P(Z_{ij}=1) = y_{ij}\lambda_j$. With z_{ij} in the complete likelihood ($1 \le i \le n$, $1 \le j \le k$), we have the observed data changing to $x_i = \max\{z_{i1}, \ldots, z_{ik}\}$, and

$$L(\lambda|Z) = \prod_{i=1}^{n}\prod_{j=1}^{k} (y_{ij}\lambda_j)^{z_{ij}} (1-y_{ij}\lambda_j)^{1-z_{ij}},$$

which has the simple binomial form.

For the *E-Step*, we find $\mathbb{E}(Z_{ij}|x_i, \lambda^{(m)})$, where $\lambda^{(m)}$ is the current estimate for $(\lambda_1, \ldots, \lambda_k)$ after m iterations of the algorithm. We need only concern ourselves with the case $x_i = 1$, so that

$$\mathbb{E}(Z_{ij}|x_i = 1) = P(y_{ij} = 1|x_i = 1) = \frac{y_{ij}\lambda_j}{1 - \prod_{j=1}^{k}(1-y_{ij}\lambda_j)}.$$

In the *M-Step*, MLEs for $(\lambda_1, \ldots, \lambda_k)$ are updated in iteration $m+1$ from ${\lambda_1}^{(m)}, \ldots, {\lambda_k}^{(m)}$ to

$${\lambda_j}^{(m+1)} = \frac{\Sigma_{i=1}^{n} y_{ij}\left[\frac{y_{ij}{\lambda_j}^{(m+1)}}{1-\prod_{j=1}^{k}\left(1-y_{ij}{\lambda_j}^{(m+1)}\right)}\right]}{\Sigma_{i=1}^{n} y_{ij}}.$$

16.6 EXERCISES

16.1. Suppose we have data generated from a mixture of two normal distributions with a common known variance. Write a MATLAB script to determine the MLE of the unknown means from an i.i.d. sample from the mixture by using the EM algorithm. Test your program using a sample of ten observations generated from an equal mixture of the two kernels $\mathcal{N}(0,1)$ and $\mathcal{N}(1,1)$.

16.2. The data in the following table come from the mixture of two Poisson random variables, $\mathcal{P}(\lambda_1)$ with probability ϵ and $\mathcal{P}(\lambda_2)$ with probability $1-\epsilon$.

Value	0	1	2	3	4	5	6	7	8	9	10
Freq.	708	947	832	635	427	246	121	51	19	6	1

(i) Develop an EM algorithm for estimating ϵ, λ_1, and λ_2.

(ii) Write MATLAB program that uses (i) in estimating ϵ, λ_1, and λ_2 for data from the table.

16.3. The following data give the numbers of occupants in 1768 cars observed on a road junction in Jakarta, Indonesia, during a certain time period on a weekday morning.

Number of occupants	1	2	3	4	5	6	7
Number of cars	897	540	223	85	17	5	1

The proposed model for number of occupants X is truncated Poisson (TP), defined as

$$P(X=i) = \frac{\lambda^i \exp\{-\lambda\}}{(1-\exp\{-\lambda\})\, i!}, \quad i = 1, 2, \ldots$$

(i) Write down the likelihood (or the log-likelihood) function. Is it straightforward to find the MLE of λ by maximizing the likelihood or log-likelihood directly?

(ii) Develop an EM algorithm for approximating the MLE of λ. *Hint:* Assume that missing data is i_0 – the number of cases when $X = 0$, so with the complete data the model is Poisson, $\mathcal{P}(\lambda)$. Estimate λ from the complete data. Update i_0 given the estimator of λ.

(iii) Write MATLAB program that will estimate the MLE of λ for Jakarta cars data using the EM procedure from (ii).

16.4. Consider the problem of right censoring in lifetime measurements in Chapter 10. Set up the EM algorithm for solving the nonparametric MLE for a sample of possibly-right censored values $X_1, \ldots, X_n$.

16.5. Write MATLAB program that will approximate the MAP estimator in Fisher's problem (Example 16.2), if the prior on ψ is $\mathcal{B}e(2,2)$. Compare the MAP and MLE solutions.

REFERENCES

Dempster, A. P., Laird, N. M., and Rubin, D. B. (1977), "Maximum Likelihood from Incomplete Data via the EM Algorithm" (with discussion), *Journal of the Royal Statistical Society*, Ser. B, 39, 1-38.

Fisher, R.A. and Balmukand, B. (1928). The estimation of linkage from the offspring of selfed heterozygotes. *Journal of Genetics,* **20**, 79-92.

Healy M. J. R., and Westmacott M. H. (1956), "Missing Values in Experiments Analysed on Automatic Computers," *Applied Statistics*,5, 203-306.

Kvam, P. H., and Samaniego, F. J. (1994) "Nonparametric Maximum Likelihood Estimation Based on Ranked Set Samples," *Journal of the American Statistical Association*, 89, 526-537.

McKendrick, A. G. (1926), "Applications of Mathematics to Medical Problems," *Proceedings of the Edinburgh Mathematical Society,* 44, 98-130.

McLachlan, G. J., and Krishnan, T. (1997), *The EM Algorithm and Extensions,* New York: Wiley.

Rao, C. R. (1973), *Linear Statistical Inference and its Applications,* 2nd ed., New York: Wiley.

Reilly, M., and Lawlor E. (1999), "A Likelihood Method of Identifying Contaminated Lots of Blood Product," *International Journal of Epidemiology*, 28, 787-792.

Slatkin, M., and Excoffier, L. (1996), "Testing for Linkage Disequilibrium in Genotypic Data Using the Expectation-Maximization Algorithm," *Heredity*, 76, 377-383.

Tsai, W. Y., and Crowley, J. (1985), A Large Sample Study of Generalized Maximum Likelihood Estimators from Incomplete Data via Self-Consistency," *Annals of Statistics*, 13, 1317-1334.

Thisted, R. A. (1988), *Elements of Statistical Computing: Numerical Computation*, New York: Chapman & Hall.

Wu, C. F. J. (1983), "On the Convergence Properties of the EM Algorithm," *Annals of Statistics*, 11, 95-103.

17

Statistical Learning

Learning is not compulsory ... neither is survival.

W. Edwards Deming (1900–1993)

A general type of artificial intelligence, called *machine learning*, refers to techniques that sift through data and find patterns that lead to optimal decision rules, such as classification rules. In a way, these techniques allow computers to "learn" from the data, adapting as trends in the data become more clearly understood with the computer algorithms. Statistical learning pertains to the data analysis in this treatment, but the field of machine learning goes well beyond statistics and into algorithmic complexity of computational methods.

In business and finance, machine learning is used to search through huge amounts of data to find structure and pattern, and this is called *data mining*. In engineering, these methods are developed for *pattern recognition*, a term for classifying images into predetermined groups based on the study of statistical classification rules that statisticians refer to as *discriminant analysis*. In electrical engineering, specifically, the study of *signal processing* uses statistical learning techniques to analyze signals from sounds, radar or other monitoring devices and convert them into digital data for easier statistical analysis.

Techniques called *neural networks* were so named because they were thought to imitate the way the human brain works. Analogous to neurons, connections between processing elements are generated dynamically in a learning system based on a large database of examples. In fact, most neural network algorithms are based on statistical learning techniques, especially nonparametric ones.

In this chapter, we will only present a brief exposition of classification and statistical learning that can be used in machine learning, discriminant analysis, pattern recognition, neural networks and data mining. Nonparametric methods now play a vital role in statistical learning. As computing power has progressed through the years, researchers have taken on bigger and more complex problems. An increasing number of these problems cannot be properly summarized using parametric models.

This research area has a large and growing knowledge base that cannot be justly summarized in this book chapter. For students who are interested in reading more about statistical learning methods, both parametric and nonparametric, we recommend books by Hastie, Tibshirani and Friedman (2001) and Duda, Hart and Stork (2001).

17.1 DISCRIMINANT ANALYSIS

Discriminant Analysis is the statistical name for categorical prediction. The goal is to predict a categorical response variable, G, from one or more predictor variables, x. For example, if there is a partition of k groups $\mathcal{G} = (G_1, \dots, G_k)$, we want to find the probability that any particular observation x belongs to group G_j, $j = 1, \dots, k$ and then use this information to classify it in one group or the other. This is called *supervised classification* or *supervised learning* because the structure of the categorical response is known, and the problem is to find out in which group each observation belongs. *Unsupervised classification*, or *unsupervised learning* on the other hand, aims to find out how many relevant classes there are and then to characterize them.

One can view this simply as a categorical extension to prediction for simple regression: using a set of data of the form (x_1, g_1) , ... , (x_n, g_n), we want to devise a rule to classify future observations $x_{n+1}, \dots, x_{n+m}$.

17.1.1 Bias Versus Variance

Recall that a loss function measures the discrepancy between the data responses and what the proposed model predicts for response values, given the corresponding set of inputs. For continuous response values y with inputs x, we are most familiar with squared error loss

$$L(y, f) = (y - f(x))^2 .$$

We want to find the predictive function f that minimizes the *expected loss*, $\mathbb{E}[L(y, f)]$, where the expectation averages over all possible response values. With the observed data set, we can estimate this as

$$\mathcal{E}_f = \frac{1}{n}\Sigma_{i=1}^{n} L\left(y_i, \hat{f}(x_i)\right).$$

The function that minimizes the squared error is the conditional mean $\mathbb{E}(Y|X = x)$, and the expected squared errors $\mathbb{E}(Y - f(Y))^2$ consists of two parts: *variance* and the square of the *bias*. If the classifier is based on a global rule, such as linear regression, it is simple, rigid, but at least stable. It has little variance, but by overlooking important nuances of the data, can be highly biased. A classifier that fits the model locally fails to garner information from as many observations and is more unstable. It has larger variance, but its adaptability to the detailed characteristics of the data ensure it has less bias. Compared to traditional statistical classification methods, most nonparametric classifiers tend to be less stable (more variable) but highly adaptable (less bias).

17.1.2 Cross-Validation

Obviously, the more local model will report less error than a global model, so instead of finding a model that simply minimizes error for the data set, it is better to put aside some of the data to test the model fit independently. The part of the data used to form the estimated model is called the *training sample*. The reserved group of data is the *test sample*.

The idea of using a training sample to develop a decision rule is paramount to empirical classification. Using the test sample to judge the method constructed from the training data is called *cross-validation*. Because data are often sparse and hard to come by, some methods use the training set to both develop the rule and to measure its misclassification rate (or error rate) as well. See the jackknife and bootstrap methods described in Chapter 15, for example.

17.1.3 Bayesian Decision Theory

There are two kinds of loss functions commonly used for categorical responses: a zero-one loss and cross-entropy loss. The zero-one loss merely counts the number of misclassified observations. Cross-entropy, on the other hand, uses the estimated class probabilities $\hat{p}_i(x) = \hat{P}(g \in G_i|x))$, and we minimize $\mathbb{E}(-2\ln\hat{p}_i(X))$.

By using zero-one loss, the estimator that minimizes risk classifies the observation to the most probable class, given the input $P(G|X)$. Because this is based on Bayes rule of probability, this is called the *Bayes Classifier*. Although, if $P(X|G_i)$ represents the distribution of observations from population G_i, it might be assumed we know a prior probability $P(G_i)$ that represents the probability any particular observation comes from population G_i. Furthermore, optimal decisions might depend on particular consequences of misclassification, which can be represented in cost variables; for example, c_{ij} = Cost of classifying an observation from population G_i into population G_j.

For example, if k=2, the *Bayes Decision Rule* which minimizes the expected cost (c_{ij}) is to classify x into G_1 if

$$\frac{P(x|G_1)}{P(x|G_2)} > \frac{(c_{21} - c_{22})P(G_2)}{(c_{12} - c_{11})P(G_1)}$$

and otherwise classify the observation into G_2.

Cross-entropy has an advantage over zero-one loss because of its continuity; in regression trees, for example, classifiers found via optimization techniques are easier to use if the loss function is differentiable.

17.2 LINEAR CLASSIFICATION MODELS

In terms of bias versus variance, a linear classification model represents a strict global model with potential for bias, but low variance that makes the classifier more stable. For example, if a categorical response depends on two ordinal inputs on the (x, y) axis, a linear classifier will draw a straight line somewhere on the graph to best separate the two groups.

The first linear rule developed was based on assuming the the underlying distribution of inputs were normally distributed with different means for the different populations. If we assume further that the distributions have an identical covariance structure ($X_i \sim \mathcal{N}(\mu_i, \Sigma)$), and the unknown parameters have MLEs $\hat{\mu}_i$ and $\hat{\Sigma}$, then the discrimination function reduces to

$$x\hat{\Sigma}^{-1}(x_1 - x_2)' - \frac{1}{2}(x_1 + x_2)\hat{\Sigma}^{-1}(x_1 - x_2) > \delta \tag{17.1}$$

for some value δ, which is a function of cost. This is called *Fisher's Linear Discrimination Function* (LDF) because with the equal variance assumption, the rule is linear in x. The LDF was developed using normal distributions, but this linear rule can also be derived using a minimal squared-error approach. This is true, you can recall, for estimating parameters in multiple linear regression as well.

If the variances are not the same, the optimization procedure is repeated with extra MLEs for the covariance matrices, and the rule is quadratic in the inputs and hence called a *Quadratic Discriminant Function* (QDF). Because the linear rule is overly simplistic for some examples, quadratic classification rules are used to extend the linear rule by including squared values of the predictors. With k predictors in the model, this begets $\binom{k+1}{2}$ additional parameters to estimate. So many parameters in the model can cause obvious problems, even in large data sets.

There have been several studies that have looked into the quality of linear and quadratic classifiers. While these rules work well if the normality assumptions are valid, the performance can be pretty lousy if they are not. There are numerous studies on the LDF and QDF robustness, for example, see Moore (1973), Marks and Dunn (1974), Randles, Bramberg, and Hogg (1978).

17.2.1 Logistic Regression as Classifier

The simple zero-one loss function makes sense in the categorical classification problem. If we relied on the squared error loss (and outputs labeled with zeroes and ones), the estimate for g is not necessarily in $[0, 1]$, and even if the large sample properties of the procedure are satisfactory, it will be hard to take such results seriously.

One of the simplest models in the regression framework is the logistic regression model, which serves as a bridge between simple linear regression and statistical classification. Logistic regression, discussed in Chapter 12 in the context of Generalized Linear Models (GLM), applies the linear model to binary response variables, relying on a *link function* that will allow the linear model to adequately describe probabilities for binary outcomes. Below we will use a simple illustration of how it can be used as a classifier. For a more comprehensive instruction on logistic regression and other models for ordinal data, Agresti's book *Categorical Data Analysis* serves as an excellent basis.

If we start with the simplest case where $k = 2$ groups, we can arbitrarily assign $g_i = 0$ or $g_i = 1$ for categories G_0 and G_1. This means we are modeling a binary response function based on the measurements on x. If we restrict our attention to a linear model $P(g = 1|x) = x'\beta$, we will be saddled with an unrefined model that can estimate probability with a value outside [0,1]. To avoid this problem, consider transformations of the linear model such as

(i) *logit:* $p(x) = P(g = 1|x) = \exp(x'\beta)/[1 + \exp(x'\beta)]$, so $x'\beta$ is estimating $\ln[p(x)/(1 - p(x))]$ which has its range on $\mathbb{R}$.

(ii) *probit:* $P(g = 1|x) = \Phi(x'\beta)$, where Φ is the standard normal CDF. In this case $x'\beta$ is estimating $\Phi^{-1}(p(x))$.

(iii) *log-log:* $p(x) = 1 - \exp(\exp(x'\beta))$ so that $x'\beta$ is estimating $\ln[-\ln(1 - p(x))]$ on $\mathbb{R}$.

Because the logit transformation is symmetric and has relation to the natural parameter in the GLM context, it is generally the default transformation in this group of three. We focus on the logit link and seek to maximize the likelihood

$$L(\beta) = \prod_{i=1}^{n} p_i(x)^{g_i}(1 - p_i(x))^{1-g_i},$$

in terms of $p(x) = 1 - \exp(\exp(x'\beta))$ to estimate β and therefore obtain MLEs for $p(x) = P(g = 1|x)$. This likelihood is rather well behaved and can be maximized in a straightforward manner. We use the MATLAB function `logistic` to perform a logistic regression in the example below.

Example 17.1 (Kutner, Nachtsheim, and Neter, 1996) A study of 25 computer programmers aims to predict task success based on the programmers' months of work experience. The MATLAB m-file `logist` computes simple ordinal logistic regressions:

```
>> x=[14 29 6 25 18 4 18 12 22 6 30 11 30 5 20 13 9 32
     24 13 19 4 28 22 8];
>> y=[0 0 0 1 1 0 0 0 1 0 1 0 1 0 1 0 0 1 0 1 0 0 1 1 1];
>> logist(y,x,1)
Number of iterations
     3
  Deviance
   25.4246
     Theta        SE
    3.0585    1.2590
     Beta         SE
    0.1614    0.0650
ans =
    0.1614
```

Here $\beta = (\beta_0, \beta_1)$ and $\hat{\beta}$=(3.0585,0.1614). The estimated logistic regression function is

$$\hat{p} = \frac{e^{-3.0585+0.1615x}}{1 + e^{-3.0585+0.1615x}}.$$

For example, in the case $x_1 = 14$, we have $\hat{p}_1 = 0.31$; i.e., we estimate that there is a 31% chance a programmer with 14 months experience will successfully complete the project.

In the logistic regression model, if we use $\hat{p}$ as a criterion for classifying observations, the regression serves as a simple linear classification model. If misclassification penalties are the same for each category, $\hat{p} = 1/2$ will be the classifier boundary. For asymmetric loss, the relative costs of the misclassification errors will determine an optimal threshold.

Example 17.2 (Fisher's Iris Data) To illustrate this technique, we use Fisher's Iris data, which is commonly used to show off classification methods. The iris data set contains physical measurements of 150 flowers – 50 for each of three types of iris (Virginica, Versicolor and Setosa). Iris flowers have three petals and three outer petal-like sepals. Figure (17.2.1a) shows a plot of petal length vs width for Versicolor (circles) and Virginica (plus signs) along with the line that best linearly categorizes them. How is this line determined?

From the logistic function $x'\beta = \ln(p/(1-p))$, $p = 1/2$ represents an observation that is half-way between the Virginica iris and the Versicolor iris. Observations with values of $p < 0.5$ are classified to be Versicolor while those with $p > 0.5$ are classified as Virginica. At $p = 1/2$, $x'\beta = \ln(p/(1-p)) = 0$, and the line is defined by $\beta_0 + \beta_1 x_1 + \beta_2 x_2 = 0$, which in this case equates to $x_2 = (42.2723 - 5.7545x_1)/10.4467$. This line is drawn in Figure (17.2.1a).

```
  >> load iris
  >> x = [PetalLength,PetalWidth];
  >> plot(PetalLength(51:100), PetalWidth(51:100),'o')
  >> hold on
  >> x2 = [PetalLength(51:150), PetalWidth(51:150)];
  >> fplot('(45.27-5.7*x)/10.4',[3,7])
  >> v2 = Variety(51:150);
  >> L2 = logist(v2,x2,1);
Number of iterations
     8
  Deviance
   20.5635
     Theta        SE
   45.2723   13.6117
     Beta         SE
    5.7545    2.3059
   10.4467    3.7556
```

While this example provides a spiffy illustration of linear classification, most populations are not so easily differentiated, and a linear rule can seem overly simplified and crude. Figure (17.2.1b) shows a similar plot of sepal width vs. length. The iris types are not so easily distinguished, and the linear classification does not help us in this example.

In the next parts of this chapter, we will look at "nonparametric" classifying methods that can be used to construct a more flexible, nonlinear classifier.

17.3 NEAREST NEIGHBOR CLASSIFICATION

Recall from Chapter 13, nearest neighbor methods can be used to create nonparametric regressions by determining the regression curve at x based on explanatory variables x_i that are considered closest to x. We will call this a k-nearest neighbor classifier if it considers the k closest points to x (using a majority vote) when constructing the rule at that point.

If we allow k to increase, the estimator eventually uses all of the data to

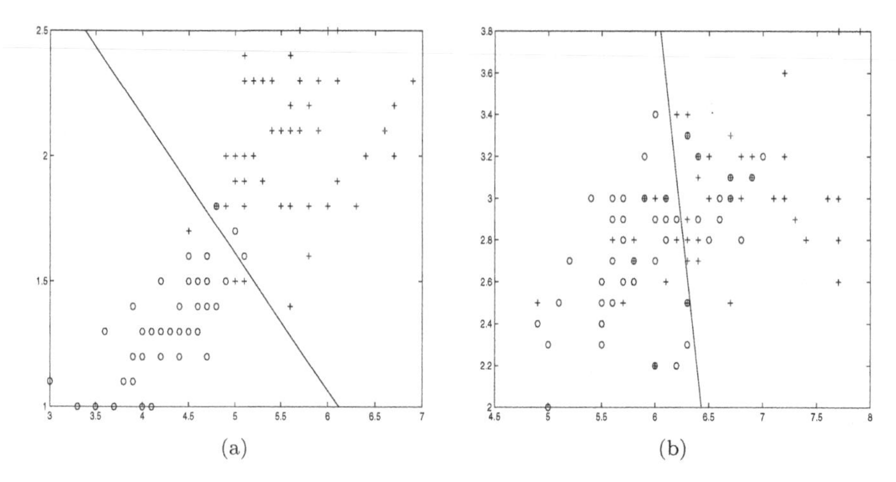

Fig. 17.1 Two types of iris classified according to (a) petal length vs. petal width, and (b) sepal length vs. sepal width. Versicolor = o, Virginica = +.

fit each local response, so the rule is a global one. This leads to a simpler model with low variance. But if the assumptions of the simple model are wrong, high bias will cause the expected mean squared error to explode. On the other hand, if we let k go down to one, the classifier will create minute neighborhoods around each observed x_i, revealing nothing from the data that a plot of the data has not already shown us. This is highly suspect as well.

The best model is likely to be somewhere in between these two extremes. As we allow k to increase, we will receive more smoothness in the classification boundary and more stability in the estimator. With small k, we will have a more jagged classification rule, but the rule will be able to identify more interesting nuances of the data. If we use a loss function to judge which is best, the 1-nearest neighbor model will fit best, because there is no penalty for over-fitting. Once we identify each estimated category (conditional on X) as the observed category in the data, there will be no error to report.

In this case, it will help to split the data into a training sample and a test sample. Even with the loss function, the idea of local fitting works well with large samples. In fact, as the input sample size n gets larger, the k-nearest neighbor estimator will be consistent as long as $k/n \to 0$. That is, it will achieve the goals we wanted without the strong model assumptions that come with parametric classification. There is an extra problem using the nonparametric technique, however. If the dimension of X is somewhat large, the amount of data needed to achieve a satisfactory answer from the nearest neighbor grows exponentially.

17.3.1 The Curse of Dimensionality

The *curse of dimensionality*, termed by Bellman (1961), describes the property of data to become sparse if the dimension of the sample space increases. For example, imagine the denseness of a data set with 100 observations distributed uniformly on the unit square. To achieve the same denseness in a 10-dimensional unit hypercube, we would require 10^{20} observations.

This is a significant problem for nonparametric classification problems including nearest neighbor classifiers and neural networks. As the dimension of inputs increase, the observations in the training set become relatively sparse. These procedures based on a large number of parameters help to handle complex problems, but must be considered inappropriate for most small or medium sized data sets. In those cases, the linear methods may seem overly simplistic or even crude, but still preferable to nearest neighbor methods.

17.3.2 Constructing the Nearest Neighbor Classifier

The classification rule is based on the ratio of the nearest-neighbor density estimator. That is, if x is from population G, then $P(x|G) \approx$ (proportion of observations in the neighborhood around x)/(volume of the neighborhood). To classify x, select the population corresponding to the largest value of

$$\frac{P(G_i)P(x|G_i)}{\Sigma_j P(G_j)P(x|G_j)}, \quad i = 1, \ldots, k.$$

This simplifies to the nearest neighbor rule; if the neighborhood around x is defined to be the closest r observations, x is classified into the population that is most frequently represented in that neighborhood.

Figure (17.4) shows the output derived from the MATLAB example below. Fifty randomly generated points are classified into one of two groups in v in a partially random way. The nearest neighbor plots reflect three different smoothing conditions of k=11, 5 and 1. As k gets smaller, the classifier acts more locally, and the rule appears more jagged.

```
>> y=rand(50,2)
>> v=round(0.3*rand(50,1)+0.3*y(:,1)+0.4*y(:,2));
>> n=100;
>> x=nby2(n);
>> m=n^2;
>> for i=1:m
     w(i,1)=nearneighbor(x(i,1:2),y,4,v);
   end
>> rr=find(w==1);
>> x2=x(rr,:);
>> plot(x2(:,1),x2(:,2),'.')
```

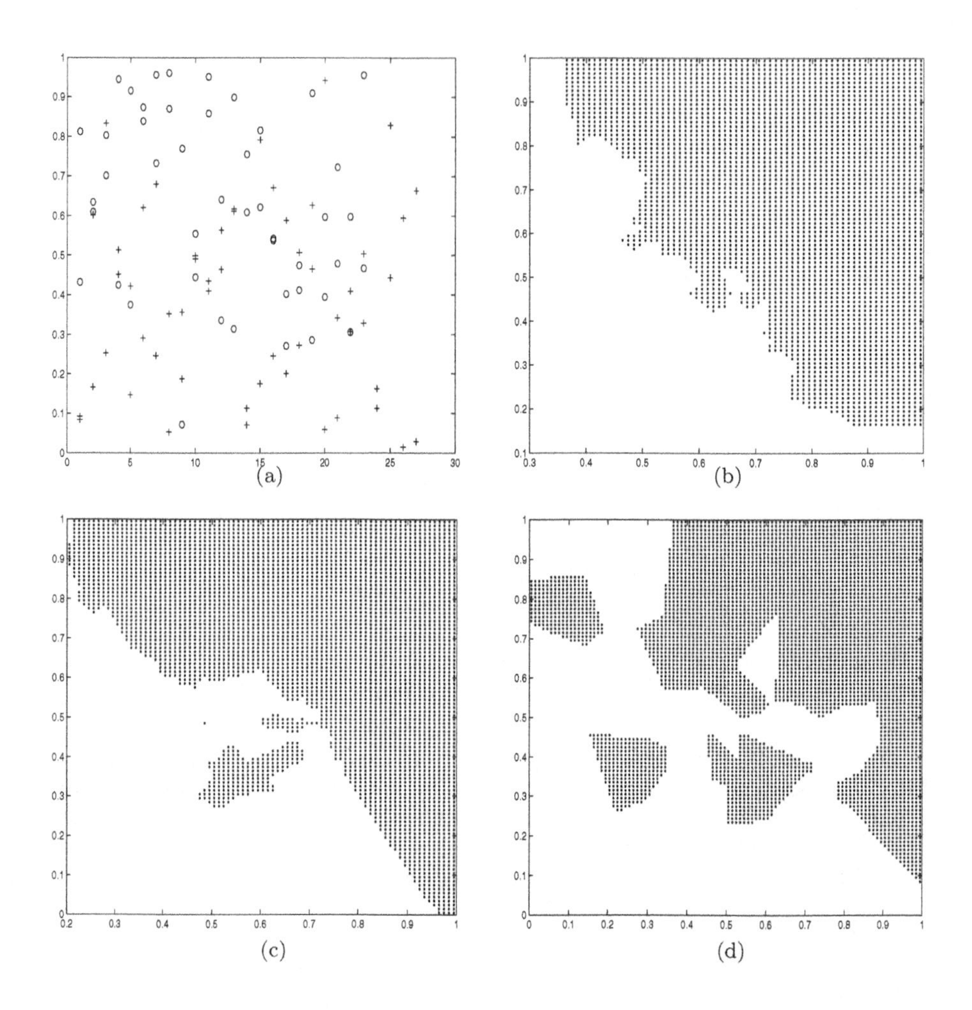

Fig. 17.2 Nearest neighbor classification of 50 observations plotted in (a) using neighborhood sizes of (b) 11, (c) 5, (d) 1.

17.4 NEURAL NETWORKS

Despite what your detractors say, you have a remarkable brain. Even with the increasing speed of computer processing, the much slower human brain has surprising ability to sort through gobs of information, disseminate some of its peculiarities and make a correct classification often several times faster than a computer. When a familiar face appears to you around a street corner, your brain has several processes working in parallel to identify this person you see, using past experience to gauge your expectation (you might not believe your eyes, for example, if you saw Elvis appear around the corner) along with all the sensory data from what you see, hear, or even smell.

The computer is at a disadvantage in this contest because despite all of the speed and memory available, the static processes it uses cannot parse through the same amount of information in an efficient manner. It cannot adapt and learn as the human brain does. Instead, the digital processor goes through sequential algorithms, almost all of them being a waste of CPU time, rather than traversing a relatively few complex neural pathways set up by our past experiences.

Rosenblatt (1962) developed a simple learning algorithm he named the *perceptron*, which consists of an input layer of several nodes that is completely connected to nodes of an output layer. The perceptron is overly simplistic and has numerous shortcomings, but it also represents the first neural network. By extending this to a two-step network which includes a *hidden layer* of nodes between the inputs and outputs, the network overcomes most of the disadvantages of the simpler map. Figure (17.4) shows a simple *feed-forward* neural network, that is, the information travels in the direction from input to output.

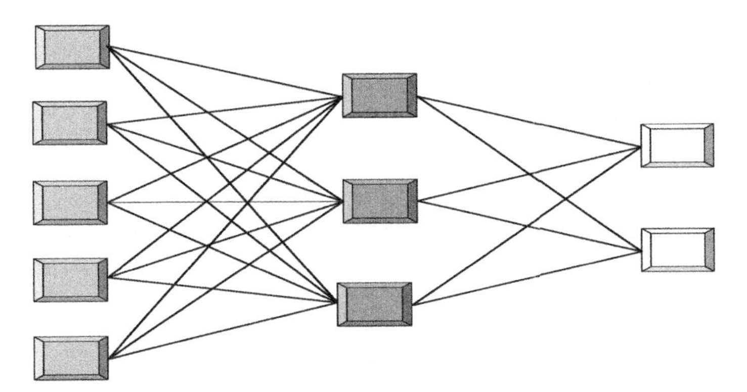

Fig. 17.3 Basic structure of feed-forward neural network.

The square nodes in Figure (17.4) represent neurons, and the connections (or *edges*) between them represent the synapses of the brain. Each connection is weighted, and this weight can be interpreted as the relative strength in the connection between the nodes. Even though the figure shows three layers, this is considered a *two-layer* network because the input layer, which does not process data or perform calculations, is not counted.

Each node in the hidden layers is characterized by an *activation function* which can be as simple as an indicator function (the binary output is similar to a computer) or have more complex nonlinear forms. A simple activation function would represent a node that would react when the weighted input surpassed some fixed threshold.

The neural network essentially looks at repeated examples (or input observations) and recalls patterns appearing in the inputs along with each subsequent response. We want to train the network to find this relationship between inputs and outputs using supervised learning. A key in training the network is to find the weights to go along with the activation functions that lead to supervised learning. To determine weights, we use a *back-propagation* algorithm.

17.4.1 Back-propagation

Before the neural network experiences any input data, the weights for the nodes are essentially random (noninformative). So at this point, the network functions like the scattered brain of a college freshman who has celebrated his first weekend on campus by drinking way too much beer.

The feed-forward neural network is represented by

$$\begin{array}{ccccc} n_I & & n_H & & n_O \\ \text{input nodes} & \Longrightarrow & \text{hidden nodes} & \Longrightarrow & \text{output nodes} \end{array}.$$

With an input vector $X = (x_1, \ldots, x_{n_I})$, each of the n_I input node codes the data and "fires" a signal across the edges to the hidden nodes. At each of the n_H hidden nodes, this message takes the form of a weighted linear combination from each attribute,

$$\mathcal{H}_j = A(\alpha_{0j} + \alpha_{1j}x_1 + \cdots + \alpha_{n_Ij}x_{n_I}), \quad j = 1, \ldots, n_H \tag{17.2}$$

where A is the activation function which is usually chosen to be the *sigmoid function*

$$A(x) = \frac{1}{1 + e^{-x}}.$$

We will discuss why A is chosen to be a sigmoid later. In the next step, the n_H hidden nodes fire this nonlinear outcome of the activation function to the output nodes, each translating the signals as a linear combination

$$\mathcal{O}_k = \beta_0 + \beta_1 \mathcal{H}_1 + \cdots + \beta_{n_H} \mathcal{H}_{nH}, \quad k = 1, \ldots, n_O. \tag{17.3}$$

Each output node is a function of the inputs, and through the steps of the neural network, each node is also a function of the weights α and β. If we observe $X_l = (x_{1l}, \ldots, x_{n_I l})$ with output $g_l(k)$ for $k = 1, \ldots, n_O$, we use the same kind of transformation used in logistic regression:

$$\hat{g}_l(k) = \frac{e^{\mathcal{H}_k}}{e^{\mathcal{H}_1} + e^{\mathcal{H}_2} + \cdots + e^{\mathcal{H}_{n_o}}}, \quad k = 1, ..., n_O.$$

For the training data $\{(X_1, g_1), \ldots, (X_n, g_n)\}$, the classification is compared to the observation's known group, which is then *back-propagated* across the network, and the network responds (learns) by adjusting weights in the cases an error in classification occurs. The loss function associated with misclassification can be squared errors, such as

$$SSQ(\alpha, \beta) = \Sigma_{l=1}^{n} \Sigma_{k=1}^{n_O} (g_l(k) - \hat{g}_l(k))^2, \tag{17.4}$$

where $g_l(k)$ is the actual response of the input X_l for output node k and $\hat{g}_l(k)$ is the estimated response.

Now we look how those weights are changed in this back-propagation. To minimize the squared error SSQ in (17.4) with respect to weights α and β from both layers of the neural net, we can take partial derivatives (with respect to weight) to find the direction the weights should go in order to decrease the error. But there are a lot of parameters to estimate: α_{ij}, with $1 \le i \le n_I$, $1 \le j \le n_H$ and β_{jk}, $1 \le j \le n_H$, $1 \le k \le n_O$. It's not helpful to think of this as a parameter set, as if they have their own intrinsic value. If you do, the network looks terribly over-parameterized and unnecessarily complicated. Remember that α and β are artificial, and our focus is on the n predicted outcomes instead of estimated parameters. We will do this iteratively using *batch learning* by updating the network after the entire data set is entered.

Actually, finding the global minimum of SSQ with respect to α and β will lead to over-fitting the model, that is, the answer will not represent the true underlying process because it is blindly mimicking every idiosyncrasy of the data. The gradient is expressed here with a constant γ called the *learning rate*:

$$\Delta\alpha_{ij} = \gamma \Sigma_{l=1}^{n} \frac{\partial(\Sigma_{k=1}^{n_O} (g_l(k) - \hat{g}_l(k))^2)}{\partial \alpha_{ij}} \tag{17.5}$$

$$\Delta\beta_{jk} = \gamma \Sigma_{l=1}^{n} \frac{\partial(\Sigma_{k=1}^{n_O} (g_l(k) - \hat{g}_l(k))^2)}{\partial \beta_{jk}} \tag{17.6}$$

and is solved iteratively with the following back-propagation equations (see

Chapter 11 of Hastie et al. (2001)) via error variables a and b:

$$a_{il} = \left[\frac{\partial A(t)}{\partial(t)} \right]_{t=\alpha_i' X_l} \Sigma_{l=1}^{n} \beta_{jk} b_{jl}. \tag{17.7}$$

Obviously, the activation function A must be differentiable. Note that if $A(x)$ is chosen as a binary function such as $I(x \geq 0)$, we end up with a regular linear model from (17.2). The sigmoid function, when scaled as $A_c(x) = A(cx)$ will look like $I(x \geq 0)$ as $c \to \infty$, but the function also has a well-behaved derivative.

In the first step, we use current values of α and β to predict outputs from (17.2) and (17.3). In the next step we compute errors b from the output layer, and use (17.7) to compute a from the hidden layer. Instead of batch processing, updates to the gradient can be made sequentially after each observation. In this case, γ is not constant, and should decrease to zero as the iterations are repeated (this is why it is called the learning rate).

The hidden layer of the network, along with the nonlinear activation function, gives it the flexibility to learn by creating convex regions for classification that need not be linearly separable like the more simple linear rules require. One can introduce another hidden layer that in effect can allow non convex regions (by combining convex regions together). Applications exist with even more hidden layers, but two hidden layers should be ample for almost every nonlinear classification problem that fits into the neural network framework.

17.4.2 Implementing the Neural Network

Implementing the steps above into a computer algorithm is not simple, nor is it free from potential errors. One popular method for processing through the back-propagation algorithm uses six steps:

1. Assign random values to the weights.
2. Input the first pattern to get outputs to the hidden layer $(\mathcal{H}_1, ..., \mathcal{H}_{n_H})$ and output layer $(\hat{g}(1), ..., \hat{g}(k))$.
3. Compute the output errors b.
4. Compute the hidden layer errors a as a function of b.
5. Update the weights using (17.5)
6. Repeat the steps for the next observation

Computing a neural network from scratch would be challenging for many of us, even if we have a good programming background. In MATLAB, there are a few modest programs that can be used for classification, such as `softmax(X,K,Prior)` that uses implements a feed-forward neural network using a training set X, a vector K for class indexing, with an optional prior argument. Instead of minimizing SSQ in (17.4), `softmax` assumes that "the outputs are a Poisson process conditional on their sum and calculates the error as the residual deviance."

MATLAB also has a Neural Networks Toolbox, see

`http://www.mathworks.com`

which features a graphical user interface (GUI) for creating, training, and running neural networks.

17.4.3 Projection Pursuit

The technique of Projection Pursuit is similar to that of neural networks, as both employ a nonlinear function that is applied only to linear combinations of the input. While the neural network is relatively fixed with a set number of hidden layer nodes (and hence a fixed number of parameters), projection pursuit seems more nonparametric because it uses unspecified functions in its transformations. We will start with a basic model

$$g(X) = \Sigma_{i=1}^{n_P} \psi\left(\theta_i' X\right), \tag{17.8}$$

where n_P represents the number of unknown parameter vectors $(\theta_1, \ldots, \theta_{n_p})$.

Note that $\theta_i' X$ is the projection of X onto the vector θ_i. If we pursue a value of θ_i that makes this projection effective, it seems logical enough to call this projection pursuit. The idea of using a linear combination of inputs to uncover structure in the data was first suggested by Kruskal (1969). Friedman and Stuetzle (1981) derived a more formal projection pursuit regression using a multi-step algorithm:

1. Define $\tau_i^{(0)} = g_i$.
2. Maximize the standardized squared errors

$$SSQ^{(j)} = 1 - \frac{\Sigma_{i=1}^{n}\left(\tau_i^{(j-1)} - \hat{g}^{(j-1)}(\hat{w}^{(j)\prime} x_i)\right)^2}{\Sigma_{i=1}^{n}\left(\tau_i^{(j-1)}\right)^2} \tag{17.9}$$

   over weights $\hat{w}^{(j)}$ (under the constraint that $\hat{w}^{(j)\prime}1 = 1$) and $\hat{g}^{(j-1)}$.

3. Update τ with $\tau_i^{(j)} = \tau_i^{(j-1)} - \hat{g}^{(j-1)}(\hat{w}^{(j)\prime} x_i)$.
4. Repeat the first step k times until $SSQ^{(k)} \leq \delta$ for some fixed $\delta > 0$.

Once the algorithm finishes, it essentially has given up trying to find other projections, and we complete the projection pursuit estimator as

$$\hat{g}(x) = \Sigma_{j=1}^{n_P} \hat{g}^{(j)}(\hat{w}^{(j)\prime}x). \tag{17.10}$$

17.5 BINARY CLASSIFICATION TREES

Binary trees offer a graphical and logical basis for empirical classification. Decisions are made sequentially through a route of branches on a tree - every time a choice is made, the route is split into two directions. Observations that are collected at the same endpoint (node) are classified into the same population. At those junctures on the route where the split is made are *nonterminal nodes*, and *terminal nodes* denote all the different endpoints where a classification of the tree. These endpoints are also called the leaves of the tree, and the starting node is called the root.

With the training set $(x_1, g_1), \ldots, (x_n, g_n)$, where x is a vector of m components, splits are based on a single variable of x, possibly a linear combination. This leads to decision rules that are fairly easy to interpret and explain, so binary trees are popular for disseminating information to a broad audience. The phases of of tree construction include

- Deciding whether to make the node a terminal node.
- Selection of splits in a nonterminal node
- Assigning classification rule at terminal nodes.

This is the essential approach of CART (*Classification and Regression Trees*). The goal is to produce a simple and effective classification tree without an excess number of nodes.

If we have k populations $G_1, \ldots, G_k$, we will use the frequencies found in the training data to estimate population frequency in the same way we constructed nearest-neighbor classification rules: the proportion of observations in training set from the ith population $= P(G_i) = n_i/n$. Suppose there are $n_i(r)$ observations from G_i that reach node r. The probability of such an observation reaching node r is estimated as

$$P_i(t) = P(G_i)P(\text{reach node } r \mid G_i) = \frac{n_i}{n} \times \frac{n_i(r)}{n_i} = \frac{n_i(r)}{n}.$$

We want to construct a perfectly pure split where we can isolate one or some of the populations into a single node that can be a terminal node (or at least split more easily into one during a later split). Figure 17.4 illustrates a

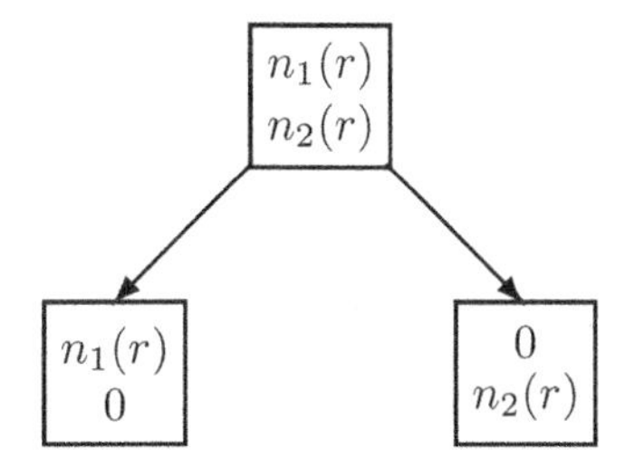

Fig. 17.4 Purifying a tree by splitting.

perfect split of node r. This, of course, is not always possible. This quality measure of a split is defined in an impurity index function

$$\mathcal{I}(r) = \psi\left(P_1(r), \ldots, P_k(r)\right),$$

where ψ is nonnegative, symmetric in its arguments, maximized at $(1/k, \ldots, 1/k)$, and minimized at any k-vector that has a one and $k-1$ zeroes.

Several different methods of impurity have been defined for constructing trees. The three most popular impurity measures are cross-entropy, Gini impurity and misclassification impurity:

1. **Cross-entropy:** $\mathcal{I}(r) = -\Sigma_{i:P_i(r)>0} P_i(r) \ln[P_i(r)]$.
2. **Gini:** $\mathcal{I}(r) = -\Sigma_{i \neq j} P_i(r) P_j(r)$.
3. **Misclassification:** $\mathcal{I}(r) = 1 - \max_j P_j(r)$

The misclassification impurity represents the minimum probability that the training set observations would be (empirically) misclassified at node r. The Gini measure and Cross-entropy measure have an analytical advantage over the discrete impurity measure by being differentiable. We will focus on the most popular index of the three, which is the cross-entropy impurity.

By splitting a node, we will reduce the impurity to

$$q(L)\mathcal{I}(r_L) + q(R)\mathcal{I}(r_R),$$

where $q(R)$ is the proportion of observations that go to node r_R, and $q(L)$ is the proportion of observations that go to node r_L. Constructed this way, the binary tree is a *recursive* classifier.

Let Q be a potential split for the input vector x. If $x = (x_1, ..., x_m)$, $Q = \{x_i > x_0\}$ would be a valid split if x_i is ordinal, or $Q = \{x_i \in \mathcal{S}\}$ if x_i is categorical and $\mathcal{S}$ is a subset of possible categorical outcomes for x_i. In either case, the split creates two additional nodes for the binary response of the data to Q. For the first split, we find the split Q_1 that will minimize the impurity measure the most. The second split will be chosen to be the Q_2 that minimizes the impurity from one of the two nodes created by Q_1.

Suppose we are the middle of constructing a binary classification tree T that has a set of terminal nodes $\mathcal{R}$. With

$$P(\text{ reach node } r) = P(r) = \Sigma P_i(r),$$

suppose the current impurity function is

$$\mathcal{I}_T = \Sigma_{r \in \mathcal{R}} \mathcal{I}(r) P(r).$$

At the next stage, then, we split the node that will most greatly decrease $\mathcal{I}_T$.

Example 17.3 The following made-up example was used in Elsner, Lehmiller, and Kimberlain (1996) to illustrate a case for which linear classification models fail and binary classification trees perform well. Hurricanes categorized according to season as "tropical only" or "baroclinically influenced". Hurricanes are classified according to location (longitude, latitude), and Figure (17.5(a)) shows that no linear rule can separate the two categories without a great amount of misclassification. The average latitude of origin for tropical-only hurricanes is 18.8°N, compared to 29.1°N for baroclinically influenced storms. The baroclinically influenced hurricane season extends from-mid May to December, while the tropical-only season is largely confined to the months of August through October.

For this problem, simple splits are considered and the ones that minimize impurity are Q_1 : Longitude ≥ 67.75, and Q_2 : Longitude ≤ 62.5 *(see homework)*. In this case, the tree perfectly separates the two types of storms with two splits and three terminal nodes in Figure 17.5(b).

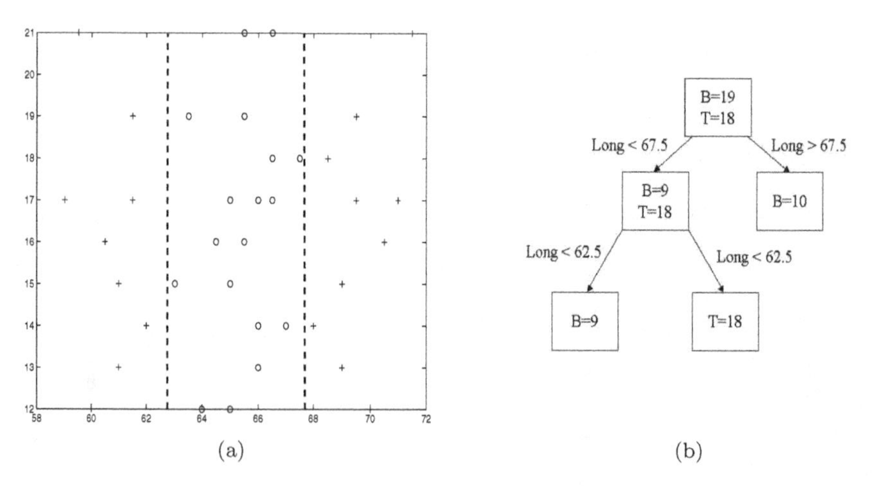

Fig. 17.5 (a) Location of 37 tropical (circles) and other (plus-signs) hurricanes from Elsner at al. (1996); (b) Corresponding separating tree.

```
>> long =[59.00 59.50 60.00 60.50 61.00 61.00 61.50 61.50 62.00 63.00...
   63.50 64.00 64.50 65.00 65.00 65.00 65.50 65.50 65.50 66.00 66.00...
   66.00 66.50 66.50 66.50 67.00 67.50 68.00 68.50 69.00 69.00 69.50...
  69.50 70.00 70.50 71.00 71.50];
>> lat = [17.00 21.00 12.00 16.00 13.00 15.00 17.00 19.00 14.00 15.00...
   19.00 12.00 16.00 12.00 15.00 17.00 16.00 19.00 21.00 13.00 14.00...
   17.00 17.00 18.00 21.00 14.00 18.00 14.00 18.00 13.00 15.00 17.00...
   19.00 12.00 16.00 17.00 21.00];
>>  trop = [0 0 0 0 0 0 0 0 0 1 1 1 1 1 1 1 1 1 1 1 1 1 1 1 1 1 1 0 0...
   0 0 0 0 0 0 0 0];
>> plot(long(find(long'.*trop'>0.5))',lat(find(long'.*trop'>0.5))','o')
>> hold on
>> plot(long(find(long'.*trop'<0.5))',lat(find(long'.*trop'<0.5))','+')
```

17.5.1 Growing the Tree

So far we have not decided how many splits will be used in the final tree; we have only determined which splits should take place first. In constructing a binary classification tree, it is standard to grow a tree that is initially too large, and to then prune it back, forming a sequence of sub-trees. This approach works well; if one of the splits made in the tree appears to have no value, it might be worth saving if there exists below it an effective split.

In this case we define a branch to be a split direction that begins at a node and includes all the subsequent nodes in the direction of that split (called a subtree or descendants). For example, suppose we consider splitting tree T at node r and T_r represents the classification tree after the split is made. The new nodes made under r will be denoted r_R and r_L. The impurity is now

$$\mathcal{I}_{T_r} = \Sigma_{s \in T, s \neq r} \mathcal{I}_T(s) P(s) + P(r_R)\mathcal{I}_T(r_R) + P(r_L)\mathcal{I}(r_L).$$

The change in impurity caused by the split is

$$\begin{aligned} \Delta\mathcal{I}_{T_r}(r) &= P(r)\mathcal{I}_T(r) - P(r_R)\mathcal{I}_T(r_R) - P(r_L)\mathcal{I}_T(r_L) \\ &= P(r)\left(\mathcal{I}_T(r) - \frac{P(r_R)}{P(r_R)}\mathcal{I}_T(r_R) - \frac{P(r_L)}{P(r_L)}\mathcal{I}_T(r_L)\right). \end{aligned}$$

Again, let $\mathcal{R}$ be the set of all terminal nodes of the tree. If we consider the potential differences for any particular split Q, say $\Delta\mathcal{I}_{T_r}(r; Q)$, then the next split should be chosen by finding the terminal node r and split Q corresponding to

$$\max_{r \in \mathcal{R}} \left(P(r) \left(\max_{Q} \Delta\mathcal{I}_{T_r}(r; Q) \right) \right).$$

To prevent the tree from splitting too much, we will have a fixed threshold level $\tau > 0$ so that splitting must stop once the change no longer exceeds τ.

We classify each terminal node according to majority vote: observations in terminal node r are classified into the population i with the highest $n_i(r)$. With this simple rule, the misclassification rate for observations arriving at node r is estimated as $1 - P_i(r)$.

17.5.2 Pruning the Tree

With a tree constructed using only a threshold value to prevent overgrowth, a large set of training data may yield a tree with an abundance of branches and terminal nodes. If τ is small enough, the tree will fit the data locally, similar to how a 1-nearest-neighbor overfits a model. If τ is too large, the tree will stop growing prematurely, and we might fail to find some interesting features of the data. The best method is to grow the tree a bit too much and then prune back unnecessary branches.

To make this efficable, there must be a penalty function $\zeta_T = \zeta_T(|\mathcal{R}|)$ for adding extra terminal nodes, where $|\mathcal{R}|$ is the cardinality, or number of terminal nodes of $\mathcal{R}$. We define our cost function to be a combination of misclassification error and penalty for over-fitting:

$$C(T) = L_T + \zeta_T,$$

where

$$L_T = \Sigma_{r \in \mathcal{R}} P(r) \left(1 - \max_j P_j(r) \right) \equiv \Sigma_{r \in \mathcal{R}} L_T(r).$$

This is called the *cost-complexity* pruning algorithm in Breiman et al (1984). Using this rule, we will always find a subtree of T that minimizes $C(T)$. If we allow $\zeta_T \to 0$, the subtree is just the original tree T, and if we allow $\zeta_T \to \infty$, the subtree is a single node that doesn't split at all. If we increase ζ_T from 0, we will get a sequence of subtrees, each one being nested in the previous one.

In deciding whether or not to prune a branch of the tree at node r, we will compare $C(T)$ of the tree to the new cost that would result from removing the branches under node r. L_T will necessarily increase, while ζ_T will decrease as the number of terminal nodes in the tree decreases.

Let T_r be the branch under node r, so the tree remaining after cutting branch T_r (we will call this $T_{(r)}$) is nested in T, i.e., $T_{(r)} \subset T$. The set of terminal nodes in the branch T_r is denoted $\mathcal{R}_r$. If another branch at node s is pruned, we will denote the remaining subtree as $T_{(r,s)} \subset T_{(r)} \subset T$. Now,

$$C(T_r) = \Sigma_{s \in \mathcal{R}_r} L_{T_r}(s) + \zeta_{T_r}$$

is equal to $C(T)$ if ζ_T is set to

$$h(r) = \frac{L_T - \Sigma_{s \in \mathcal{R}_r} L_{T_r}(s)}{|\mathcal{R}_r|}.$$

Using this approach, we want to trim the node r that minimizes $h(r)$. Obviously, only non-terminal nodes $r \in \mathcal{R}^C$ because terminal nodes have no branches. If we repeat this procedure after recomputing $h(r)$ for the resulting subtree, this pruning will create another sequence of nested trees

$$T \supset T_{(r_1)} \supset T_{(r_1, r_2)} \supset \cdots \supset r_0,$$

where r_0 is the the first node of the original tree T. Each subtree has an associated cost $(C(T), C(T_{r_1}), \cdots, C(r_0))$ which can be used to determine at what point the pruning should finish. The problem with this procedure is that the misclassification probability is based only on the training data.

A better estimator can be constructed by cross-validation. If we divide the training data into ν subsets $S_1, \ldots, S_\nu$, we can form ν artificial training sets as

$$S_{(j)} \equiv \bigcup_{i \neq j} S_i$$

and constructing a binary classification tree based on each of the ν sets $S_{(1)}, \cdots, S_{(\nu)}$. This type of cross-validation is analogous to the jackknife "leave-one-out" resampling procedure. If we let $L^{(j)}$ be the estimated misclassification probability based on the subtree chosen in the jth step of the cross validation (i.e., leaving out S_j), and let $\zeta^{(j)}$ be the corresponding penalty function, then

$$L^{CV} \equiv \frac{1}{n} \Sigma_{j=1}^{\nu} L^{(j)}$$

provides a bona fide estimator for misclassification error. The corresponding penalty function for L^{CV} is estimated as the geometric mean of the penalty functions in the cross validation. That is,

$$\zeta^{CV} = \sqrt[\nu]{\prod_{j=1}^{\nu} \zeta^{(j)}}.$$

To perform a binary tree search in MATLAB, the function `treefit` creates a decision tree based on an input an nxm matrix of inputs and a n-vector y of outcomes. The function `treedisp` creates a graphical representation of the tree using the same inputs. Several options are available to control tree growth, tree pruning, and misclassification costs (see Chapter 23). The function `treeprune` produces a sequence of trees by pruning according to a threshold rule. For example, if T is the output of a treeprune function, then treeprune(T) generates an unpruned tree of T and adds information about optimal pruning.

```
>> numobs = size(meas,1);
>> tree = treefit(meas(:,1:2), species);
>> [dtnum,dtnode,dtclass] = treeval(tree, meas(:,1:2));
>> bad = ~strcmp(dtclass,species);
>> sum(bad) / numobs

ans =
    0.1333
%The decision tree misclassifies 13.3% or 20 of the specimens.

>> [grpnum,node,grpname] = treeval(tree, [x y]);
>> gscatter(x,y,grpname,'grb','sod')
>> treedisp(tree,'name',{'SL' 'SW'})
>> resubcost = treetest(tree,'resub');
>> [cost,secost,ntermnodes,bestlevel] = ...
>>      treetest(tree,'cross',meas(:,1:2),species);
>> plot(ntermnodes,cost,'b-', ntermnodes,resubcost,'r--')
>> xlabel('Number of terminal nodes')
>> ylabel('Cost(misclassification error)')
>> legend('Cross-validation','Resubstitution')
```

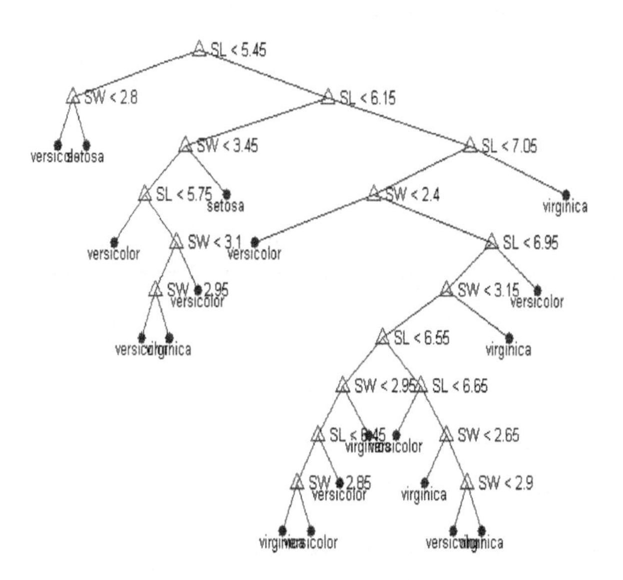

Fig. 17.6 MATLAB function **treedisp** applied to Fisher's Iris Data.

17.5.3 General Tree Classifiers

Classification and regression trees can be conveniently divided to five different families.

(i) The CART family : Simple versions of CART have been emphasized in this chapter. This method is characterized by its use of two branches from each nonterminal node. Cross-validation and pruning are used to determine size of tree. Response variable can be quantitative or nominal. Predictor variables can be nominal or ordinal, and continuous predictors are supported. *Motivation*: statistical prediction.

(ii) The CLS family: These include ID3, originally developed by Quinlan (1979), and off-shoots such as CLS and C4.5. For this method, the number of branches equals the number of categories of the predictor. Only nominal response and predictor variables are supported in early versions, so continuous inputs had to be binned. However, the latest version of C4.5 supports ordinal predictors. *Motivation*: concept learning.

(iii) The AID family: Methods include AID, THAID, CHAID, MAID, XAID, FIRM, and TREEDISC. The number of branches varies from two to the number of categories of the predictor. Statistical significance tests (with multiplicity adjustments in the later versions) are used to determine the size of tree. AID, MAID, and XAID are for quantitative responses. THAID, CHAID, and TREEDISC are for nominal responses, although the version of CHAID from Statistical Innovations, distributed by SPSS, can handle a quantitative categorical response. FIRM comes in two varieties for categorical or continuous response. Predictors can be nominal or ordinal and there is usually provision for a missing-value category. Some versions can handle continuous predictors, others cannot. *Motivation*: detecting complex statistical relationships.

(iv) Linear combinations: Methods include OC1 and SE-Trees. The Number of branches varies from two to the number of categories of predictor. *Motivation*: Detecting linear statistical relationships combined to concept learning.

(v) Hybrid models: IND is one example. IND combines CART and C4 as well as Bayesian and minimum encoding methods. Knowledge Seeker combines methods from CHAID and ID3 with a novel multiplicity adjustment.*Motivation*: Combines methods from other families to find optimal algorithm.

17.6 EXERCISES

17.1. Create a simple nearest-neighbor program using MATLAB. It should input a training set of data in $m+1$ columns; one column should contain the population identifier 1, ..., k and the others contain the input vectors that can have length m. Along with this training set, also input another m column matrix representing the classification set. The output should contain n, m, k and the classifications for the input set.

17.2. For the Example 17.3, show the optimal splits, using the cross-entropy measure, in terms of intervals { longitude $\geq l_0$} and { latitude $\geq l_1$}

17.3. In this exercise the goal is to discriminate between observations coming from two different normal populations, using logistic regression.

Simulate a training data set, $\{(X_i, Y_i), i = 1, \dots, n\}$, (take n even) as follows: For the first half of data, X_i, $i = 1, \dots, n/2$ are sampled from the standard normal distribution and $Y_i = 0$, $i = 1, \dots, n/2$. For the second half, X_i, $i = n/2+1, \dots, n$ are sampled from normal distribution with mean 2 and variance 1, while $Y_i = 1$, $i = n/2 + 1, \dots, n$. Fit the logistic regression to this data, $\hat{P}(Y = 1) = f(X)$.

Simulate a validation set $\{(X_j^*, Y_j^*), j = 1, \dots, m\}$ the same way, and classify these new Y_j^*'s as 0 or 1 depending whether $f(X_j^*) < 0.5$ or ≥ 0.5.

(a) Calculate the error of this logistic regression classifier,

$$L_n(m) = \frac{1}{m} \Sigma_{j=1}^{m} \mathbf{1} \left(\mathbf{1}(f(X_j^*) > 0.5) \neq Y_j^* \right).$$

In your simulations use $n = 60, 200$, and 2000 and $m = 100$.

(b) Can the error $L_n(m)$ be made arbitrarily small by increasing n?

REFERENCES

Agresti, A. (1990), *Categorical Data Analysis*, New York: Wiley.

Bellman, R. E. (1961), *Adaptive Control Processes,* Princeton, NJ: Princeton University Press.

Breiman, L., Friedman, J., Olshen, R., and Stone, C. (1984), *Classification and Regression Trees*, Belmont, CA: Wadsworth.

Duda, R. O., Hart, P. E. and Stork, D. G. (2001), *Pattern Classification*, New York: Wiley.

Fisher, R. A. (1936), "The Use of Multiple Measurements in Taxonomic Problems," *Annals of Eugenics,* 7, 179-188.

Elsner, J. B., Lehmiller, G. S., and Kimberlain, T. B. (1996), "Objective Classification of Atlantic Basin Hurricanes," *Journal of Climate,* 9, 2880-2889.

Friedman, J., and Stuetzle, W. (1981), "Projection Pursuit Regression," *Journal of the American Statistical Association,* 76, 817-823.

Hastie, T., Tibshirani, R., and Friedman, J. (2001), *The Elements of Statistical Learning,* New York: Springer Verlag.

Kutner, M. A., Nachtsheim, C. J., and Neter, J. (1996), *Applied Linear Regression Models,* 4th ed., Chicago: Irwin.

Kruskal J. (1969), "Toward a Practical Method which Helps Uncover the Structure of a Set of Multivariate Observations by Finding the Linear Transformation which Optimizes a New Index of Condensation," *Statistical Computation,* New York: Academic Press, pp. 427-440.

Marks, S., and Dunn, O. (1974), "Discriminant Functions when Covariance Matrices are Unequal," *Journal of the American Statistical Association,* 69, 555-559.

Moore, D. H. (1973), "Evaluation of Five Discrimination Procedures for Binary Variables," *Journal of the American Statistical Association,* 68, 399-404.

Quinlan, J. R. (1979), "Discovering Rules from Large Collections of Examples: A Case Study." in *Expert Systems in the Microelectronics Age,* Ed. D. Michie, Edinburgh: Edinburgh University Press.

Randles, R. H., Broffitt, J.D., Ramberg, J. S., and Hogg, R. V. (1978), "Generalized Linear and Quadratic Discriminant Functions Using Robust Estimates," *Journal of the American Statistical Association,* 73, 564-568.

Rosenblatt, R. (1962), *Principles of Neurodynamics: Perceptrons and the Theory of Brain Mechanisms,* Washington, DC: Spartan.

18

Nonparametric Bayes

> **Bayesian** (bey' -zh*uh*n) *n.* **1.** Result of breeding a statistician with a clergyman to produce the much sought honest statistician.
>
> Anonymous

This chapter is about nonparametric Bayesian inference. Understanding the computational machinery needed for non-conjugate Bayesian analysis in this chapter can be quite challenging and it is beyond the scope of this text. Instead, we will use specialized software, WinBUGS, to implement complex Bayesian models in a user-friendly manner. Some applications of WinBUGS have been discussed in Chapter 4 and an overview of WinBUGS is given in the Appendix B.

Our purpose is to explore the useful applications of the nonparametric side of Bayesian inference. At first glance, the term *nonparametric Bayes* might seem like an oxymoron; after all, Bayesian analysis is all about introducing prior distributions on parameters. Actually, nonparametric Bayes is often seen as a synonym for Bayesian models with process priors on the spaces of densities and functions. Dirichlet process priors are the most popular choice. However, many other Bayesian methods are nonparametric in spirit. In addition to Dirichlet process priors, Bayesian formulations of contingency tables and Bayesian models on the coefficients in atomic decompositions of functions will be discussed later in this chapter.

18.1 DIRICHLET PROCESSES

The central idea of traditional nonparametric Bayesian analysis is to draw inference on an unknown distribution function. This leads to models on function spaces, so that the Bayesian nonparametric approach to modeling requires a dramatic shift in methodology. In fact, a commonly used technical definition of nonparametric Bayes models involves infinitely many parameters, as mentioned in Chapter 10.

Results from Bayesian inference are comparable to classical nonparametric inference, such as density and function estimation, estimation of mixtures and smoothing. There are two main groups of nonparametric Bayes methodologies: (1) methods that involve prior/posterior analysis on distribution spaces, and (2) methods in which standard Bayes analysis is performed on a vast number of parameters, such as atomic decompositions of functions and densities. Although the these two methodologies can be presented in a unified way (see Mueller and Quintana, 2005), because of simplicity we present them separately.

Recall a Dirichlet random variable can be constructed from gamma random variables. If $X_1, \dots, X_n$ are i.i.d. $\mathcal{G}amma(a_i, 1)$, then for $Y_i = X_i/\Sigma_{j=1}^n X_j$, the vector $(Y_1, \dots, Y_n)$ has Dirichlet $\mathcal{D}ir(a_1, \dots, a_n)$ distribution. The Dirichlet distribution represents a multivariate extension of the beta distribution: $\mathcal{D}ir(a_1, a_2) \equiv \mathcal{B}e(a_1, a_2)$. Also, from Chapter 2, $\mathbb{E}Y_i = a_i/\Sigma_{j=1}^n a_j$, $\mathbb{E}Y_i^2 = a_i(a_i+1)/\Sigma_{j=1}^n a_j(1+\Sigma_{j=1}^n a_j)$, and $\mathbb{E}(Y_i Y_j) = a_i a_j/\Sigma_{j=1}^n a_j(1+\Sigma_{j=1}^n a_j)$.

The Dirichlet process (DP), with precursors in the work of Freedman (1963) and Fabius (1964), was formally developed by Ferguson (1973, 1974). It is the first prior developed for spaces of distribution functions. The DP is, formally, a probability measure (distribution) on the space of probability measures (distributions) defined on a common probability space $\mathcal{X}$. Hence, a realization of DP is a random distribution function.

The DP is characterized by two parameters: (i) Q_0, a specific probability measure on $\mathcal{X}$ (or equivalently, G_0 a specified distribution function on $\mathcal{X}$); (ii) α, a positive scalar parameter.

Definition 18.1 *(Ferguson, 1973) The DP generates random probability measures (random distributions) Q on $\mathcal{X}$ such that for any finite partition $B_1, \dots, B_k$ of $\mathcal{X}$,*

$$(Q(B_1), \dots, Q(B_k)) \sim \mathcal{D}ir(\alpha Q_0(B_1), \dots, \alpha Q_0(B_k)),$$

where, $Q(B_i)$ (a random variable) and $Q_0(B_i)$ (a constant) denote the probability of set B_i under Q and Q_0, respectively. Thus, for any B,

$$Q(B) \sim \mathcal{B}e(\alpha Q_0(B), \alpha(1 - Q_0(B)))$$

and

$$\mathbb{E}(Q(B)) = Q_0(B), \quad \mathbb{V}ar(Q(B)) = \frac{Q_0(B)(1 - Q_0(B))}{\alpha + 1}.$$

The probability measure Q_0 plays the role of the center of the DP, while α can be viewed as a *precision* parameter. Large α implies small variability of DP about its center Q_0.

The above can be expressed in terms of CDFs, rather than in terms of probabilities. For $B = (-\infty, x]$ the probability $Q(B) = Q((-\infty, x]) = G(x)$ is a distribution function. As a result, we can write

$$G(x) \sim \mathcal{B}e(\alpha G_0(x), \alpha(1 - G_0(x)))$$

and

$$\mathbb{E}(G(x)) = G_0(x)), \quad \mathbb{V}\text{ar}(G(x)) = \frac{G_0(x)(1 - G_0(x))}{\alpha + 1}.$$

The notation $G \sim DP(\alpha G_0)$ indicates that the DP prior is placed on the distribution G.

Example 18.1 Let $G \sim DP(\alpha G_0)$ and $x_1 < x_2 < \cdots < x_n$ are arbitrary real numbers from the support of G. Then

$$\begin{aligned} (G(x_1), G(x_2) - G(x_1), \ldots, G(x_n) - G(x_{n-1})) &\sim \\ Dir(\alpha G_0(x_1), \alpha(G_0(x_2) - G_0(x_1)), \ldots, \alpha(G_0(x_n) - G_0(x_{n-1}))), & \quad (18.1) \end{aligned}$$

which suggests a way to generate a realization of density from DP at discrete points.

If $(d_1, \ldots, d_n)$ is a draw from (18.1), then $(d_1, d_1 + d_2, \ldots, \Sigma_{i=1}^n d_i)$ is a draw from $(G(x_1), G(x_2), \ldots, G(x_n))$. The MATLAB program `dpgen.m` generates 15 draws form $DP(\alpha G_0)$ for the base CDF $G_0 \equiv \mathcal{B}e(2, 2)$ and the precision parameter $\alpha = 20$. In Figure 18.1 the base CDF $\mathcal{B}e(2, 2)$ is shown as a dotted line. Fifteen random CDF's from DP(20, $\mathcal{B}e(2, 2)$) are scattered around the base CDF.

```
>> n = 30;  %generate random CDF's at 30 equispaced points
>> a = 2; %a, b are parameters of the
>>          %BASE distribution G_0 = Beta(2,2)
>> b = 2;
>> alpha = 20; %The precision parameter alpha = 20 describes
>> % scattering about the BASE distribution.
>> % Higher alpha, less variability.
>> %-------------------
>> x = linspace(1/n,1,n); %the equispaced points at which
>> % random CDF's are evaluated.
>> y = CDF_beta(x, a, b);   % find CDF's of BASE
```

```
>> par = [y(1) diff(y)];      % and form a Dirichlet parameter
>> %-----------------------
>> for i = 1:15  % Generate 15 random CDF's.
>> yy = rand_dirichlet(alpha * par,1);
>> plot( x, cumsum(yy),'-','linewidth',1) %cummulative sum
>> % of Dirichlet vector is a random CDF
>> hold on
>> end
>> yyy = 6 .* (x.^2/2 - x.^3/3);  %Plot BASE CDF as reference
>> plot( x, yyy, ':', 'linewidth',3)
```

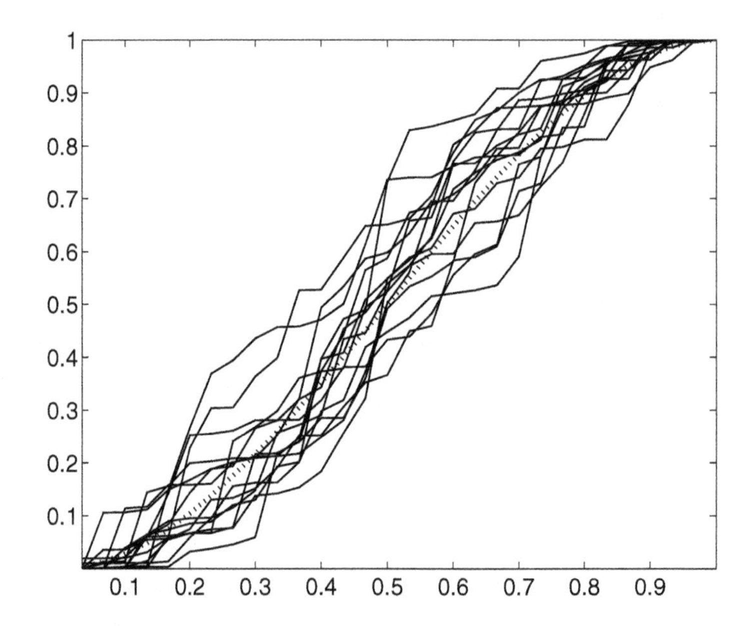

Fig. 18.1 The base CDF $\mathcal{B}e(2,2)$ is shown as a dotted line. Fifteen random CDF's from DP(20, $\mathcal{B}e(2,2)$) are scattered around the base CDF.

An alternative definition of DP, due to Sethuraman and Tiwari (1982) and Sethuraman (1994), is known as the *stick-breaking algorithm.*

Definition 18.2 *Let* $U_i \sim \mathcal{B}e(1,\alpha),\ i = 1,2,\dots$ *and* $V_i \sim G_0, i = 1,2,\dots$ *be two independent sequences of i.i.d. random variables. Define weights* $\omega_1 = U_1$ *and* $\omega_i = U_i \prod_{j=1}^{i-1}(1 - U_j), i > 1$. *Then,*

$$G = \Sigma_{k=1}^{\infty} \omega_k \delta(V_k) \sim DP(\alpha G_0),$$

where $\delta(V_k)$ *is a point mass at* V_k.

The distribution G is discrete, as a countable mixture of point masses, and from this definition one can see that with probability 1 only discrete distributions fall in the support of DP. The name stick-breaking comes from the fact that $\Sigma \omega_i = 1$ with probability 1, that is, the unity is broken on infinitely many random weights. The Definition 18.2 suggests another way to generate approximately from a given DP.

Let $G_K = \Sigma_{k=1}^{K} \omega_k \delta(V_k)$ where the weights $\omega_1, \ldots, \omega_{K-1}$ are as in Definition 18.2 and the last weight ω_K is modified as $1 - \omega_1 - \cdots - \omega_{K-1}$, so that the sum of K weights is 1. In practical applications, K is selected so that $(1 - (\alpha/(1+\alpha))^K)$ is small.

18.1.1 Updating Dirichlet Process Priors

The critical step in any Bayesian inference is the transition from the prior to the posterior, that is, updating a prior when data are available. If $Y_1, Y_2, \ldots, Y_n$ is a random sample from G, and G has Dirichlet prior $DP(\alpha G_0)$, the posterior is remains Dirichlet, $G|Y_1, \ldots, Y_n \sim DP(\alpha^* G_0^*)$, with $\alpha^* = \alpha + n$, and

$$G_0^*(t) = \frac{\alpha}{\alpha+n} G_0(t) + \frac{n}{\alpha+n} \left(\frac{1}{n} \Sigma_{i=1}^{n} I(Y_i \leq t \leq \infty) \right). \tag{18.2}$$

Notice that the DP prior and the EDF constitute a *conjugate pair* because the posterior is also a DP. The posterior estimate of distribution is $E(G|Y_1, \ldots, Y_n) = G_0^*(t)$ which is, as we saw in several examples with conjugate priors, a weighted average of the "prior mean" and the maximum likelihood estimator (the EDF).

Example 18.2 In the spirit of classical nonparametrics, the problem of estimating the CDF at a fixed value x, has a simple nonparametric Bayes solution. Suppose the sample $X_1, \ldots, X_n \sim F$ is observed and that one is interested in estimating $F(x)$. Suppose the $F(x)$ is assigned a Dirichlet process prior with a center F_0 and a small precision parameter α. The posterior distribution for $F(x)$ is $\mathcal{B}e(\alpha F_0(x) + \ell_x, \alpha(1 - F_0(x)) + n - \ell_x)$ where ℓ_x is the number of observations in the sample smaller than or equal to x. As $\alpha \to 0$, the posterior tends to a $\mathcal{B}e(\ell_x, n - \ell_x)$. This limiting posterior is often called *noninformative.* By inspecting the $\mathcal{B}e(\ell_x, n - \ell_x)$ distribution, or generating from it, one can find a posterior probability region for the CDF at any value x. Note that the posterior expectation of $F(x)$ is equal to the classical estimator ℓ_x/n, which makes sense because the prior is noninformative.

Example 18.3 The underground train at Hartsfield-Jackson airport arrives at its starting station every four minutes. The number of people Y entering a single car of the train is random variable with a Poisson distribution,

$$Y|\lambda \sim \mathcal{P}(\lambda).$$

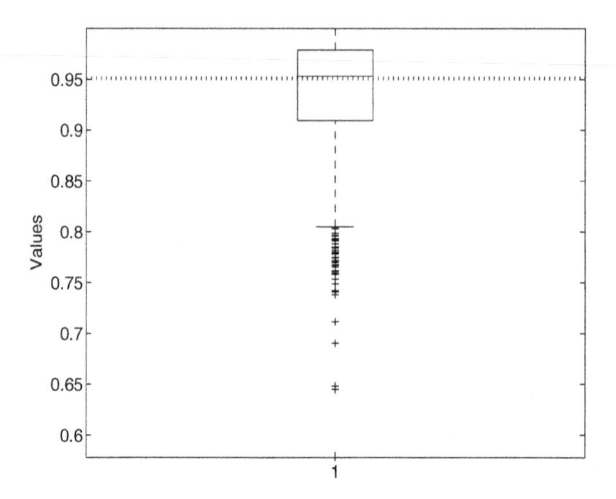

Fig. 18.2 For a sample $n = 15$ Beta(2,2) observations a boxplot of "noninformative" posterior realizations of $P(X \leq 1)$ is shown. Exact value $F(1)$ for Beta(2,2) is shown as dotted line.

A sample of size $N = 20$ for Y is obtained below.

9	7	7	8	8	11	8	7	5	7
13	5	7	14	4	6	18	9	8	10

The prior on λ is *any* discrete distribution supported on integers [1, 17],

$$\lambda | P \sim \mathcal{D}iscr\left((1, 2, \ldots, 17), P = (p_1, p_2, \ldots, p_{17})\right),$$

where $\Sigma_i \, p_i = 1$. The hyperprior on probabilities P is Dirichlet,

$$P \sim \mathcal{D}ir(\alpha G_0(1), \alpha G_0(2), \ldots, \alpha G_0(17)).$$

We can assume that the prior on λ is a Dirichlet process with

$$G_0 = [1, 1, 1, 2, 2, 3, 3, 4, 4, 5, 6, 5, 4, 3, 2, 1, 1]/48$$

and $\alpha = 48$. We are interested in posterior inference on the rate parameter λ.

```
model
{
for (i in 1:N)
    {
     y[i] ~ dpois(lambda)
```

```
    }
    lambda ~ dcat(P[])
    P[1:bins] ~ ddirch(alphaG0[])
}
#data
list(bins=17, alphaG0=c(1,1,1,2,2,3,3,4,4,5,6,5,4,3,2,1,1),
 y=c(9,7,7,8,8,11,8,7,5,7,13,5,7,14,4,6,18,9,8,10),  N=20
)

#inits
list(lambda=12,
P=c(0,0,0,0,0,0,0,0,0,0,0,1,0,0,0,0,0))
)
```

The summary posterior statistics were found directly from within WinBUGS:

node	mean	sd	MC error	2.5%	median	97.5%
lambda	8.634	0.6687	0.003232	8	9	10
P[1]	0.02034	0.01982	8.556E-5	5.413E-4	0.01445	0.07282
P[2]	0.02038	0.01995	78.219E-5	5.374E-4	0.01423	0.07391
P[3]	0.02046	0.02004	8.752E-5	5.245E-4	0.01434	0.07456
P[4]	0.04075	0.028	1.179E-4	0.004988	0.03454	0.1113
P[5]	0.04103	0.028	1.237E-4	0.005249	0.03507	0.1107
P[6]	0.06142	0.03419	1.575E-4	0.01316	0.05536	0.143
P[7]	0.06171	0.03406	1.586E-4	0.01313	0.05573	0.1427
P[8]	0.09012	0.04161	1.981E-4	0.02637	0.08438	0.1859
P[9]	0.09134	0.04163	1.956E-4	0.02676	0.08578	0.1866
P[10]	0.1035	0.04329	1.85E-4	0.03516	0.09774	0.2022
P[11]	0.1226	0.04663	2.278E-4	0.04698	0.1175	0.2276
P[12]	0.1019	0.04284	1.811E-4	0.03496	0.09649	0.1994
P[13]	0.08173	0.03874	1.71E-4	0.02326	0.07608	0.1718
P[14]	0.06118	0.03396	1.585E-4	0.01288	0.05512	0.1426
P[15]	0.04085	0.02795	1.336E-4	0.005309	0.03477	0.1106
P[16]	0.02032	0.01996	9.549E-5	5.317E-4	0.01419	0.07444
P[17]	0.02044	0.01986	8.487E-5	5.475E-4	0.01445	0.07347

The main parameter of interest is the arrival rate, λ. The posterior mean of λ is 8.634. The median is 9 passengers every four minutes. Either number could be justified as an estimate of the passenger arrival rate per four minute interval. WinBUGS provides an easy way to save the simulated parameter values, in order, to a text file. This then enables the data to be easily imported into another environment, such as **R** or MATLAB, for data analysis and graphing. In this example, MATLAB was used to provide the histograms for λ and p_{10}. The histograms in Figure 18.3 illustrate that λ is pretty much confined to the five integers 7, 8, 9, 10, and 11, with the mode 9.

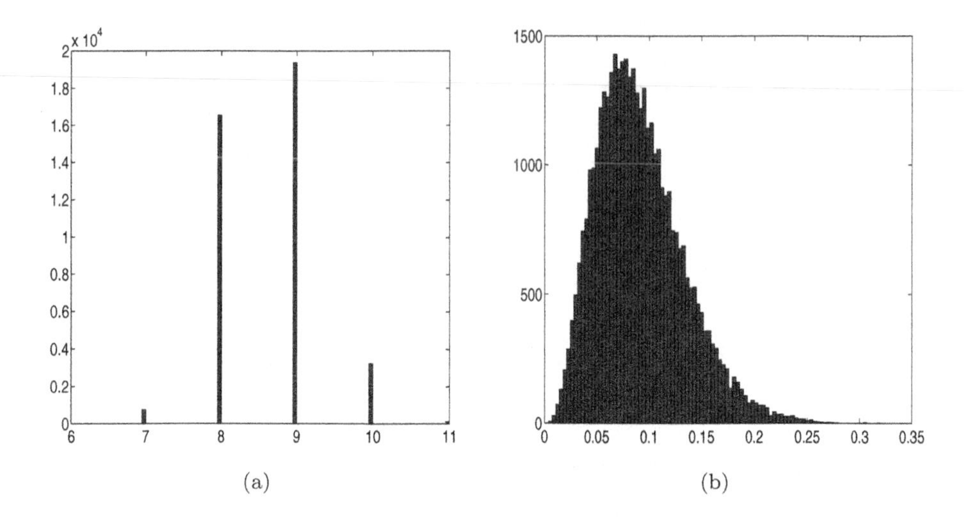

Fig. 18.3 Histograms of 40,000 samples from the posterior for λ and $P[10]$.

18.1.2 Generalizing Dirichlet Processes

Some popular NP Bayesian models employ a mixture of Dirichlet processes. The motivation for such models is their extraordinary modeling flexibility. Let $X_1, X_2, \ldots, X_n$ be the observations modeled as

$$\begin{aligned} X_i|\theta_i &\sim \mathcal{B}in(n_i, \theta_i), \\ \theta_i|F &\sim F, \ i = 1, \ldots, n \\ F &\sim \mathcal{D}ir(\alpha). \end{aligned} \tag{18.3}$$

If α assigns mass to every open interval on $[0, 1]$ then the support of the distributions on F is the class of *all* distributions on $[0, 1]$. This model allows for pooling information across the samples. For example, observation X_i will have an effect on the posterior distribution of θ_j, $j \neq i$, via the hierarchical stage of the model involving the common Dirichlet process.

The model (18.3) is used extensively in the applications of Bayesian non-parametrics. For example, Berry and Christensen (1979) use the model for the quality of welding material submitted to a naval shipyard, implying an interest in posterior distributions of θ_i. Liu (1996) uses the model for results of flicks of thumbtacks and focusses on distribution of $\theta_{n+1}|X_1, \ldots, X_n$. McEarchern, Clyde, and Liu (1999) discuss estimation of the posterior predictive $X_{n+1}|X_1, \ldots, X_n$, and some other posterior functionals.

The DP is the most popular nonparametric Bayes model in the literature (for a recent review, see MacEachern and Mueller, 2000). However, limiting the prior to discrete distributions may not be appropriate for some applications. A simple extension to remove the constraint of discrete measures is to use a convoluted DP:

$$\begin{aligned} X|F &\sim F \\ F(x) &= \int f(x|\theta)dG(\theta), \\ G &\sim DP(\alpha G_0). \end{aligned}$$

This model is called *Dirichlet Process Mixture* (DPM), because the mixing is done by the DP. Posterior inference for DMP models is based on MCMC posterior simulation. Most approaches proceed by introducing latent variables θ as $X_i|\theta_i \sim f(x|\theta_i)$,$\theta_i|G \sim G$ and $G \sim DP(\alpha G_0)$. Efficient MCMC simulation for general MDP models is discussed, among others, in Escobar (1994), Escobar and West (1995), Bush and MacEachern (1996) and MacEachern and Mueller (1998). Using a Gaussian kernel, $f(x|\mu, \Sigma) \propto \exp\{(x-\mu)')\Sigma(x-\mu)/2\}$, and mixing with respect to $\theta = (\mu, \Sigma)$, a density estimate resembling traditional kernel density estimation is obtained. Such approaches have been studied in Lo (1984) and Escobar and West (1995).

A related generalization of Dirichlet Processes is the *Mixture of Dirichlet Processes* (MDP). The MDP is defined as a DP with a center CDF which depends on random θ,

$$\begin{aligned} F &\sim DP(\alpha G_\theta) \\ \theta &\sim \pi(\theta). \end{aligned}$$

Antoniak (1974) explored theoretical properties of MDP's and obtained posterior distribution for θ.

18.2 BAYESIAN CONTINGENCY TABLES AND CATEGORICAL MODELS

In contingency tables, the cell counts N_{ij} can be modeled as realizations from a count distribution, such as Multinomial $\mathcal{M}n(n, p_{ij})$ or Poisson $\mathcal{P}(\lambda_{ij})$. The hypothesis of interest is independence of row and column factors, H_0 : $p_{ij} = a_i b_j$, where a_i and b_j are marginal probabilities of levels of two factors satisfying $\Sigma_i a_i = \Sigma_j b_j = 1$.

The expected cell count for the multinomial distribution is $\mathbb{E}N_{ij} = np_{ij}$. Under H_0, this equals $na_i b_j$, so by taking the logarithm on both sides, one obtains

$$\begin{aligned} \log \mathbb{E}N_{ij} &= \log n + \log a_i + \log b_j \\ &= \text{const} + \alpha_i + \beta_j, \end{aligned}$$

for some parameters α_i and β_j. This shows that testing the model for additivity in parameters α and β is equivalent to testing the original independence hypothesis H_0. For the Poisson counts, the situation is analogous; one uses $\log \lambda_{ij} = \text{const} + \alpha_i + \beta_j$.

Example 18.4 Activities of Dolphin Groups Revisited. We revisit the Dolphin's Activity example from p. 162. Groups of dolphins were observed off the coast of Iceland and the table providing group counts is given below. The counts are listed according to the time of the day and the main activity of the dolphin group. The hypothesis of interest is independence of the type of activity from the time of the day.

	Travelling	Feeding	Socializing
Morning	6	28	38
Noon	6	4	5
Afternoon	14	0	9
Evening	13	56	10

The WinBUGS program implementing the additive model is quite simple. We assume the cell counts are assumed distributed Poisson and the logarithm of intensity (expectation) is represented in an additive manner. The model parts (intercept, α_i, and β_j) are assigned normal priors with mean zero and precision parameter `xi`. The precision parameter is given a gamma prior with mean 1 and variance 10. In addition to the model parameters, the WinBUGS program will calculate the deviance and chi-square statistics that measure goodness of fit for this model.

```
model {
     for (i in 1:nrow) {
          for (j in 1:ncol) {
          groups[i,j]  ~ dpois(lambda[i,j])
          log(lambda[i,j]) <-  c + alpha[i] + beta[j]
          } }
#
         c ~ dnorm(0, xi)
         for (i in 1:nrow) { alpha[i] ~ dnorm(0, xi)  }
         for (j in 1:ncol) {  beta[j] ~ dnorm(0, xi)   }
         xi ~ dgamma(0.01, 0.01)
#
  for (i in 1:nrow) {
          for (j in 1:ncol) {
             devG[i,j] <- groups[i,j] * log((groups[i,j]+0.5)/
             (lambda[i,j]+0.5))-(groups[i,j]-lambda[i,j]);
             devX[i,j] <- (groups[i,j]-lambda[i,j])
                    *(groups[i,j]-lambda[i,j])/lambda[i,j];} }
         G2 <- 2 * sum( devG[,] );
         X2 <- sum( devX[,] )}
```

The data are imported as

```
list(nrow=4, ncol=3,
            groups = structure(
                  .Data = c(  6, 28,  38, 6, 4, 5,
                  14, 0, 9, 13, 56, 10), .Dim=c(4,3)) ) )
```

and initial parameters are

```
list(xi=0.1, c = 0, alpha=c(0,0,0,0), beta=c(0,0,0) )
```

The following output gives Bayes estimators of the parameters, and measures of fit. This additive model conforms poorly to the observations; under the hypothesis of independence, the test statistic is χ^2 with $3 \times 4 - 6 = 6$ degrees of freedom, and the observed value $X^2 = 77.73$ has a p-value (`1-chi2cdf(77.73, 6)`) that is essentially zero.

node	mean	sd	MC error	2.5%	median	97.5%
c	1.514	0.7393	0.03152	−0.02262	1.536	2.961
alpha[1]	1.028	0.5658	0.0215	−0.07829	1.025	2.185
alpha[2]	−0.5182	0.5894	0.02072	−1.695	−0.5166	0.6532
alpha[3]	−0.1105	0.5793	0.02108	−1.259	−0.1113	1.068
alpha[4]	1.121	0.5656	0.02158	0.02059	1.117	2.277
beta[1]	0.1314	0.6478	0.02492	−1.134	0.1101	1.507
beta[2]	0.9439	0.6427	0.02516	−0.3026	0.9201	2.308
beta[3]	0.5924	0.6451	0.02512	−0.6616	0.5687	1.951
c	1.514	0.7393	0.03152	−0.02262	1.536	2.961
G2	77.8	3.452	0.01548	73.07	77.16	86.2
X2	77.73	9.871	0.03737	64.32	75.85	102.2

Example 18.5 Cæsarean Section Infections Revisited. We now consider the Bayesian solution to the Cæsarean section birth problem from p. 236. The model for probability of infection in a birth by Cæsarean section was given in terms of the *logit* link as,

$$\log \frac{P(\texttt{infection})}{P(\texttt{no infection})} = \beta_0 + \beta_1 \texttt{ noplan} + \beta_2 \texttt{ riskfac} + \beta_3 \texttt{ antibio}.$$

The WinBUGS program provided below implements the model in which the number of infections is $\mathcal{B}in(n,p)$ with p connected to covariates `noplan` `riskfac` and `antibio` via the logit link. Priors on coefficients in the linear predictor are set to be a vague Gaussian (small precision parameter).

```
model{
 for(i in 1:N){
   inf[i] ~ dbin(p[i],total[i])
   logit(p[i]) <- beta0 + beta1*noplan[i] +
```

```
            beta2*riskfac[i] + beta3*antibio[i]
   }
 beta0 ~dnorm(0, 0.00001)
 beta1 ~dnorm(0, 0.00001)
 beta2 ~dnorm(0, 0.00001)
 beta3 ~dnorm(0, 0.00001)
}

#DATA
list( inf=c(1, 11, 0, 0, 28, 23, 8, 0),
      total = c(18, 98, 2, 0, 58, 26, 40, 9),
     noplan = c(0,1,0,1,0,1,0,1),
     riskfac = c(1,1, 0, 0, 1,1, 0, 0),
     antibio =c(1,1,1,1,0,0,0,0), N=8)

#INITS
list(beta0 =0, beta1=0,
       beta2=0, beta3=0)
```

The Bayes estimates of the parameters $\beta_0 - \beta_3$ are given in the WinBUGS output below.

node	mean	sd	MC error	2.5%	median	97.5%
beta0	−1.962	0.4283	0.004451	−2.861	−1.941	−1.183
beta1	1.115	0.4323	0.003004	0.29	1.106	1.988
beta2	2.101	0.4691	0.004843	1.225	2.084	3.066
beta3	−3.339	0.4896	0.003262	−4.338	−3.324	−2.418

Note that Bayes estimators are close to the estimators obtained in the frequentist solution in Chapter 12: $(\beta_0, \beta_1, \beta_2, \beta_3) = (-1.89,\ 1.07,\ 2.03,\ -3.25)$ and that in addition to the posterior means, posterior medians and 95% credible sets for the parameters are provided. WinBUGS can provide various posterior location and precision measures. From the table, the 95% credible set for β_0 is $[-2.861, -1.183]$.

18.3 BAYESIAN INFERENCE IN INFINITELY DIMENSIONAL NONPARAMETRIC PROBLEMS

Earlier in the book we argued that many statistical procedures classified as nonparametric are, in fact, infinitely parametric. Examples include wavelet regression, orthogonal series density estimators and nonparametric MLEs (Chapter 10). In order to estimate such functions, we rely on shrinkage, tapering or truncation of coefficient estimators in a potentially infinite expansion class. (Chencov's orthogonal series density estimators, Fourier and wavelet shrinkage, and related.) The benefits of shrinkage estimation in statistics were first explored in the mid-1950's by C. Stein. In the 1970's and 1980's, many statisticians were active in research on statistical properties of classical and Bayesian shrinkage estimators.

Bayesian methods have become popular in shrinkage estimation because Bayes rules are, in general, "shrinkers". Most Bayes rules shrink large coefficients slightly, whereas small ones are more heavily shrunk. Furthermore, interest for Bayesian methods is boosted by the possibility of incorporating prior information about the function to model wavelet coefficients in a realistic way.

Wavelet transformations W are applied to noisy measurements $y_i = f_i + \epsilon_i, \ \ i = 1, \dots, n$, or, in vector notation, $\boldsymbol{y} = \boldsymbol{f} + \boldsymbol{\epsilon}$. The linearity of W implies that the transformed vector $\boldsymbol{d} = W(\boldsymbol{y})$ is the sum of the transformed signal $\boldsymbol{\theta} = W(\boldsymbol{f})$ and the transformed noise $\boldsymbol{\eta} = W(\boldsymbol{\epsilon})$. Furthermore, the orthogonality of W implies that ϵ_i, i.i.d. normal $\mathcal{N}(0, \sigma^2)$ components of the noise vector $\boldsymbol{\epsilon}$, are transformed into components of $\boldsymbol{\eta}$ with the same distribution.

Bayesian methods are applied in the wavelet domain, that is, after the wavelet transformation has been applied and the model $d_i \sim \mathcal{N}(\theta_i, \sigma^2), i = 1, \dots, n$, has been obtained. We can model coefficient-by-coefficient because wavelets decorrelate and d_i's are approximately independent.

Therefore we concentrate just on a single typical wavelet coefficient and one model: $d = \theta + \epsilon$. Bayesian methods are applied to estimate the location parameter θ. As θ's correspond to the function to be estimated, back-transforming an estimated vector $\boldsymbol{\theta}$ will give the estimator of the function.

18.3.1 BAMS Wavelet Shrinkage

BAMS (stands for *Bayesian Adaptive Multiscale Shrinkage*) is a simple efficient shrinkage in which the shrinkage rule is a Bayes rule for properly selected prior and hyperparameters of the prior. Starting with $[d|\theta, \sigma^2] \sim \mathcal{N}(\theta, \sigma^2)$ and the prior $\sigma^2 \sim \mathcal{E}(\mu), \ \ \mu > 0$, with density $f(\sigma^2|\mu) = \mu e^{-\mu\sigma^2}$, we obtain the marginal likelihood

$$d|\theta \sim \mathcal{DE}\left(\theta, \sqrt{2\mu}\right), \qquad \text{with density } f(d|\theta) = \frac{1}{2}\sqrt{2\mu} e^{-\sqrt{2\mu}|d-\theta|}.$$

If the prior on θ is a mixture of a point mass δ_0 at zero, and a double-exponential distribution,

$$\theta|\epsilon \sim \epsilon\delta_0 + (1-\epsilon)\mathcal{DE}(0, \tau), \tag{18.4}$$

then the posterior mean of θ (from Bayes rule) is:

$$\delta^*(d) = \frac{(1-\epsilon)\ m(d)\ \delta(d)}{(1-\epsilon)\ m(d) + \epsilon\ f(d|0)}, \tag{18.5}$$

where

$$m(d) = \frac{\frac{1}{\tau} e^{-\tau|d|} - \frac{1}{\sqrt{2\mu}} e^{-\sqrt{2\mu}|d|}}{2/\tau^2 - 1/\mu}, \tag{18.6}$$

and

$$\delta(d) = \frac{(1/\tau^2 - 1/(2\mu)) d e^{-\tau|d|}/\tau + (e^{-\sqrt{2\mu}|d|} - e^{-\tau|d|})/(\mu\tau^2)}{(1/\tau^2 - 1/(2\mu))(e^{-\tau|d|}/\tau - e^{-\sqrt{2\mu}|d|}/\sqrt{2\mu})}. \tag{18.7}$$

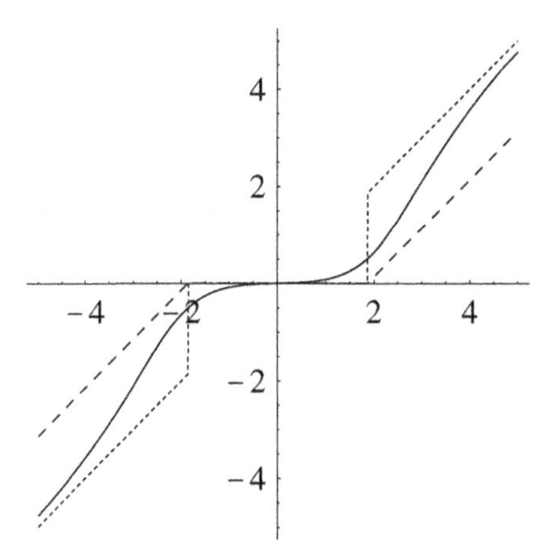

Fig. 18.4 Bayes rule (18.7) and comparable hard and soft thresholding rules.

As evident from Figure 18.4, the Bayes rule (18.5) falls between comparable hard- and soft-thresholding rules. To apply the shrinkage in (18.5) on a specific problem, the hyperparameters μ, τ, and ϵ have to be specified. A default choice for the parameters is suggested in Vidakovic and Ruggeri (2001); see also Antoniadis, Bigot, and Sapatinas (2001) for a comparative study of many shrinkage rules, including BAMS. Their analysis is accompanied by MATLAB routines and can be found at

`http://www-lmc.imag.fr/SMS/software/GaussianWaveDen/.`

Figure 18.5(a) shows a noisy `doppler` function of size $n = 1024$, where the signal-to-noise ratio (defined as a ratio of variances of signal and noise) is 7. Panel (b) in the same figure shows the smoothed function by BAMS. The graphs are based on default values for the hyperparameters.

Example 18.6 Bayesian Wavelet Shrinkage in WinBUGS. Because of the decorrelating property of wavelet transforms, the wavelet coefficients are modeled independently. A selected coefficient d is assumed to be normal

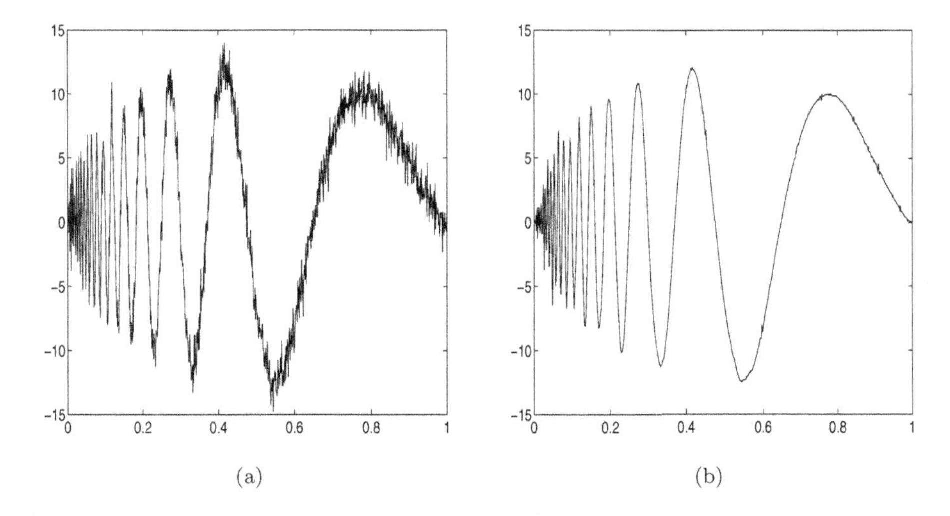

Fig. 18.5 (a) A noisy **doppler** signal [SNR=7, n=1024, noise variance $\sigma^2 = 1$]. (b) Signal reconstructed by BAMS.

$d \sim \mathcal{N}(\theta, \xi)$ where θ is the coefficient corresponding to the underlying signal in data and ξ is the precision, reciprocal of variance. The signal component θ is modeled as a mixture of two double-exponential distributions with zero mean and different precisions, because WinBUGS will not allow a point mass prior. The precision of one part of the mixture is large (so the variance is small) indicating coefficients that could be ignored as negligible. The corresponding precision of the second part is small (so the variance is large) indicating important coefficients of possibly large magnitude. The densities in the prior mixture are taken in proportion $p : (1-p)$ where p is Bernoulli. For all other parameters and hyperparameters, appropriate prior distributions are adopted.

We are interested in the posterior means for θ. Here is the WinBUGS implementation of the described model acting on some imaginary wavelet coefficients ranging from -50 to 50, as an illustration. Figure 18.6 shows the Bayes rule. Note a desirable shape close to that of the thresholding rules.

```
model{
  for (j in 1:N){
    DD[j] ~ dnorm(theta[j], tau);
    theta[j] <- p[j] * mu1[j] + (1-p[j]) * mu2[j];
    mu1[j] ~ ddexp(0, tau1);
    mu2[j] ~ ddexp(0, tau2);
    p[j] ~ dbern(r);
  }
  r ~ dbeta(1,10);
  tau ~ dgamma(0.5, 0.5);
```

```
   tau1 ~ dgamma(0.005, 0.5);
   tau2 ~dgamma(0.5, 0.005);
}

#data
list( DD=c(-50, -10, -5,-4,-3,-2,-1,-0.5, -0.1, 0,
        0.1, 0.5,  1, 2,3,4,5, 10, 50),  N=19);

#inits
list(tau=1, tau1=0.1, tau2=10);
```

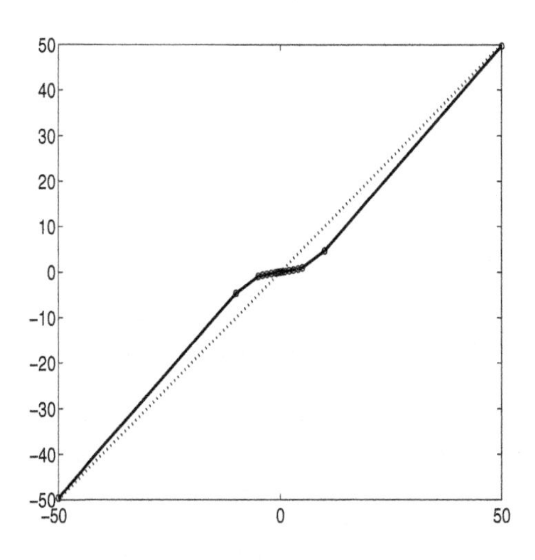

Fig. 18.6 Approximation of Bayes shrinkage rule calculated by WinBUGS.

18.4 EXERCISES

18.1. Show that in the DP Definition 18.2, $\mathbb{E}(\Sigma_{i=1}^{T}\omega_i) = 1 - [\alpha/(1-\alpha)]^T$.

18.2. Let $\mu = \int_{-\infty}^{\infty} y dG(y)$ and let G be a random CDF with Dirichlet process prior $DP(\alpha G_0)$. Let $\mathbf{y}$ be a sample of size n from G. Using (18.2), show that

$$\mathbb{E}(\mu|\mathbf{y}) = \frac{\alpha}{\alpha+n}\mathbb{E}\mu + \frac{n}{\alpha+n}\bar{y}.$$

In other words, show that the expected posterior mean is a weighted average of the expected prior mean and the sample mean.

18.3. Redo Exercise 9.13, where the results for 148 survey responses are broken down by program choice and by race. Test the fit of the properly set additive Bayesian model. Use WinBUGS for model analysis.

18.4. Show that $m(d)$ and $\delta(d)$ from (18.6) and (18.7) are marginal distributions and the Bayes rule for the model is

$$d|\theta \sim \mathcal{DE}\left(\theta, \sqrt{2\mu}\right), \quad \theta \sim \mathcal{DE}(0, \tau),$$

where μ and τ are the hyperparameters.

18.5. This is an open-ended question. Select a data set with noise present in it (a noisy signal), transform the data to the wavelet domain, apply shrinkage on wavelet coefficients by the Bayes procedure described below, and back-transform the shrunk coefficients to the domain of original data.

(i) Prove that for $[d|\theta] \sim \mathcal{N}(\theta, 1)$, $[\theta|\tau^2] \sim \mathcal{N}(0, \tau^2)$, and $\tau^2 \sim (\tau^2)^{-3/4}$, the posterior is unimodal at 0 if $0 < d^2 < 2$ and bimodal otherwise with the second mode

$$\delta(d) = \left(1 - \frac{1 - \sqrt{1 - 2/d^2}}{2}\right) d.$$

(ii) Generalize to $[d|\theta] \sim \mathcal{N}(\theta, \sigma^2)$, σ^2 known, and apply *the larger mode shrinkage*. Is this shrinkage of the thresholding type?

(iii) Use the approximation $(1-u)^\alpha \sim (1-\alpha u)$ for u small to argue that the largest mode shrinkage is close to a James-Stein-type rule $\delta^*(d) = \left(1 - \frac{1}{2d^2}\right)_+ d$, where $(f)_+ = \max\{0, f\}$.

18.6. Chipman, Kolaczyk, and McCulloch (1997) propose the following model for Bayesian wavelet shrinkage (ABWS) which we give in a simplified form,

$$d|\theta \sim \mathcal{N}(\theta, \sigma^2).$$

The prior on θ is defined as a mixture of two normals with a hyperprior on the mixing proportion,

$$\begin{aligned} \theta|\gamma &\sim \gamma\mathcal{N}(0, (c\tau)^2) + (1-\gamma)\mathcal{N}(0, \tau^2), \\ \gamma &\sim \mathcal{B}in(1, p). \end{aligned}$$

Variance σ^2 is considered known, and $c \gg 1$.

i) Show that the Bayes rule (posterior expectation) for θ has the explicit form of

$$\delta(d) = \left[P(\gamma = 1|d)\frac{(c\tau)^2}{\sigma^2 + (c\tau)^2} + P(\gamma = 0|d)\frac{\tau^2}{\sigma^2 + \tau^2}\right] d,$$

where

$$P(\gamma = 1|d) = \frac{p\pi(d|\gamma = 1)}{p\pi(d|\gamma = 1) + (1-p)\pi(d|\gamma = 0)}$$

and $\pi(d|\gamma = 1)$ and $\pi(d|\gamma = 0)$ are densities of $\mathcal{N}(0, \sigma^2 + (c\tau)^2)$ and $\mathcal{N}(0, \sigma^2 + \tau^2)$ distributions, respectively, evaluated at d.

(ii) Plot the Bayes rule from (i) for selected values of parameters and hyperparameters $(\sigma^2, \tau^2, \gamma, c)$ so that the shape of the rule is reminiscent of thresholding.

REFERENCES

Antoniadis, A., Bigot, J., and Sapatinas, T. (2001), "Wavelet Estimators in Nonparametric Regression: A Comparative Simulation Study," *Journal of Statistical Software*, 6, 1-83.

Antoniak, C. E. (1974), "Mixtures of Dirichlet Processes with Applications to Bayesian Nonparametric Problems," *Annals of Statistics*, 2, 1152-1174.

Berry, D. A., and Christensen, R. (1979), " Empirical Bayes Estimation of a Binomial Parameter Via Mixtures of Dirichlet Processes," *Annals of Statistics,* 7, 558-568.

Bush, C. A., and MacEachern S. N. (1996), "A Semi-Parametric Bayesian Model for Randomized Block Designs," *Biometrika,* 83, 275-286.

Chipman, H. A., Kolaczyk, E. D., and McCulloch, R. E. (1997), "Adaptive Bayesian Wavelet Shrinkage," *Journal of American Statistical Association,* 92, 1413-1421.

Escobar, M. D. (1994), "Estimating Normal Means with a Dirichlet Process Prior," *Journal of American Statistical Association,* 89, 268-277.

Escobar, M. D., and West, M. (1995), "Bayesian Density Estimation and Inference Using Mixtures, *Journal of American Statistical Association,* 90, 577-588.

Fabius, J. (1964), "Asymptotic Behavior of Bayes' Estimates," *Annals of Mathematical Statistics,* 35, 846-856.

Ferguson, T. S. (1973), "A Bayesian Analysis of Some Nonparametric Problems," *Annals of Statistics,* 1, 209-230.

_______ (1974), "Prior Distributions on Spaces of Probability Measures," *Annals of Statistics,* 2, 615-629.

Freedman, D. A. (1963), "On the Asymptotic Behavior of Bayes' Estimates in the Discrete Case," *Annals of Mathematical Statistics,* 34, 1386-1403.

Liu, J. S. (1996), " Nonparametric Hierarchical Bayes via Sequential Imputations," *Annals of Statistics,* 24, 911-930.

Lo, A. Y. (1984), "On a Class of Bayesian Nonparametric Estimates, I. Density Estimates," *Annals of Statistics,* 12, 351-357.

MacEachern, S. N., and Mueller, P. (1998), "Estimating Mixture of Dirichlet Process Models, *Journal of Computational and Graphical Statistics,* 7, 223-238.

_______ (2000), "Efficient MCMC Schemes for Robust Model Extensions Using Encompassing Dirichlet Process Mixture Models," in *Robust Bayesian Analysis,* Eds. F. Ruggeri and D. Rios-Insua, New York: Springer Verlag.

MacEachern, S. N., Clyde, M., and Liu, J. S. (1999), "Sequential Importance Sampling for Nonparametric Bayes Models: The Next Generation," *Canadian Journal of Statistics,* 27, 251-267.

Mueller, P., and Quintana, F. A. (2004), "Nonparametric Bayesian Data Analysis," *Statistical Science,* 19, 95-110.

Sethuraman, J., and Tiwari, R. C. (1982), "Convergence of Dirichlet Measures and the Interpretation of their Parameter," in *Statistical Decision Theory and Related Topics III,* eds. S. Gupta and J. O. Berger, New York: Springer Verlag, 2, pp. 305-315.

Sethuraman, J. (1994), "A Constructive Definition of Dirichlet Priors," *Statistica Sinica,* 4, 639-650.

Vidakovic, B., and Ruggeri, F. (2001), "BAMS Method: Theory and Simulations." *Sankhyā,* Ser. B, 63, 234-249.

Appendix A: MATLAB

The combination of some data and an aching desire for an answer does not ensure that a reasonable answer can be extracted from a given body of data.

J. W. Tukey (1915–2000)

A.1 USING MATLAB

MATLAB is a interactive environment that allows the user to perform computational tasks and create graphical output. The user types in expressions and commands in a Command Window where numerical results of the commands are displayed with the user input. Graphical output will be produced in a new (graphics) window that can usually be printed or stored.

When MATLAB is launched, several windows are available to the user as you can see in Fig. A.7. Their uses are listed below:

- **Command Window:** Typing commands and expressions – this is the main interactive window in the user interface
- **Launch Pad Window:** Allows user to run demos
- **Workspace Window:** List of variables entered or created during session

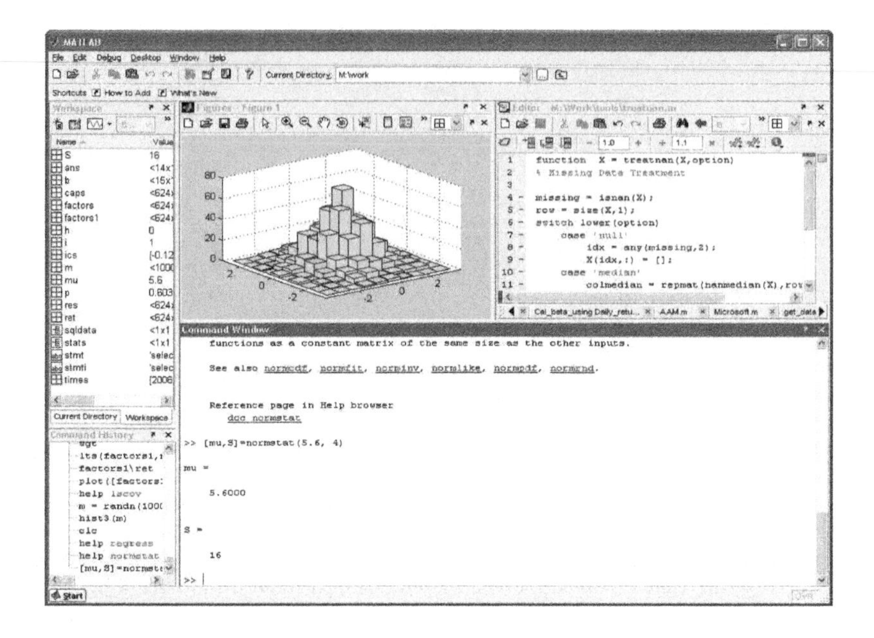

Fig. A.7 Interactive environment of MATLAB.

- **Command History Window:** List of recent commands used
- **Array Editor Window:** Allows user to manipulate arrays variables using spreadsheet
- **Current Directory Window:** To specify directory where MATLAB will search for or store files

MATLAB is a high-level technical computing language for algorithm development, data visualization, data analysis, and numeric computation. Some highlight features of MATLAB can be summarized as

- High-level language for technical computing, which are easy to learn
- Development environment for managing code, files, and data
- Mathematical functions for linear algebra, statistics, Fourier analysis, filtering, optimization, and numerical integration
- 2-D and 3-D graphics functions for visualizing data
- Tools for building custom graphical user interfaces
- Functions to communicate with other statistical software, such as R, WinBUGS

To get started, you can type doc in the command window. This will bring you to an HTML help window and you can search keyword or browse topics therein.

```
>> doc
```

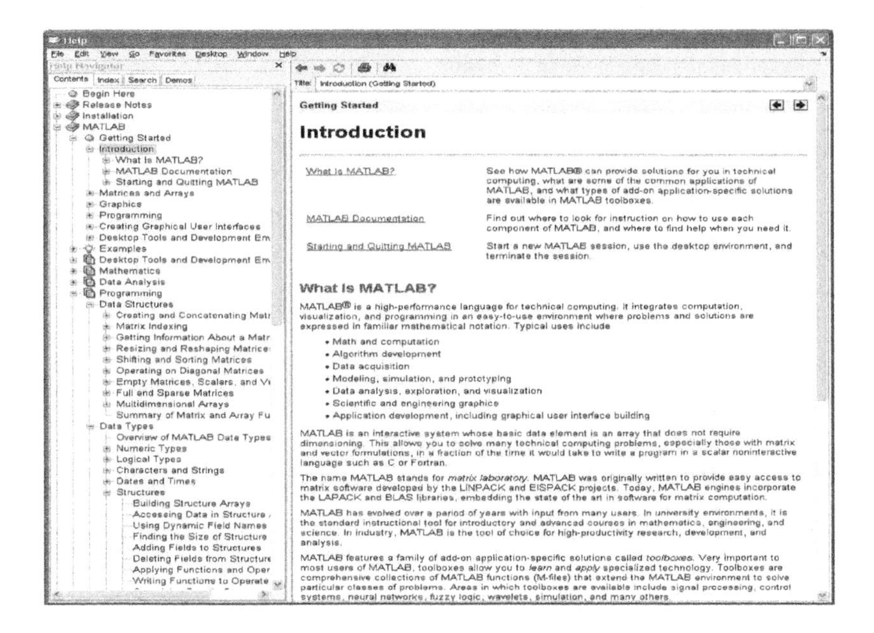

Fig. A.8 Help window of MATLAB.

If you know the function name, but do not know how to use it, it is often useful to type "help *function name*" in command window. For example, if you want to know how to use function randg or find out what randg does.

```
>> help randg

RANDG Gamma random numbers (unit scale).
Note: To generate gamma random numbers with specified shape and
scale parameters, you should call GAMRND.

R = RANDG returns a scalar random value chosen from a gamma
distribution with unit scale and shape.

R = RANDG(A) returns a matrix of random values chosen from gamma
distributions with unit scale.  R is the same size as A, and RANDG
generates each element of R using a shape parameter equal to the
corresponding element of A.
....
```

A.1.1 Toolboxes

Serving as extensions to the basic MATLAB programming environment, toolboxes are available for specific research interests. Toolboxes available include

```
Communications Toolbox
Control System Toolbox
DSP Blockset
Extended Symbolic Math Toolbox
Financial Toolbox
Frequency Domain System Identification
Fuzzy Logic Toolbox
Higher-Order Spectral Analysis Toolbox
Image Processing Toolbox
LMI Control Toolbox
Mapping Toolbox
Model Predictive Control Toolbox
Mu-Analysis and Synthesis Toolbox
NAG Foundation Blockset
Neural Network Toolbox
Optimization Toolbox
Partial Differential Equation Toolbox
QFT Control Design Toolbox
Robust Control Toolbox
Signal Processing Toolbox
Spline Toolbox
Statistics Toolbox
System Identification Toolbox
Wavelet Toolbox
```

For the most part we use functions in the base MATLAB product, but where necessary we also use functions from the Statistics Toolbox. There are numerous procedures from other toolboxes that can be helpful in nonparametric data analysis (e.g., Neural Network Toolbox, Wavelet Toolbox) but we restrict routine applications to basic and fundamental computational algorithms to avoid making the book depend on any pre-written software code.

A.2 MATRIX OPERATIONS

MATLAB was originally written to provide easy interaction with matrix software developed by the NASA[1]-sponsored LINPACK and EISPACK projects. Today, MATLAB engines incorporate the LAPACK and BLAS libraries, embedding the state of the art in software for matrix computation. Instead of

[1]National Aeronautics and Space Administration.

relying on do loops to perform repeated tasks, MATLAB is better suited to using arrays because MATLAB is an interpreted language.

MATLAB was originally written to provide easy access to matrix software developed by the LINPACK and EISPACK projects, (these projects were sponsored by NASA and much of the source code is in public domain) which together represent the state-of-the-art in software for matrix computation.

A.2.1 Entering a Matrix

There are a few basic conventions of entering a matrix in MATLAB, which include

- Separating the elements of a row with blanks or commas.
- Using a semicolon ';' to indicate the end of each row.
- Surrounding the entire list of elements with square brackets, [ ].

```
   >> A = [3 0 1; 1 2 1; 1 1 1] % columns separated by a space
                                 % rows separated by ";"
A =
     3     0     1
     1     2     1
     1     1     1
```

A.2.2 Arithmetic Operations

MATLAB uses familiar arithmetic operators and precedence rules, but unlike most programming languages, these expressions involve entire matrices. The common matrix operators used in MATLAB are listed as follows:

+	addition	−	subtraction
*	multiplication	^	power
'	transpose	.'	transpose
\	left division	/	right division
.*	element-wise multiplication	.^	element-wise power
./	element-wise right division		

```
   >> X=[10 10 20]';       % semicolon suppresses output of X
   >> A*X                  % A is 3x3, X is 3x1 and X' is 1x
                           % so A*X is 3x1
 ans =
    50
    50
    40
```

```
>> y=A\X                % y is the solution of Ay=X

y =
  -10.0000
  -10.0000
   40.0000

>> A.*A                 % ".*" multiplies corresponding elements of
                        % matching matrices; this is equivalent to A.^2
ans =
     9     0     1
     1     4     1
     1     1     1
```

A.2.3 Logical Operations

The relational operators in MATLAB are

$<$	less than	$>$	greater than
$<=$	less-than-or-equal	$==$	equal
$>=$	greater-than-or-equal	$\sim=$	not equal
&	(logical) and	|	(logical) or
$\sim$	(logical) not		

When relational operators are applied to scalars, 0 represents false and 1 represents true.

A.2.4 Matrix Functions

These extra matrix functions are helpful in creating and manipulating arrays:

`eye`	identity matrix	`ones`	matrix of ones
`zeros`	matrix of zeros	`diag`	diagonal matrix
`rand`	matrix of random U(0,1)	`inv`	matrix inverse
`det`	matrix determinant	`rank`	rank of matrix
`find`	indices of nonzero entries	`norm`	normalized matrix

A.3 CREATING FUNCTIONS IN MATLAB

Along with the extensive collection of existing MATLAB functions, you can create your own problem-specific function using input variables and generating array or graphical output. Once you look at a simple example, you can easily see how a function is constructed. For example, here is a way to compute the PDF of a triangular distribution, centered at zero with the support $[-c, c]$:

```
function y = tripdf(x,c)
 y1 = max(0,c-abs(x)) / c^2;
 y = y1
```

The function starts with the `function y = function_name(input)` where y is just a dummy variable assigned as function output at the end of the function. Local variables (such as y1) can be defined and combined with input variables (x,c) and the output can be scalar or matrix form. Once the function is named, it will override any previous function with the same name (so try not to call your function "sort", "inv" or any other known MATLAB function you might want to use later).

The function can be typed and saved as an m-file (i.e., `tripdf.m`) because that is how MATLAB recognizes an external file with executable code. Alternatively, you can type the entire function (line by line) directly into the program, but it won't be automatically saved after you finish. Then you can "call" the new function as

```
 >> v = tripdf(0:4,3)

v = [0.3333 0.2222 0.1111 0 0]

 >> tripdf(-1,2) <= 0.5      % =1 if statement is true
 ans =
      1
```

It also possible to define a function as a variable. For example, if you want to define a truncated (and unnormalized) normal PDF, use the following command

```
 >> tnormpdf = @(x, mu, sig, left, right) ...
                normpdf(x,mu,sig).*(x>left & x <right);
 >> tnormpdf(-3:3,0,1,-2,2)

 ans =
          0         0    0.2420    0.3989    0.2420         0         0
```

The `tnormpdf` function does not integrate to 1. To normalize it, one can divide the result by `(normcdf(right,mu,sigma) - normcdf(left,mu,sigma))`.

A.4 IMPORTING AND EXPORTING DATA

As a first step of data analysis, we may need to import data from external sources. The most common types of files used in the MATLAB statistical computing are MATLAB data files, Text files, and Spreadsheet files. The MATLAB data file has the extension name *.mat. Here is an example of importing such data to MATLAB workspace.

A.4.1 MAT Files

You can use the command whos to look what variables are in the data file.

```
>> whos -file dataexample

  Name         Size                        Bytes  Class

  Sigma        2x2                            32  double array
  ans          1x1                             8  double array
  mu           1x2                            16  double array
  xx         500x2                          8000  double array

Grand total is 1007 elements using 8056 bytes
```

Then you can use the command load to load all variables in this data file.

```
>> clear    % clear variables in the workspace
>> load dataexample
>> whos     % check what variables are in the workspace

  Name         Size
Bytes  Class

  Sigma        2x2                            32  double array
  ans          1x1                             8  double array
  mu           1x2                            16  double array
  xx         500x2                          8000  double array

Grand total is 1007 elements using 8056 bytes
```

In some cases, you may only want to load some variables in the MAT file to the workspace. Here is how you can do it.

```
>> clear
>> varlist = {'Sigma','mu'};    % Created a list of variables
>> load('dataexample.mat',varlist{:})
>> clear varlist                % remove varlist from workspace
>> whos                         % see what is in the workspace

  Name         Size                          Bytes  Class
```

```
Sigma       2x2                         32  double array
mu          1x2                         16  double array

Grand total is 6 elements using 48 bytes
```

Another way of creating variables of interest is to use an index.

```
>> clear
>> vars = whos('-file', 'dataexample.mat');
>> load('dataexample.mat',vars([1,3]).name)
```

If you do not want to use full variable names, but want to use some patterns in these names, the `load` command can be used with a '-regexp' option. The following command will load the same variable as the previous one.

```
>> load('dataexample.mat', '-regexp', '^S|^m');
```

Text files usually have the extension name `*.txt`, `*.dat`, `*.csv`, and so forth.

A.4.2 Text Files

If the data in the text file are organized as a matrix, you can still use `load` to import the data into the workspace.

```
>> load mytextdata.dat
>> mytextdata

mytextdata =
   -0.3097    0.2950   -0.1681   -1.4250
   -1.5219   -0.3927   -0.6873    0.4615
    0.8265    0.5759   -0.9907    1.0915
   -0.6130   -1.1414   -0.0498   -1.0443
    0.9597    0.0611    0.7193   -2.8428
    1.9730    0.0123   -0.2831    0.9968
```

You can also assign the loading data to be stored in a new variable.

```
>> x = load('mytextdata.dat');
```

The command `load` will not work if the text file is not organized in matrix form. For example, if you have a text file `mydata.txt`

```
>> type mydata.txt

var1        var2        var3        var4      name
-0.3097     0.2950     -0.1681     -1.4250   Olive
-1.5219    -0.3927     -0.6873      0.4615   Richard
```

```
 0.8265    0.5759   -0.9907    1.0915  Dwayne
-0.6130   -1.1414   -0.0498   -1.0443  Edwin
 0.9597    0.0611    0.7193   -2.8428  Sheryl
 1.9730    0.0123   -0.2831    0.9968  Frank
```

You should use a new function **txtread** to import variables to workspace.

```
>> [var1,var2,var3,var4,str] = ...
textread('mydata.txt','%f%f%f%f%s','headerlines',1 );
```

Alternatively, you can use **textscan** to finish the import.

```
>> fid = fopen('mydata.txt');
>> C = textscan(fid, '%f%f%f%f%s','headerLines',1);
>> fclose(fid);
>> [C{1:4}] % var1 - var4

ans =
   -0.3097    0.2950   -0.1681   -1.4250
   -1.5219   -0.3927   -0.6873    0.4615
    0.8265    0.5759   -0.9907    1.0915
   -0.6130   -1.1414   -0.0498   -1.0443
    0.9597    0.0611    0.7193   -2.8428
    1.9730    0.0123   -0.2831    0.9968

>> C{5}     % name

ans =
    'Olive'
    'Richard'
    'Dwayne'
    'Edwin'
    'Sheryl'
    'Frank'
```

Comma-separated values files are useful when exchanging data. Given the file `data.csv` that contains the comma-separated values

```
>> type data.csv

02, 04, 06, 08, 10, 12
03, 06, 09, 12, 15, 18
05, 10, 15, 20, 25, 30
07, 14, 21, 28, 35, 42
11, 22, 33, 44, 55, 66
```

You can use **csvread** to read the entire file into workspace

```
>> csvread('data.csv')
```

```
ans =
     2     4     6     8    10    12
     3     6     9    12    15    18
     5    10    15    20    25    30
     7    14    21    28    35    42
    11    22    33    44    55    66
```

A.4.3 Spreadsheet Files

Data from a spreadsheet can be imported into the workspace using the function `xlsread`.

```
>> [NUMERIC,TXT,RAW]=xlsread('data.xls');
>> NUMERIC

NUMERIC =
    1.0000    0.3000       NaN
    2.0000    0.4500       NaN
    3.0000    0.3000   12.0000
    4.0000    0.3500    5.0000
    5.0000    0.3500    5.0000
    6.0000    0.3500   10.0000
    7.0000    0.3500   13.0000
    8.0000    0.3500    5.0000
    9.0000    0.3500   23.0000


>> TXT

TXT =
    'Date '         'var1 '    'var2'    'var3'    'name'
    '1/1/2001'           ''        ''        ''    'Frank'
    '1/2/2001'           ''        ''        ''         ''
    '1/3/2001'           ''        ''        ''    'Sheryl'
    '1/4/2001'           ''        ''        ''         ''
    '1/5/2001'           ''        ''        ''    'Richard'
    '1/6/2001'           ''        ''        ''    'Olive'
    '1/7/2001'           ''        ''        ''    'Dwayne'
    '1/8/2001'           ''        ''        ''    'Edwin'
    '1/9/2001'           ''        ''        ''    'Stan'

>> RAW

RAW =
    'Date '         'var1 '    'var2'        'var3'      'name'
    '1/1/2001'      [    1]    [0.3000]      [ NaN]      'Frank'
    '1/2/2001'      [    2]    [0.4500]      [ NaN]      [  NaN]
```

```
'1/3/2001'    [    3]    [0.3000]    [  12]    'Sheryl'
'1/4/2001'    [    4]    [0.3500]    [   5]    [  NaN]
'1/5/2001'    [    5]    [0.3500]    [   5]    'Richard'
'1/6/2001'    [    6]    [0.3500]    [  10]    'Olive'
'1/7/2001'    [    7]    [0.3500]    [  13]    'Dwayne'
'1/8/2001'    [    8]    [0.3500]    [   5]    'Edwin'
'1/9/2001'    [    9]    [0.3500]    [  23]    'Stan'
```

It is also possible to specify the sheet name of xls file as the source of the data.

```
>> NUMERIC = xlsread('data.xls','rnd'); % read data from
                                        % a sheet named as rnd
```

From an xls file, you can get data from a specified region in a named sheet:

```
>> NUMERIC = xlsread('data.xls','data','b2:c10');
```

The following command also allows you do interactive region selection:

```
>> NUMERIC = xlsread('data.xls',-1);
```

The simplest way to save the variables from a workspace to a permanent file in the format of a MAT file is to use the command `save`. If you have a single matrix to save, `save filename varname -ascii` will save export the result to text file. You can also save numeric array or cell array in an Excel workbook using `xlswrite`.

A.5 DATA VISUALIZATION

A.5.1 Scatter Plot

A scatterplot is a useful summary of a set of bivariate data (two variables), usually drawn before working out a linear correlation coefficient or fitting a regression line. It gives a good visual picture of the relationship between the two variables, and aids the interpretation of the correlation coefficient or regression model.

In MATLAB, a simple way of make a plot matrix is to use the command `plot`. Fig. A.9 gives the result of the following MATLAB commands:

```
>> x = rand(1000,1);
>> y = .5*x + 5*x.^2 + .3*randn(1000,1);
>> plot(x,y,'.')
```

However, this is is not enough if you are dealing with more than two variables. In this case, the function `plotmatrix` should used in stead (Fig.A.10).

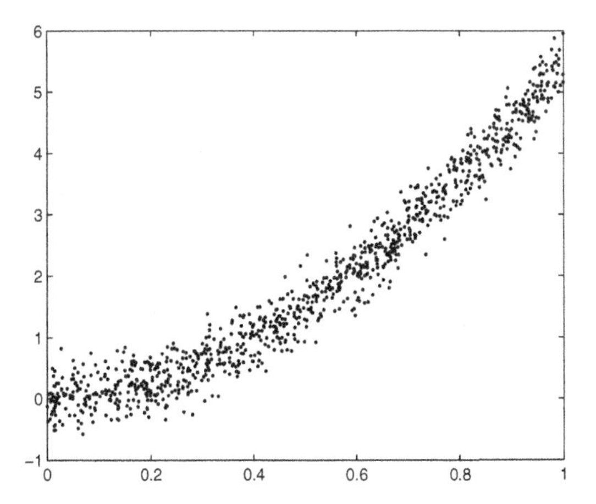

Fig. A.9 Scatterplot of (x, y) for `x = rand(1000,1)` and `y = .5*x + 5*x.2 + .3*randn(1000,1)`.

```
>> x = randn(50,3);
>> y = x*[-1 2 1;2 0 1;1 -2 3;]';
>> plotmatrix(y)
```

In classification problems, it is also useful to look at scatter plot matrix with grouping variable (Fig.A.11).

```
>> load carsmall;
>> X = [MPG,Acceleration,Displacement,Weight,Horsepower];
>> varNames = {'MPG' 'Acceleration' 'Displacement' ...
                                    'Weight' 'Horsepower'};
>> gplotmatrix(X,[],Cylinders,'bgrcm',[],[],'on','hist',varNames);
>> set(gcf,'color','white')
```

A.5.2 Box Plot

Box plot is an excellent tool for conveying location and variation information in data sets, particularly for detecting and illustrating location and variation changes between different groups of data. Here is an example of how MATLAB makes a boxplot (Fig. A.12).

```
>> load carsmall
>> boxplot(MPG, Origin, 'grouporder', ...
      {'France' 'Germany' 'Italy' 'Japan' 'Sweden' 'USA'})
>> set(gcf,'color','white')
```

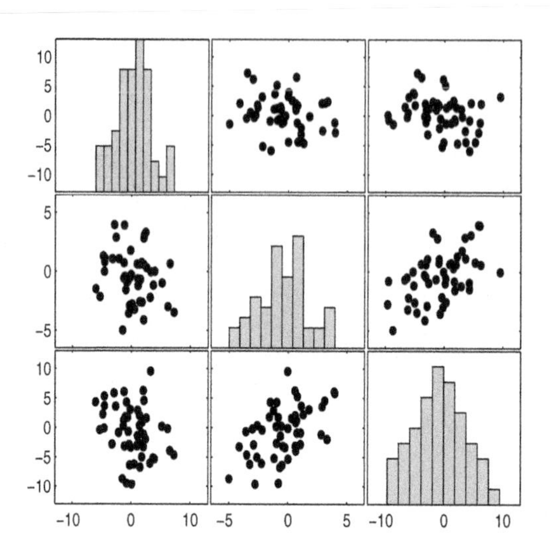

Fig. A.10 Simulated data visualized by `plotmatrix`.

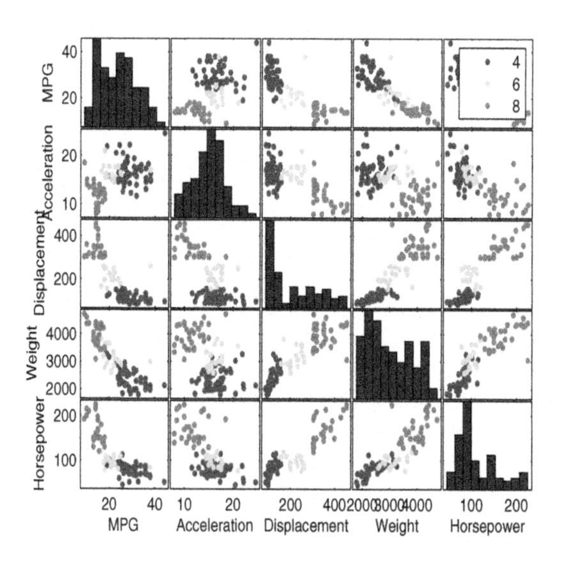

Fig. A.11 Scatterplot matrix for Car Data.

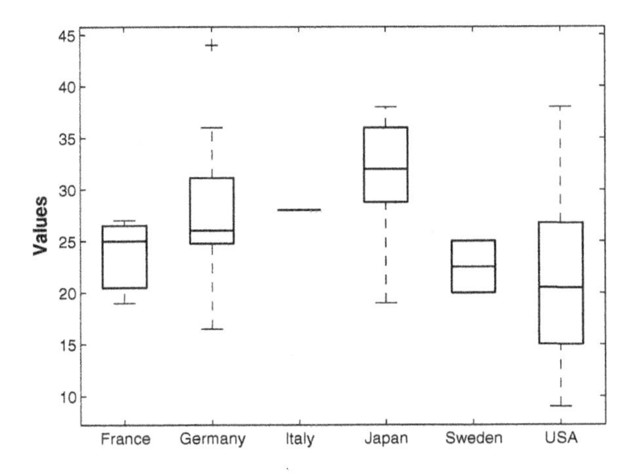

Fig. A.12 Boxplot for Car Data.

A.5.3 Histogram and Density Plot

A histogram of univariate data can be plotted using **hist** (Fig.A.13),

```
>> hist(randn(100,1))
```

while a three-dimensional histogram of bivariate data is plotted using **hist3**, (Fig.A.14),

```
>> mu = [1 -1]; Sigma = [.9 .4; .4 .3];
>> r = mvnrnd(mu, Sigma, 500);
>> hist3(r)
```

If you like a smoother density plot, you may turn to a kernel density or distribution estimate implemented in **ksdensity** (Fig.A.15). Also, in recent versions of MATLAB you have the option of not asking for outputs from the **ksdensity**, and the function plots the results directly.

```
>> [y,x] = ksdensity(randn(100,1));
>> plot(x,y)
```

A.5.4 Plotting Function List

Here is a complete list of statistical plotting functions available in MATLAB

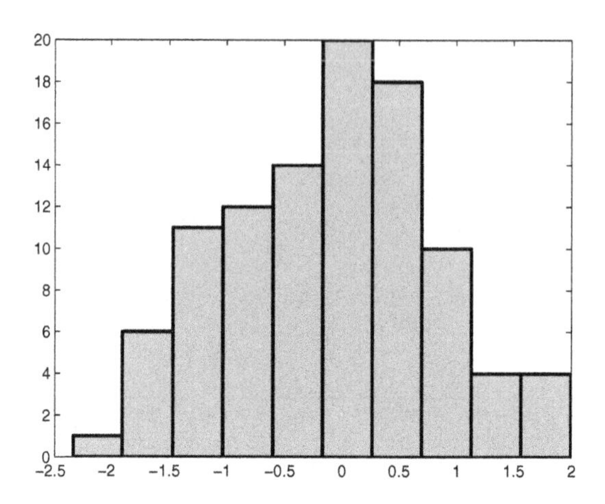

Fig. A.13 Histogram for simulated random normal data.

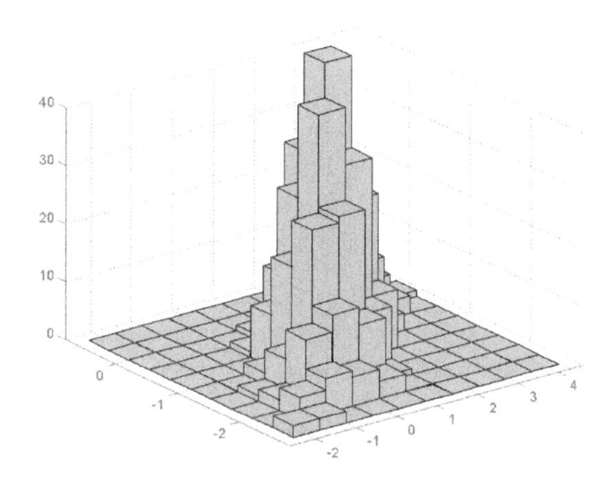

Fig. A.14 Spatial histogram for simulated two-dimensional random normal data.

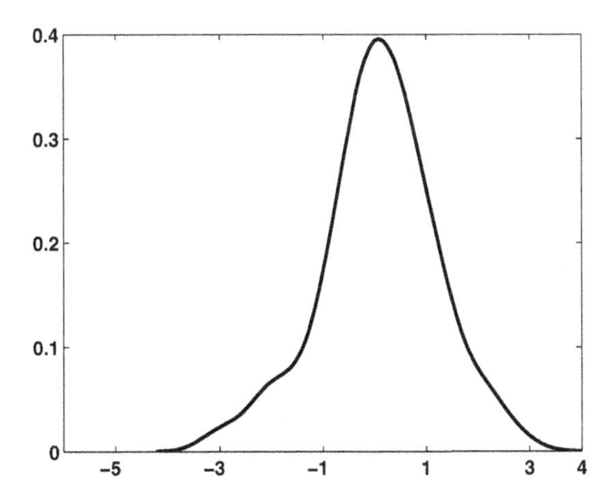

Fig. A.15 Kernel density estimator for simulated random normal data.

```
andrewsplot - Andrews plot for multivariate data.
bar         - Bar graph.
biplot      - Biplot of variable/factor coefficients and scores.
boxplot     - Boxplots of a data matrix (one per column).
cdfplot     - Plot of empirical cumulative distribution function.
contour     - Contour plot.
ecdf        - Empirical CDF (Kaplan-Meier estimate).
ecdfhist    - Histogram calculated from empirical CDF.
fplot       - Plots scalar function $f(x)$ at values of $x$.
fsurfht     - Interactive contour plot of a function.
gline       - Point, drag and click line drawing on figures.
glyphplot   - Plot stars or Chernoff faces for multivariate data.
gname       - Interactive point labeling in x-y plots.
gplotmatrix - Matrix of scatter plots grouped by a common variable.
gscatter    - Scatter plot of two variables grouped by a third.
hist        - Histogram (in MATLAB toolbox).
hist3       - Three-dimensional histogram of bivariate data.
ksdensity   - Kernel smoothing density estimation.
lsline      - Add least-square fit line to scatter plot.
normplot    - Normal probability plot.
parallelcoords - Parallel coordinates plot for multivariate data.
probplot    - Probability plot.
qqplot      - Quantile-Quantile plot.
refcurve    - Reference polynomial curve.
refline     - Reference line.
stairs      - Stair-step of y with jumps at points x.
surfht      - Interactive contour plot of a data grid.
```

```
wblplot      - Weibull probability plot.
```

A.6 STATISTICS

For your convenience, let's look at a list of functions that can be used to compute summary statistics from data.

```
corr         - Linear or rank correlation coefficient.
corrcoef     - Correlation coefficient with confidence intervals.
cov          - Covariance.
crosstab     - Cross tabulation.
geomean      - Geometric mean.
grpstats     - Summary statistics by group.
harmmean     - Harmonic mean.
iqr          - Interquartile range.
kurtosis     - Kurtosis.
mad          - Median Absolute Deviation.
mean         - Sample average (in MATLAB toolbox).
median       - 50th percentile of a sample.
moment       - Moments of a sample.
nancov       - Covariance matrix ignoring NaNs.
nanmax       - Maximum ignoring NaNs.
nanmean      - Mean ignoring NaNs.
nanmedian    - Median ignoring NaNs.
nanmin       - Minimum ignoring NaNs.
nanstd       - Standard deviation ignoring NaNs.
nansum       - Sum ignoring NaNs.
nanvar       - Variance ignoring NaNs.
partialcorr  - Linear or rank partial correlation coefficient.
prctile      - Percentiles.
quantile     - Quantiles.
range        - Range.
skewness     - Skewness.
std          - Standard deviation (in MATLAB toolbox).
tabulate     - Frequency table.
trimmean     - Trimmed mean.
var          - Variance
```

A.6.1 Distributions

Distribution	CDF	PDF	Inverse CDF	RNG
Beta	betacdf	betapdf	betainv	betarnd
Binomial	binocdf	binopdf	binoinv	binornd
Chi square	chi2cdf	chi2pdf	chi2inv	chi2rnd
Exponential	expcdf	exppdf	expinv	exprnd
Extreme value	evcdf	evpdf	evinv	evrnd
F	fcdf	fpdf	finv	frnd
Gamma	gamcdf	gampdf	gaminv	gamrnd
Geometric	geocdf	geopdf	geoinv	geornd
Hypergeometric	hygecdf	hygepdf	hygeinv	hygernd
Lognormal	logncdf	lognpdf	logninv	lognrnd
Multivariate normal	mvncdf	mvnpdf	mvninv	mvnrnd
Negative binomial	nbincdf	nbinpdf	nbininv	nbinrnd
Normal (Gaussian)	normcdf	normpdf	norminv	normrnd
Poisson	poisscdf	poisspdf	poissinv	poissrnd
Rayleigh	raylcdf	raylpdf	raylinv	raylrnd
t	tcdf	tpdf	tinv	trnd
Discrete uniform	unidcdf	unidpdf	unidinv	unidrnd
Uniform distribution	unifcdf	unifpdf	unifinv	unifrnd
Weibull	wblcdf	wblpdf	wblinv	wblrnd

A.6.2 Distribution Fitting

```
betafit     - Beta parameter estimation.
binofit     - Binomial parameter estimation.
evfit       - Extreme value parameter estimation.
expfit      - Exponential parameter estimation.
gamfit      - Gamma parameter estimation.
gevfit      - Generalized extreme value parameter estimation.
gpfit       - Generalized Pareto parameter estimation.
lognfit     - Lognormal parameter estimation.
mle         - Maximum likelihood estimation (MLE).
mlecov      - Asymptotic covariance matrix of MLE.
lognfit     - Negative binomial parameter estimation.
normfit     - Normal parameter estimation.
poissfit    - Poisson parameter estimation.
raylfit     - Rayleigh parameter estimation.
unifit      - Uniform parameter estimation.
wblfit      - Weibull parameter estimation.
```

In addition to the command line function listed above, there is also a GUI to used for distribution fitting. You can use the command `dfittool` to invoke this tool (Fig.A.16).

```
>> dfittool
```

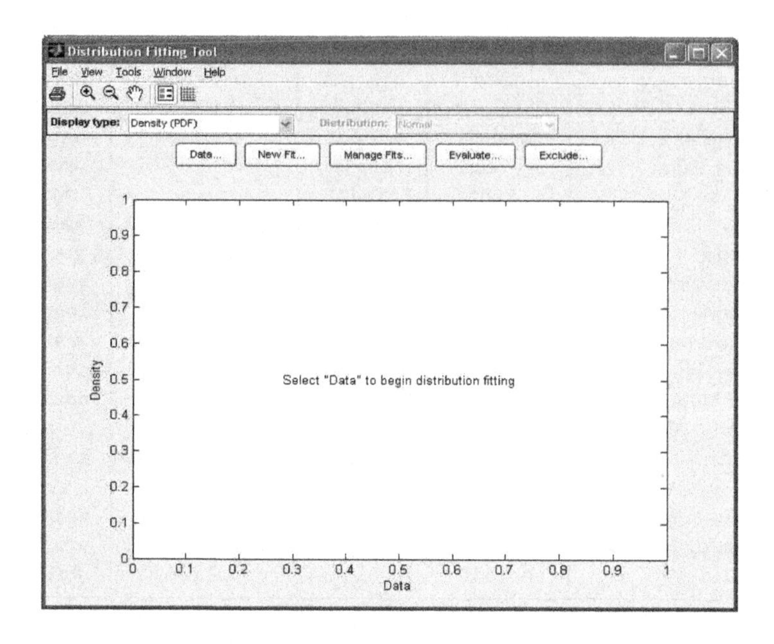

Fig. A.16 GUI for **dfittool**.

A.6.3 Nonparametric Procedures

```
kstest          - Kolmogorov-Smirnov two-sample test.
kstest2         - Kolmogorov-Smirnov one or two-sample test
mtest           - Cramer Von Mises test for normality
dagosptest      - D'Agostino-Pearson's test for normality
runs_test       - Runs test
sign_test1      - Two-sample sign test.
kruskal_wallis  - Kruskal-Wallis rank test.
friedman        - Friedman randomized block design test
kendall         - Computes Kendall's tau correlation statistic
spear           - Spearman correlation coefficient.
wmw             - Wilcoxon-Mann-Whitney two-sample test.
tablerxc        - test of independence for $r$x$c$ table.
mantel_haenszel- Mantel-Haenszel statistic for $2$x$2$ tables.
```

The listed nonparametric functions that are not distributed with MATLAB or its Statistics Toolbox can be downloaded from the book home page.

A.6.4 Regression Models

A.6.4.1 Ordinary Least Squares (OLS) The most straightforward way of implementing OLS is based on normal equations.

```
>> x = rand(20,1);
>> y = 2 + 3*x + randn(size(x));
>> X = [ones(length(x),1),x];
>> b = inv(X'*X)*X'*y   % normal equation

b =
    1.8778
    3.4689
```

A better solution uses backslash because it is more numerically stable than `inv`.

```
>> b2 = X\y

b2 =
    1.8778
    3.4689
```

The pseudo inverse function **pinv** is also an option. It too is numerically stable, but it will yield subtly different results when your matrix is singular or nearly so. Is **pinv** better? There are arguments for both backslash and **pinv**. The difference really lies in what happens on singular or nearly singular matrixes. **pinv** will not work on sparse problems, and because **pinv** relies on the singular value decomposition, it may be slower for large problems.

```
>> b3 = pinv(X)*y

b3 =
    1.8778
    3.4689
```

Large-scale problems where X is sparse may sometimes benefit from a sparse iterative solution. **lsqr** is an iterative solver.

```
>> b4 = lsqr(X,y,1.e-13,10)

lsqr converged at iteration 2 to  a solution with relative residual 0.33
```

```
b4 =
     1.8778
     3.4689
```

There is another option, `lscov`. `lscov` is designed to handle problems where the data covariance matrix is known. It can also solve a weighted regression problem.

```
>> b5 = lscov(X,y)

b5 =
     1.8778
     3.4689
```

Directly related to the backslash solution is one based on the QR factorization. If our over-determined system of equations to solve is $Xb = y$, then a quick look at the normal equations,

$$b = (X'X)^{-1}X'y$$

combined with the qr factorization of X,

$$X = QR$$

yields

$$b = (R'Q'QR)^{-1}R'Q'y.$$

Of course, we know that Q is an orthogonal matrix, so $Q'Q$ is an identity matrix.

$$b = (R'R)^{-1}R'Q'y$$

If R is non-singular, then $(R'R)^{-1} = R^{-1}R'^{-1}$, so we can further reduce to

$$b = R^{-1}Q'y$$

This solution is also useful for computing confidence intervals on the parameters.

```
>> [Q,R] = qr(X,0);
>> b6 = R\(Q'*y)

b6 =
     1.8778
     3.4689
```

A.6.4.2 Weighted Least Squares (WLS) Weighted Least Squares (WLS) is special case of Generalized Least Squares (GLS). It should be applied when there is heteroscedasticity in the regression, i.e. the variance of the error term is not a constant across observations. The optimal weights should be inversely proportional to the error variances.

```
>> x = (1:10)';
>> wgts = 1./rand(size(x));
>> y = 2 + 3*x + wgts.*randn(size(x));
>> X = [ones(length(x),1),x];
>> b7 = lscov(M,y,wgts)

b7 =
  -89.6867
   27.9335
```

Another alternative way of doing WLS is to transform the independent and dependent variables so that we apply OLS to the transformed data.

```
>> yw = y.*sqrt(wgts);
>> Xw = X.*repmat(sqrt(wgts),1,size(M,2));
>> b8 = Xw\yw

coef8 =
  -89.6867
   27.9335
```

A.6.4.3 Iterative Reweighted Least Squares (IRLS) IRLS can be used for multiple purposes. One is to get robust estimates by reducing the effect of outliers. Another is to fit a generalized linear model, as described in Section A.6.6. MATLAB provides a function `robustfit` which performs iterative reweighted least squares estimation which yield robust coefficient estimates.

```
>> x = (1:10)';
>> y = 10 - 2*x + randn(10,1); y(10) = 0;
>> brob = robustfit(x,y)

brob =
   10.5208
   -2.0902
```

A.6.4.4 Nonlinear Least Squares MATLAB provides a function `nlinfit` which performs nonlinear least squares estimation.

```
>> mymodel = @(beta, x) (beta(1)*x(:,2) - x(:,3)/beta(5)) ./ ...
             (1+beta(2)*x(:,1)+beta(3)*x(:,2)+beta(4)*x(:,3));
>> load reaction;
```

```
>> beta = nlinfit(reactants,rate,mymodel,ones(5,1))

beta =
    1.2526
    0.0628
    0.0400
    0.1124
    1.1914
```

A.6.4.5 Other Regression Functions

```
coxphfit    - Cox proportional hazards regression.
nlintool    - Graphical tool for prediction in nonlinear models.
nlpredci    - Confidence intervals for prediction in nonlinear models.
nlparci     - Confidence intervals for parameters in nonlinear models.
polyconf    - Polynomial evaluation and with confidence intervals.
polyfit     - Least-squares polynomial fitting.
polyval     - Predicted values for polynomial functions.
rcoplot     - Residuals case order plot.
regress     - Multivariate linear regression, also return the
              R-square statistic, the F statistic and p value for
              the full model, and an estimate of the error variance.
regstats    - Regression diagnostics for linear regression.
ridge       - Ridge regression.
rstool      - Multidimensional response surface visualization (RSM).
stepwise    - Interactive tool for stepwise regression.
stepwisefit - Non-interactive stepwise regression.
```

A.6.5 ANOVA

The following function set can be used to perform ANOVA in a parametric or nonparametric fashion.

```
anova1        - One-way analysis of variance.
anova2        - Two-way analysis of variance.
anovan        - n-way analysis of variance.
aoctool       - Interactive tool for analysis of covariance.
friedman      - Friedman's test (nonparametric two-way anova).
kruskalwallis - Kruskal-Wallis test (nonparametric one-way anova).
```

A.6.6 Generalized Linear Models

MATLAB provides the `glmfit` and `glmval` functions to fit generalized linear models. These models include Poisson regression, gamma regression, and binary probit or logistic regression. The functions allow you to specify a link function that relates the distribution parameters to the predictors. It is also possible to fit a weighted generalized linear model. Fig. A.17 is a result of the following MATLAB commands:

```
>> x = [2100 2300 2500 2700 2900 3100 3300 3500 3700 3900 4100 4300]';
>> n = [48 42 31 34 31 21 23 23 21 16 17 21]';
>> y = [1 2 0 3 8 8 14 17 19 15 17 21]';
>> b = glmfit(x, [y n], 'binomial', 'link', 'probit');
>> yfit = glmval(b, x, 'probit', 'size', n);
>> plot(x, y./n, 'o', x, yfit./n, '-')
```

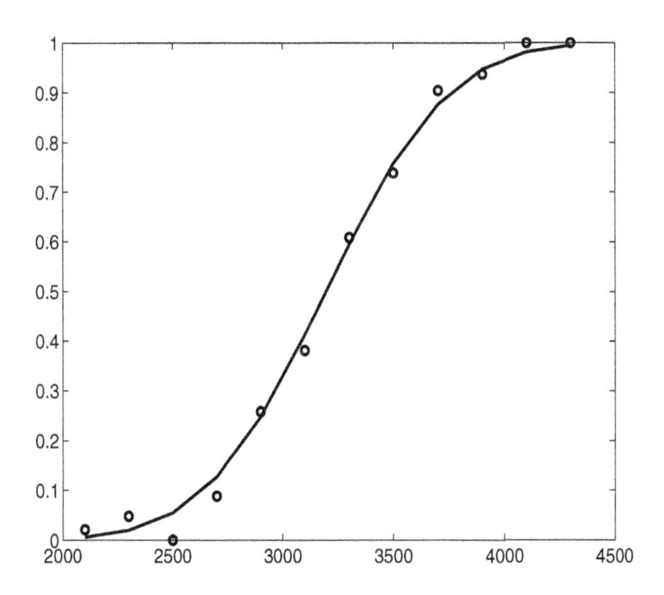

Fig. A.17 Probit regression example.

A.6.7 Hypothesis Testing

MATLAB also provide a set of functions to perform some important statistical tests. These tests include tests on location or dispersion. For example, `ttest` and `ttest2` can be used to do a t test.

```
Hypothesis Tests.
  ansaribradley - Ansari-Bradley two-sample test for equal dispersions.
  dwtest        - Durbin-Watson test for autocorrelation in regression.
  ranksum       - Wilcoxon rank sum test (independent samples).
  runstest      - Runs test for randomness.
  signrank      - Wilcoxon sign rank test (paired samples).
  signtest      - Sign test (paired samples).
```

```
  ztest        - Z test.
  ttest        - One sample t test.
  ttest2       - Two sample t test.
  vartest      - One-sample test of variance.
  vartest2     - Two-sample F test for equal variances.
  vartestn     - Test for equal variances across multiple groups.
```

Distribution tests, sometimes called goodness of fit tests, are also included. For example, `kstest` and `kstest2` are functions to perform a Kolmogorov-Smirnov test.

```
Distribution Testing.
  chi2gof      - Chi-square goodness-of-fit test.
  jbtest       - Jarque-Bera test of normality.
  kstest       - Kolmogorov-Smirnov test for one sample.
  kstest2      - Kolmogorov-Smirnov test for two samples.
  lillietest   - Lilliefors test of normality.
```

A.6.8 Statistical Learning

The following function provide tools to develop data mining/machine learning programs.

```
Factor Models
  factoran     - Factor analysis.
  pcacov       - Principal components from covariance matrix.
  pcares       - Residuals from principal components.
  princomp     - Principal components analysis from raw data.
  rotatefactors - Rotation of FA or PCA loadings.

Decision Tree Techniques.
  treedisp     - Display decision tree.
  treefit      - Fit data using a classification or regression tree.
  treeprune    - Prune decision tree or create optimal pruning sequence.
  treetest     - Estimate error for decision tree.
  treeval      - Compute fitted values using decision tree.

Discrimination Models
  classify     - Discriminant analysis with  'linear', 'quadratic',
 'diagLinear', 'diagQuadratic', or 'mahalanobis' discriminant function
```

A.6.9 Bootstrapping

In MATLAB, `boot` and `bootci` are used to obtain boostrap estimates. The former is used to draw bootstrapped samples from data and compute the bootstrapped statistics based on these samples. The latter computes the improved bootstrap confidence intervals, including the BCa interval.

```
>> load lawdata gpa lsat
>> se = std(bootstrp(1000,@corr,gpa,lsat))
>> bca = bootci(1000,{@corr,gpa,lsat})

se =
    0.1322

bca =
    0.3042
    0.9407
```

Appendix B: WinBUGS

Beware: MCMC sampling can be dangerous! (Disclaimer from WinBUGS User Manual)

BUGS is freely available software for constructing Bayesian statistical models and evaluating them using MCMC methodology.

BUGS and WinBUGS are distributed freely and are the result of many years of development by a team of statisticians and programmers at the Medical research Council Biostatistics Research Unit in Cambridge (BUGS and WinBUGS), and from recently by a team at University of Helsinki (OpenBUGS) see the project pages: `http://www.mrc-bsu.cam.ac.uk/bugs/` and `http://mathstat.helsinki.fi/openbugs/`.

Models are represented by a flexible language, and there is also a graphical feature, DOODLEBUGS, that allows users to specify their models as directed graphs. For complex models the DOODLEBUGS can be very useful. As of May 2007, the latest version of WinBUGS is 1.4.1 and OpenBUGS 3.0.

B.1 USING WINBUGS

We start the introduction to WinBUGS with a simple regression example. Consider the model

$$
\begin{aligned}
y_i|\mu_i,\tau &\sim \mathcal{N}(\mu_i,\tau),\ i=1,\dots,n \\
\mu_i &= \alpha+\beta(x_i-\bar{x}), \\
\alpha &\sim \mathcal{N}(0,10^{-4}) \\
\beta &\sim \mathcal{N}(0,10^{-4}) \\
\tau &\sim \mathcal{G}amma(0.001,0.001).
\end{aligned}
$$

The scale in normal distributions here is parameterized in terms of a *precision* parameter τ which is the reciprocal of variance, $\tau = 1/\sigma^2$. Natural distributions for the precision parameters are gamma and small values of the precision reflect the flatness (noninformativeness) of the priors. The parameters α and β are less correlated if predictors $x_i - \bar{x}$ are used instead of x_i. Assume that (x, y)-pairs $(1,1), (2,3), (3,3), (4,3)$, and $(5,5)$ are observed.

Estimators in classical, Least Square regression of y on $x - \bar{x}$, are given in the following table.

```
Coef      LSEstimate  SE Coef     t      p
ALPHA         3.0000   0.3266  9.19  0.003
BETA          0.8000   0.2309  3.46  0.041
S = 0.730297    R-Sq = 80.0%   R-Sq(adj) = 73.3%
```

How about Bayesian estimators? We will find the estimators by MCMC calculations as means on the simulated posteriors. Assume that the initial values of parameters are $\alpha_0 = 0.1$, $\beta_0 = 0.6$, and $\tau = 1$. Start BUGS and input the following code in [**File** > **New**].

```
# A simple regression
model{
  for (i in 1:N) {
  Y[i] ~ dnorm(mu[i],tau);
  mu[i] <- alpha + beta * (x[i] - x.bar);
  }
x.bar <- mean(x[]);
alpha ~ dnorm(0, 0.0001);
beta ~ dnorm(0, 0.0001);
tau ~ dgamma(0.001, 0.001);
sigma <- 1.0/sqrt(tau);
}
#-----------------------------
#these are observations
list( x=c(1,2,3,4,5), Y=c(1,3,3,3,5), N=5);
#-----------------------------
#the initial values
```

```
list(alpha = 0.1, beta = 0.6, tau = 1);
```

Next, put the cursor at an arbitrary position within the scope of `model` which delimited by wiggly brackets. Select the **Model** menu and open **Specification.** The **Specification Tool** window will pop-out. If your model is highlighted, you may **check model** in the specification tool window. If the model is correct, the response on the lower bar of the BUGS window should be: **model is syntactically correct.** Next, highlight the "list" statement in the data-part of your code. In the Specification Tool window select **load data.** If the data are in correct format, you should receive response on the bottom bar of BUGS window: **data loaded.** You will need to compile your model on order to activate **inits**-buttons. Select **compile** in the Specification Tool window. The response should be: **model compiled**, and the buttons **load inits** and **gen inits** become active. Finally, highlight the "list" statement in the initials-part of your code and in the Specification Tool window select **load inits.** The response should be: **model is initialized**, and this finishes reading in the model. If the response is **initial values loaded but this or other chain contain uninitialized variables**, click on the **gen inits** button. The response should be: **initial values generated, model initialized.**

Now, you are ready to Burn-in some simulations and at the same time check that the program is working. In the **Model** menu, choose **Update...** and open **Update Tool** to check if your model updates.

From the **Inference** menu, open **Samples....** A window titled **Sample Monitor Tool** will pop out. In the **node** sub-window input the names of the variables you want to monitor. In this case, the variables are `alpha, beta,` and `tau.` If you correctly input the variable the **set** button becomes active and you should set the variable. Do this for all 3 variables of interest. In fact, `sigma` as transformation of `tau` is available, as well.

Now choose `alpha` from the subwindow in **Sample Monitor Tool**. All of the buttons (**clear, set, trace, history, density, stats, coda, quantiles, bgr diag, auto cor**) are now active. Return to **Update Tool** and select the desired number of simulations, say 10000, in the **updates** subwindow. Press the **update** button.

Return to **Sample Monitor Tool** and check **trace** for the part of MC trace for α, **history** for the complete trace, **density** for a density estimator of α, etc. For example, pressing **stats** button will produce something like the following table

	mean	sd	MCerror	val2.5pc	median	val97.5pc	start	sample
`alpha`	3.003	0.549	0.003614	1.977	3.004	4.057	10000	20001

The mean 3.003 is the Bayes estimator (as the mean from the sample from the posterior for α. There are two precision outputs, `sd` and `MCerror`. The former is an estimator of the standard deviation of the posterior and can be improved by increasing the sample size but not the number of simulations. The later one is the error of simulation and can be improved by additional simulations. The 95% credible set is bounded by `val2.5pc` and `val97.5pc`, which are the 0.025 and 0.975 (empirical) quantiles from the posterior. The empirical median of the posterior is given by `median`. The outputs `start` and `sample` show the starting index for the simulations (after burn-in) and the available number of simulations.

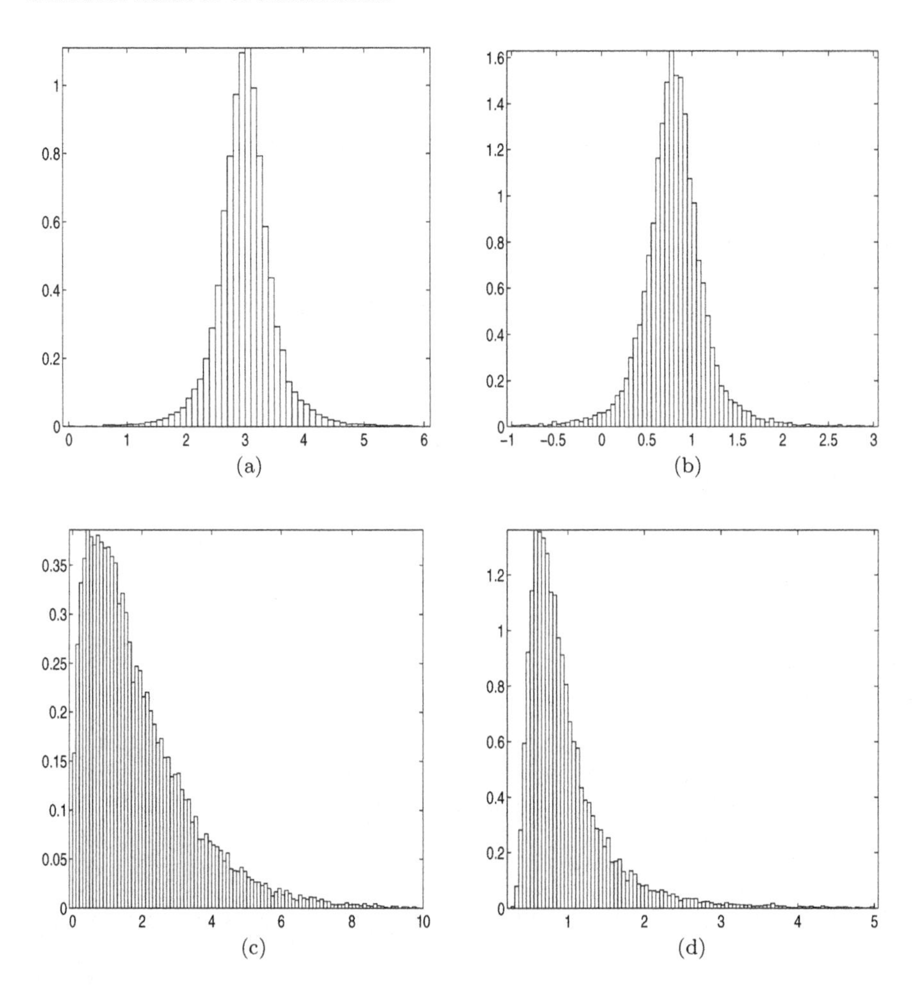

Fig. B.18 Traces of the four parameters from simple example: (a) α, (b) β, (c) τ, and (d) σ from WinBUGS. Data are plotted in MATLAB after being exported from WinBUGS.

For all parameters a comparative table is

	mean	sd	MCerror	val2.5pc	median	val97.5pc	start	sample
alpha	3.003	0.549	0.003614	1.977	3.004	4.057	10000	20001
beta	0.7994	0.3768	0.002897	0.07088	0.7988	1.534	10000	20001
tau	1.875	1.521	0.01574	0.1399	1.471	5.851	10000	20001
sigma	1.006	0.7153	0.009742	0.4134	0.8244	2.674	10000	20001

If you want to save the trace for α in a file and process it in MATLAB, say, select **coda** and the data window will open with an information window as well. Keep the data window active and select **Save As** from the **File** menu. Save the αs in `alphas.txt` where it will be ready to be imported to MATLAB.

Kevin Murphy lead the project for communication between WinBUGS and MATLAB:

`http://www.cs.ubc.ca/~murphyk/Software/MATBUGS/matbugs.html.`

His suite MATBUGS, maintained by several researchers, communicates with WinBUGS directly from MATLAB.

B.2 BUILT-IN FUNCTIONS AND COMMON DISTRIBUTIONS IN BUGS

This section contains two tables: one with the list of built-in functions and the second with the list of available distributions.

The first-time WinBUGS user may be disappointed by the selection of built in functions – the set is minimal but sufficient. The full list of distributions in WinBUGS can be found in **Help>WinBUGS User Manual** under **The_BUGS_language:_stochastic_nodes>Distributions**. BUGS also allows for construction of distributions for which are not in default list. In Table B.23 a list of important continuous and discrete distributions, with their BUGS syntax and parametrization, is provided. BUGS has the capability to define custom distributions, both as likelihood or as a prior, via the so called *zero-Poisson device.*

Table B.22 Built-in Functions in WinBUGS

BUGS Code	function
`abs(y)`	$\|y\|$
`cloglog(y)`	$\ln(-\ln(1-y))$
`cos(y)`	$\cos(y)$
`equals(y, z)`	1 if $y = z$; 0 otherwise
`exp(y)`	$\exp(y)$
`inprod(y, z)`	$\Sigma_i y_i z_i$
`inverse(y)`	y^{-1} for symmetric positive-definite matrix y
`log(y)`	$\ln(y)$
`logfact(y)`	$\ln(y!)$
`loggam(y)`	$\ln(\Gamma(y))$
`logit(y)`	$\ln(y/(1-y))$
`max(y, z)`	y if $y > z$; y otherwise
`mean(y)`	$n^{-1}\Sigma_i y_i, \quad n = dim(y)$
`min(y, z)`	y if $y < z$; z otherwise
`phi(y)`	standard normal CDF $\Phi(y)$
`pow(y, z)`	y^z
`sin(y)`	$\sin(y)$
`sqrt(y)`	$\sqrt{y}$
`rank(v, s)`	number of components of v less than or equal to v_s
`ranked(v, s)`	the sth smallest component of v
`round(y)`	nearest integer to y
`sd(y)`	standard deviation of components of y
`step(y)`	1 if $y \geq 0$; 0 otherwise
`sum(y)`	$\Sigma_i y_i$
`trunc(y)`	greatest integer less than or equal to y

Distribution	BUGS Code	Density
Bernoulli	`x ~ dbern(p)`	$p^x(1-p)^{1-x},\ x=0,1;\ 0 \le p \le 1$
Binomial	`x ~ dbin(p, n)`	$\binom{n}{x}p^x(1-p)^{n-x},\ x=0,\ldots,n;\ 0 \le p \le 1$
Categorical	`x ~ dcat(p[])`	$p[x],\ x=1,2,\ldots,\dim(p)$
Poisson	`x ~ dpois(lambda)`	$\frac{\lambda^x}{x!}\exp\{-\lambda\}, x=0,1,2,\ldots,\ \lambda>0$
Beta	`x ~ dbeta(a,b)`	$\frac{1}{B(a,b)}x^{a-1}(1-x)^{b-1},\ 0\ =\ x \le 1,\ a,b>-1$
Chi-square	`x ~ dchisqr(k)`	$\frac{x^{k/1-1}\exp\{-x/2\}}{2^{k/2}\Gamma(k/2)},\ x \ge 0,\ k>0$
Double Exponential	`x ~ ddexp(mu, tau)`	$\frac{\tau}{2}\exp\{-\tau\lvert x-\mu\rvert\}, x \in R,\ \tau>0,\ \mu \in R$
Exponential	`x ~ dexp(lambda)`	$\lambda\exp\{-\lambda x\},\ x \ge 0,\ \lambda \ge 0$
Flat	`x ~ dflat()`	constant; not a proper density
Gamma	`x ~ dgamma(a, b)`	$\frac{b^a x^{a-1}}{\Gamma(a)}\exp(-bx),\ x,a,b>0$
Normal	`x ~ dnorm(mu, tau)`	$\sqrt{\tau/(2\pi)}\exp\{-\frac{\tau}{2}(x-\mu)^2\},\ x,\mu \in R, \tau>0$
Pareto	`x ~ dpar)alpha,c)`	$\alpha c^\alpha x^{-(\alpha+1)}, x>c$
Student-t	`x ~ dt(mu, tau, k)`	$\frac{\Gamma((k+1)/2)}{\Gamma(k/2)}\sqrt{\frac{\tau}{k\pi}}[1+\frac{\tau}{k}(x-\mu)^2]^{-(k+1)/2}, x \in \mathbb{R}, k \ge 2$
Uniform	`x ~ dunif(a, b)`	$\frac{1}{b-a},\ a \le x \le b$
Weibull	`x ~ dweib(v, lambda)`	$v\lambda x^{v-1}\exp\{-\lambda x^v\},\ x,v,\lambda>0,$
Multinomial	`x[] ~ dmulti(p[], N)`	$\frac{(\Sigma_i x_i)!}{\prod_i x_i!}\prod_i p_i^{x_i}, \Sigma_i x_i=N, 0<p_i<1, \Sigma_i p_i=1$
Dirichlet	`p[] ~ ddirch(alpha[])`	$\frac{\Gamma(\Sigma_i\alpha_i)}{\prod_i\Gamma(\alpha_i)}\prod_i p_i^{\alpha_i-1},\ 0<p_i<1, \Sigma_i p_i=1$
Multivariate Normal	`x[] ~ dmnorm(mu[], T[,])`	$(2\pi)^{-d/2}\lvert T\rvert^{1/2}\exp\{-1/2(x-\mu)'T(x-\mu)\},\ x \in R^d$
Multivariate Student-t	`x[] ~ dmt(mu[], T[,], k)`	$\frac{\Gamma((k+d)/2)}{\Gamma(k/2)}\frac{\lvert T\rvert^{1/2}}{k^{d/2}\pi^{d/2}}\left[1+\frac{1}{k}(x-\mu)'T(x-\mu)\right]^{-(k+d)/2},\ x \in \mathbb{R}^d, k \ge 2$
Wishart	`x[,] ~ dwish(R[,], k)`	$\lvert R\rvert^{k/2}\lvert x\rvert^{(k-p-1)/2}\exp\{-1/2Tr(Rx)\}$

Table B.23 Built-in distributions with BUGS names and their parametrizations.

MATLAB Index

Author Index

Subject Index

WILEY SERIES IN PROBABILITY AND STATISTICS

ESTABLISHED BY WALTER A. SHEWHART AND SAMUEL S. WILKS

The ***Wiley Series in Probability and Statistics*** is well established and authoritative. It covers many topics of current research interest in both pure and applied statistics and probability theory. Written by leading statisticians and institutions, the titles span both state-of-the-art developments in the field and classical methods.

Reflecting the wide range of current research in statistics, the series encompasses applied, methodological and theoretical statistics, ranging from applications and new techniques made possible by advances in computerized practice to rigorous treatment of theoretical approaches.

This series provides essential and invaluable reading for all statisticians, whether in academia, industry, government, or research.

† ABRAHAM and LEDOLTER · Statistical Methods for Forecasting
AGRESTI · Analysis of Ordinal Categorical Data
AGRESTI · An Introduction to Categorical Data Analysis, *Second Edition*
AGRESTI · Categorical Data Analysis, *Second Edition*
ALTMAN, GILL, and McDONALD · Numerical Issues in Statistical Computing for the Social Scientist
AMARATUNGA and CABRERA · Exploration and Analysis of DNA Microarray and Protein Array Data
ANDĚL · Mathematics of Chance
ANDERSON · An Introduction to Multivariate Statistical Analysis, *Third Edition*
* ANDERSON · The Statistical Analysis of Time Series
ANDERSON, AUQUIER, HAUCK, OAKES, VANDAELE, and WEISBERG · Statistical Methods for Comparative Studies
ANDERSON and LOYNES · The Teaching of Practical Statistics
ARMITAGE and DAVID (editors) · Advances in Biometry
ARNOLD, BALAKRISHNAN, and NAGARAJA · Records
* ARTHANARI and DODGE · Mathematical Programming in Statistics
* BAILEY · The Elements of Stochastic Processes with Applications to the Natural Sciences
BALAKRISHNAN and KOUTRAS · Runs and Scans with Applications
BALAKRISHNAN and NG · Precedence-Type Tests and Applications
BARNETT · Comparative Statistical Inference, *Third Edition*
BARNETT · Environmental Statistics
BARNETT and LEWIS · Outliers in Statistical Data, *Third Edition*
BARTOSZYNSKI and NIEWIADOMSKA-BUGAJ · Probability and Statistical Inference
BASILEVSKY · Statistical Factor Analysis and Related Methods: Theory and Applications
BASU and RIGDON · Statistical Methods for the Reliability of Repairable Systems
BATES and WATTS · Nonlinear Regression Analysis and Its Applications

*Now available in a lower priced paperback edition in the Wiley Classics Library.
†Now available in a lower priced paperback edition in the Wiley–Interscience Paperback Series.

BECHHOFER, SANTNER, and GOLDSMAN · Design and Analysis of Experiments for Statistical Selection, Screening, and Multiple Comparisons
BELSLEY · Conditioning Diagnostics: Collinearity and Weak Data in Regression
† BELSLEY, KUH, and WELSCH · Regression Diagnostics: Identifying Influential Data and Sources of Collinearity
BENDAT and PIERSOL · Random Data: Analysis and Measurement Procedures, *Third Edition*
BERRY, CHALONER, and GEWEKE · Bayesian Analysis in Statistics and Econometrics: Essays in Honor of Arnold Zellner
BERNARDO and SMITH · Bayesian Theory
BHAT and MILLER · Elements of Applied Stochastic Processes, *Third Edition*
BHATTACHARYA and WAYMIRE · Stochastic Processes with Applications
BILLINGSLEY · Convergence of Probability Measures, *Second Edition*
BILLINGSLEY · Probability and Measure, *Third Edition*
BIRKES and DODGE · Alternative Methods of Regression
BLISCHKE AND MURTHY (editors) · Case Studies in Reliability and Maintenance
BLISCHKE AND MURTHY · Reliability: Modeling, Prediction, and Optimization
BLOOMFIELD · Fourier Analysis of Time Series: An Introduction, *Second Edition*
BOLLEN · Structural Equations with Latent Variables
BOLLEN and CURRAN · Latent Curve Models: A Structural Equation Perspective
BOROVKOV · Ergodicity and Stability of Stochastic Processes
BOULEAU · Numerical Methods for Stochastic Processes
BOX · Bayesian Inference in Statistical Analysis
BOX · R. A. Fisher, the Life of a Scientist
BOX and DRAPER · Response Surfaces, Mixtures, and Ridge Analyses, *Second Edition*
* BOX and DRAPER · Evolutionary Operation: A Statistical Method for Process Improvement
BOX and FRIENDS · Improving Almost Anything, *Revised Edition*
BOX, HUNTER, and HUNTER · Statistics for Experimenters: Design, Innovation, and Discovery, *Second Editon*
BOX and LUCEÑO · Statistical Control by Monitoring and Feedback Adjustment
BRANDIMARTE · Numerical Methods in Finance: A MATLAB-Based Introduction
BROWN and HOLLANDER · Statistics: A Biomedical Introduction
BRUNNER, DOMHOF, and LANGER · Nonparametric Analysis of Longitudinal Data in Factorial Experiments
BUCKLEW · Large Deviation Techniques in Decision, Simulation, and Estimation
CAIROLI and DALANG · Sequential Stochastic Optimization
CASTILLO, HADI, BALAKRISHNAN, and SARABIA · Extreme Value and Related Models with Applications in Engineering and Science
CHAN · Time Series: Applications to Finance
CHARALAMBIDES · Combinatorial Methods in Discrete Distributions
CHATTERJEE and HADI · Regression Analysis by Example, *Fourth Edition*
CHATTERJEE and HADI · Sensitivity Analysis in Linear Regression
CHERNICK · Bootstrap Methods: A Practitioner's Guide
CHERNICK and FRIIS · Introductory Biostatistics for the Health Sciences
CHILÈS and DELFINER · Geostatistics: Modeling Spatial Uncertainty
CHOW and LIU · Design and Analysis of Clinical Trials: Concepts and Methodologies, *Second Edition*
CLARKE and DISNEY · Probability and Random Processes: A First Course with Applications, *Second Edition*
* COCHRAN and COX · Experimental Designs, *Second Edition*
CONGDON · Applied Bayesian Modelling
CONGDON · Bayesian Models for Categorical Data
CONGDON · Bayesian Statistical Modelling

*Now available in a lower priced paperback edition in the Wiley Classics Library.
†Now available in a lower priced paperback edition in the Wiley–Interscience Paperback Series.

CONOVER · Practical Nonparametric Statistics, *Third Edition*
COOK · Regression Graphics
COOK and WEISBERG · Applied Regression Including Computing and Graphics
COOK and WEISBERG · An Introduction to Regression Graphics
CORNELL · Experiments with Mixtures, Designs, Models, and the Analysis of Mixture Data, *Third Edition*
COVER and THOMAS · Elements of Information Theory
COX · A Handbook of Introductory Statistical Methods
* COX · Planning of Experiments
CRESSIE · Statistics for Spatial Data, *Revised Edition*
CSÖRGŐ and HORVÁTH · Limit Theorems in Change Point Analysis
DANIEL · Applications of Statistics to Industrial Experimentation
DANIEL · Biostatistics: A Foundation for Analysis in the Health Sciences, *Eighth Edition*
* DANIEL · Fitting Equations to Data: Computer Analysis of Multifactor Data, *Second Edition*
DASU and JOHNSON · Exploratory Data Mining and Data Cleaning
DAVID and NAGARAJA · Order Statistics, *Third Edition*
* DEGROOT, FIENBERG, and KADANE · Statistics and the Law
DEL CASTILLO · Statistical Process Adjustment for Quality Control
DEMARIS · Regression with Social Data: Modeling Continuous and Limited Response Variables
DEMIDENKO · Mixed Models: Theory and Applications
DENISON, HOLMES, MALLICK and SMITH · Bayesian Methods for Nonlinear Classification and Regression
DETTE and STUDDEN · The Theory of Canonical Moments with Applications in Statistics, Probability, and Analysis
DEY and MUKERJEE · Fractional Factorial Plans
DILLON and GOLDSTEIN · Multivariate Analysis: Methods and Applications
DODGE · Alternative Methods of Regression
* DODGE and ROMIG · Sampling Inspection Tables, *Second Edition*
* DOOB · Stochastic Processes
DOWDY, WEARDEN, and CHILKO · Statistics for Research, *Third Edition*
DRAPER and SMITH · Applied Regression Analysis, *Third Edition*
DRYDEN and MARDIA · Statistical Shape Analysis
DUDEWICZ and MISHRA · Modern Mathematical Statistics
DUNN and CLARK · Basic Statistics: A Primer for the Biomedical Sciences, *Third Edition*
DUPUIS and ELLIS · A Weak Convergence Approach to the Theory of Large Deviations
EDLER and KITSOS · Recent Advances in Quantitative Methods in Cancer and Human Health Risk Assessment
* ELANDT-JOHNSON and JOHNSON · Survival Models and Data Analysis
ENDERS · Applied Econometric Time Series
† ETHIER and KURTZ · Markov Processes: Characterization and Convergence
EVANS, HASTINGS, and PEACOCK · Statistical Distributions, *Third Edition*
FELLER · An Introduction to Probability Theory and Its Applications, Volume I, *Third Edition,* Revised; Volume II, *Second Edition*
FISHER and VAN BELLE · Biostatistics: A Methodology for the Health Sciences
FITZMAURICE, LAIRD, and WARE · Applied Longitudinal Analysis
* FLEISS · The Design and Analysis of Clinical Experiments
FLEISS · Statistical Methods for Rates and Proportions, *Third Edition*
† FLEMING and HARRINGTON · Counting Processes and Survival Analysis
FULLER · Introduction to Statistical Time Series, *Second Edition*
† FULLER · Measurement Error Models

*Now available in a lower priced paperback edition in the Wiley Classics Library.
†Now available in a lower priced paperback edition in the Wiley–Interscience Paperback Series.

GALLANT · Nonlinear Statistical Models
GEISSER · Modes of Parametric Statistical Inference
GELMAN and MENG · Applied Bayesian Modeling and Causal Inference from Incomplete-Data Perspectives
GEWEKE · Contemporary Bayesian Econometrics and Statistics
GHOSH, MUKHOPADHYAY, and SEN · Sequential Estimation
GIESBRECHT and GUMPERTZ · Planning, Construction, and Statistical Analysis of Comparative Experiments
GIFI · Nonlinear Multivariate Analysis
GIVENS and HOETING · Computational Statistics
GLASSERMAN and YAO · Monotone Structure in Discrete-Event Systems
GNANADESIKAN · Methods for Statistical Data Analysis of Multivariate Observations, *Second Edition*
GOLDSTEIN and LEWIS · Assessment: Problems, Development, and Statistical Issues
GREENWOOD and NIKULIN · A Guide to Chi-Squared Testing
GROSS and HARRIS · Fundamentals of Queueing Theory, *Third Edition*
* HAHN and SHAPIRO · Statistical Models in Engineering
HAHN and MEEKER · Statistical Intervals: A Guide for Practitioners
HALD · A History of Probability and Statistics and their Applications Before 1750
HALD · A History of Mathematical Statistics from 1750 to 1930
† HAMPEL · Robust Statistics: The Approach Based on Influence Functions
HANNAN and DEISTLER · The Statistical Theory of Linear Systems
HEIBERGER · Computation for the Analysis of Designed Experiments
HEDAYAT and SINHA · Design and Inference in Finite Population Sampling
HEDEKER and GIBBONS · Longitudinal Data Analysis
HELLER · MACSYMA for Statisticians
HINKELMANN and KEMPTHORNE · Design and Analysis of Experiments, Volume 1: Introduction to Experimental Design
HINKELMANN and KEMPTHORNE · Design and Analysis of Experiments, Volume 2: Advanced Experimental Design
HOAGLIN, MOSTELLER, and TUKEY · Exploratory Approach to Analysis of Variance
* HOAGLIN, MOSTELLER, and TUKEY · Exploring Data Tables, Trends and Shapes
* HOAGLIN, MOSTELLER, and TUKEY · Understanding Robust and Exploratory Data Analysis
HOCHBERG and TAMHANE · Multiple Comparison Procedures
HOCKING · Methods and Applications of Linear Models: Regression and the Analysis of Variance, *Second Edition*
HOEL · Introduction to Mathematical Statistics, *Fifth Edition*
HOGG and KLUGMAN · Loss Distributions
HOLLANDER and WOLFE · Nonparametric Statistical Methods, *Second Edition*
HOSMER and LEMESHOW · Applied Logistic Regression, *Second Edition*
HOSMER and LEMESHOW · Applied Survival Analysis: Regression Modeling of Time to Event Data
† HUBER · Robust Statistics
HUBERTY · Applied Discriminant Analysis
HUBERTY and OLEJNIK · Applied MANOVA and Discriminant Analysis, *Second Edition*
HUNT and KENNEDY · Financial Derivatives in Theory and Practice, *Revised Edition*
HUSKOVA, BERAN, and DUPAC · Collected Works of Jaroslav Hajek—with Commentary
HUZURBAZAR · Flowgraph Models for Multistate Time-to-Event Data
IMAN and CONOVER · A Modern Approach to Statistics

*Now available in a lower priced paperback edition in the Wiley Classics Library.
†Now available in a lower priced paperback edition in the Wiley–Interscience Paperback Series.

† JACKSON · A User's Guide to Principle Components
JOHN · Statistical Methods in Engineering and Quality Assurance
JOHNSON · Multivariate Statistical Simulation
JOHNSON and BALAKRISHNAN · Advances in the Theory and Practice of Statistics: A Volume in Honor of Samuel Kotz
JOHNSON and BHATTACHARYYA · Statistics: Principles and Methods, *Fifth Edition*
JOHNSON and KOTZ · Distributions in Statistics
JOHNSON and KOTZ (editors) · Leading Personalities in Statistical Sciences: From the Seventeenth Century to the Present
JOHNSON, KOTZ, and BALAKRISHNAN · Continuous Univariate Distributions, Volume 1, *Second Edition*
JOHNSON, KOTZ, and BALAKRISHNAN · Continuous Univariate Distributions, Volume 2, *Second Edition*
JOHNSON, KOTZ, and BALAKRISHNAN · Discrete Multivariate Distributions
JOHNSON, KEMP, and KOTZ · Univariate Discrete Distributions, *Third Edition*
JUDGE, GRIFFITHS, HILL, LÜTKEPOHL, and LEE · The Theory and Practice of Econometrics, *Second Edition*
JUREČKOVÁ and SEN · Robust Statistical Procedures: Aymptotics and Interrelations
JUREK and MASON · Operator-Limit Distributions in Probability Theory
KADANE · Bayesian Methods and Ethics in a Clinical Trial Design
KADANE AND SCHUM · A Probabilistic Analysis of the Sacco and Vanzetti Evidence
KALBFLEISCH and PRENTICE · The Statistical Analysis of Failure Time Data, *Second Edition*
KARIYA and KURATA · Generalized Least Squares
KASS and VOS · Geometrical Foundations of Asymptotic Inference
† KAUFMAN and ROUSSEEUW · Finding Groups in Data: An Introduction to Cluster Analysis
KEDEM and FOKIANOS · Regression Models for Time Series Analysis
KENDALL, BARDEN, CARNE, and LE · Shape and Shape Theory
KHURI · Advanced Calculus with Applications in Statistics, *Second Edition*
KHURI, MATHEW, and SINHA · Statistical Tests for Mixed Linear Models
KLEIBER and KOTZ · Statistical Size Distributions in Economics and Actuarial Sciences
KLUGMAN, PANJER, and WILLMOT · Loss Models: From Data to Decisions, *Second Edition*
KLUGMAN, PANJER, and WILLMOT · Solutions Manual to Accompany Loss Models: From Data to Decisions, *Second Edition*
KOTZ, BALAKRISHNAN, and JOHNSON · Continuous Multivariate Distributions, Volume 1, *Second Edition*
KOVALENKO, KUZNETZOV, and PEGG · Mathematical Theory of Reliability of Time-Dependent Systems with Practical Applications
KVAM and VIDAKOVIC · Nonparametric Statistics with Applications to Science and Engineering
LACHIN · Biostatistical Methods: The Assessment of Relative Risks
LAD · Operational Subjective Statistical Methods: A Mathematical, Philosophical, and Historical Introduction
LAMPERTI · Probability: A Survey of the Mathematical Theory, *Second Edition*
LANGE, RYAN, BILLARD, BRILLINGER, CONQUEST, and GREENHOUSE · Case Studies in Biometry
LARSON · Introduction to Probability Theory and Statistical Inference, *Third Edition*
LAWLESS · Statistical Models and Methods for Lifetime Data, *Second Edition*
LAWSON · Statistical Methods in Spatial Epidemiology
LE · Applied Categorical Data Analysis
LE · Applied Survival Analysis

*Now available in a lower priced paperback edition in the Wiley Classics Library.
†Now available in a lower priced paperback edition in the Wiley–Interscience Paperback Series.

LEE and WANG · Statistical Methods for Survival Data Analysis, *Third Edition*
LEPAGE and BILLARD · Exploring the Limits of Bootstrap
LEYLAND and GOLDSTEIN (editors) · Multilevel Modelling of Health Statistics
LIAO · Statistical Group Comparison
LINDVALL · Lectures on the Coupling Method
LIN · Introductory Stochastic Analysis for Finance and Insurance
LINHART and ZUCCHINI · Model Selection
LITTLE and RUBIN · Statistical Analysis with Missing Data, *Second Edition*
LLOYD · The Statistical Analysis of Categorical Data
LOWEN and TEICH · Fractal-Based Point Processes
MAGNUS and NEUDECKER · Matrix Differential Calculus with Applications in Statistics and Econometrics, *Revised Edition*
MALLER and ZHOU · Survival Analysis with Long Term Survivors
MALLOWS · Design, Data, and Analysis by Some Friends of Cuthbert Daniel
MANN, SCHAFER, and SINGPURWALLA · Methods for Statistical Analysis of Reliability and Life Data
MANTON, WOODBURY, and TOLLEY · Statistical Applications Using Fuzzy Sets
MARCHETTE · Random Graphs for Statistical Pattern Recognition
MARDIA and JUPP · Directional Statistics
MASON, GUNST, and HESS · Statistical Design and Analysis of Experiments with Applications to Engineering and Science, *Second Edition*
McCULLOCH and SEARLE · Generalized, Linear, and Mixed Models
McFADDEN · Management of Data in Clinical Trials
* McLACHLAN · Discriminant Analysis and Statistical Pattern Recognition
McLACHLAN, DO, and AMBROISE · Analyzing Microarray Gene Expression Data
McLACHLAN and KRISHNAN · The EM Algorithm and Extensions
McLACHLAN and PEEL · Finite Mixture Models
McNEIL · Epidemiological Research Methods
MEEKER and ESCOBAR · Statistical Methods for Reliability Data
MEERSCHAERT and SCHEFFLER · Limit Distributions for Sums of Independent Random Vectors: Heavy Tails in Theory and Practice
MICKEY, DUNN, and CLARK · Applied Statistics: Analysis of Variance and Regression, *Third Edition*
* MILLER · Survival Analysis, *Second Edition*
MONTGOMERY, PECK, and VINING · Introduction to Linear Regression Analysis, *Fourth Edition*
MORGENTHALER and TUKEY · Configural Polysampling: A Route to Practical Robustness
MUIRHEAD · Aspects of Multivariate Statistical Theory
MULLER and STOYAN · Comparison Methods for Stochastic Models and Risks
MURRAY · X-STAT 2.0 Statistical Experimentation, Design Data Analysis, and Nonlinear Optimization
MURTHY, XIE, and JIANG · Weibull Models
MYERS and MONTGOMERY · Response Surface Methodology: Process and Product Optimization Using Designed Experiments, *Second Edition*
MYERS, MONTGOMERY, and VINING · Generalized Linear Models. With Applications in Engineering and the Sciences
† NELSON · Accelerated Testing, Statistical Models, Test Plans, and Data Analyses
† NELSON · Applied Life Data Analysis
NEWMAN · Biostatistical Methods in Epidemiology
OCHI · Applied Probability and Stochastic Processes in Engineering and Physical Sciences
OKABE, BOOTS, SUGIHARA, and CHIU · Spatial Tesselations: Concepts and Applications of Voronoi Diagrams, *Second Edition*

*Now available in a lower priced paperback edition in the Wiley Classics Library.
†Now available in a lower priced paperback edition in the Wiley–Interscience Paperback Series.

OLIVER and SMITH · Influence Diagrams, Belief Nets and Decision Analysis
PALTA · Quantitative Methods in Population Health: Extensions of Ordinary Regressions
PANJER · Operational Risk: Modeling and Analytics
PANKRATZ · Forecasting with Dynamic Regression Models
PANKRATZ · Forecasting with Univariate Box-Jenkins Models: Concepts and Cases
* PARZEN · Modern Probability Theory and Its Applications
PEÑA, TIAO, and TSAY · A Course in Time Series Analysis
PIANTADOSI · Clinical Trials: A Methodologic Perspective
PORT · Theoretical Probability for Applications
POURAHMADI · Foundations of Time Series Analysis and Prediction Theory
PRESS · Bayesian Statistics: Principles, Models, and Applications
PRESS · Subjective and Objective Bayesian Statistics, *Second Edition*
PRESS and TANUR · The Subjectivity of Scientists and the Bayesian Approach
PUKELSHEIM · Optimal Experimental Design
PURI, VILAPLANA, and WERTZ · New Perspectives in Theoretical and Applied Statistics
† PUTERMAN · Markov Decision Processes: Discrete Stochastic Dynamic Programming
QIU · Image Processing and Jump Regression Analysis
* RAO · Linear Statistical Inference and Its Applications, *Second Edition*
RAUSAND and HØYLAND · System Reliability Theory: Models, Statistical Methods, and Applications, *Second Edition*
RENCHER · Linear Models in Statistics
RENCHER · Methods of Multivariate Analysis, *Second Edition*
RENCHER · Multivariate Statistical Inference with Applications
* RIPLEY · Spatial Statistics
* RIPLEY · Stochastic Simulation
ROBINSON · Practical Strategies for Experimenting
ROHATGI and SALEH · An Introduction to Probability and Statistics, *Second Edition*
ROLSKI, SCHMIDLI, SCHMIDT, and TEUGELS · Stochastic Processes for Insurance and Finance
ROSENBERGER and LACHIN · Randomization in Clinical Trials: Theory and Practice
ROSS · Introduction to Probability and Statistics for Engineers and Scientists
ROSSI, ALLENBY, and McCULLOCH · Bayesian Statistics and Marketing
† ROUSSEEUW and LEROY · Robust Regression and Outlier Detection
* RUBIN · Multiple Imputation for Nonresponse in Surveys
RUBINSTEIN · Simulation and the Monte Carlo Method
RUBINSTEIN and MELAMED · Modern Simulation and Modeling
RYAN · Modern Experimental Design
RYAN · Modern Regression Methods
RYAN · Statistical Methods for Quality Improvement, *Second Edition*
SALEH · Theory of Preliminary Test and Stein-Type Estimation with Applications
* SCHEFFE · The Analysis of Variance
SCHIMEK · Smoothing and Regression: Approaches, Computation, and Application
SCHOTT · Matrix Analysis for Statistics, *Second Edition*
SCHOUTENS · Levy Processes in Finance: Pricing Financial Derivatives
SCHUSS · Theory and Applications of Stochastic Differential Equations
SCOTT · Multivariate Density Estimation: Theory, Practice, and Visualization
† SEARLE · Linear Models for Unbalanced Data
† SEARLE · Matrix Algebra Useful for Statistics
† SEARLE, CASELLA, and McCULLOCH · Variance Components
SEARLE and WILLETT · Matrix Algebra for Applied Economics
SEBER and LEE · Linear Regression Analysis, *Second Edition*
† SEBER · Multivariate Observations
† SEBER and WILD · Nonlinear Regression

*Now available in a lower priced paperback edition in the Wiley Classics Library.
†Now available in a lower priced paperback edition in the Wiley–Interscience Paperback Series.

SENNOTT · Stochastic Dynamic Programming and the Control of Queueing Systems
* SERFLING · Approximation Theorems of Mathematical Statistics
SHAFER and VOVK · Probability and Finance: It's Only a Game!
SILVAPULLE and SEN · Constrained Statistical Inference: Inequality, Order, and Shape Restrictions
SMALL and McLEISH · Hilbert Space Methods in Probability and Statistical Inference
SRIVASTAVA · Methods of Multivariate Statistics
STAPLETON · Linear Statistical Models
STAUDTE and SHEATHER · Robust Estimation and Testing
STOYAN, KENDALL, and MECKE · Stochastic Geometry and Its Applications, *Second Edition*
STOYAN and STOYAN · Fractals, Random Shapes and Point Fields: Methods of Geometrical Statistics
STREET and BURGESS · The Construction of Optimal Stated Choice Experiments: Theory and Methods
STYAN · The Collected Papers of T. W. Anderson: 1943–1985
SUTTON, ABRAMS, JONES, SHELDON, and SONG · Methods for Meta-Analysis in Medical Research
TAKEZAWA · Introduction to Nonparametric Regression
TANAKA · Time Series Analysis: Nonstationary and Noninvertible Distribution Theory
THOMPSON · Empirical Model Building
THOMPSON · Sampling, *Second Edition*
THOMPSON · Simulation: A Modeler's Approach
THOMPSON and SEBER · Adaptive Sampling
THOMPSON, WILLIAMS, and FINDLAY · Models for Investors in Real World Markets
TIAO, BISGAARD, HILL, PEÑA, and STIGLER (editors) · Box on Quality and Discovery: with Design, Control, and Robustness
TIERNEY · LISP-STAT: An Object-Oriented Environment for Statistical Computing and Dynamic Graphics
TSAY · Analysis of Financial Time Series, *Second Edition*
UPTON and FINGLETON · Spatial Data Analysis by Example, Volume II: Categorical and Directional Data
VAN BELLE · Statistical Rules of Thumb
VAN BELLE, FISHER, HEAGERTY, and LUMLEY · Biostatistics: A Methodology for the Health Sciences, *Second Edition*
VESTRUP · The Theory of Measures and Integration
VIDAKOVIC · Statistical Modeling by Wavelets
VINOD and REAGLE · Preparing for the Worst: Incorporating Downside Risk in Stock Market Investments
WALLER and GOTWAY · Applied Spatial Statistics for Public Health Data
WEERAHANDI · Generalized Inference in Repeated Measures: Exact Methods in MANOVA and Mixed Models
WEISBERG · Applied Linear Regression, *Third Edition*
WELSH · Aspects of Statistical Inference
WESTFALL and YOUNG · Resampling-Based Multiple Testing: Examples and Methods for p-Value Adjustment
WHITTAKER · Graphical Models in Applied Multivariate Statistics
WINKER · Optimization Heuristics in Economics: Applications of Threshold Accepting
WONNACOTT and WONNACOTT · Econometrics, *Second Edition*
WOODING · Planning Pharmaceutical Clinical Trials: Basic Statistical Principles
WOODWORTH · Biostatistics: A Bayesian Introduction
WOOLSON and CLARKE · Statistical Methods for the Analysis of Biomedical Data, *Second Edition*

*Now available in a lower priced paperback edition in the Wiley Classics Library.
†Now available in a lower priced paperback edition in the Wiley–Interscience Paperback Series.

WU and HAMADA · Experiments: Planning, Analysis, and Parameter Design Optimization
WU and ZHANG · Nonparametric Regression Methods for Longitudinal Data Analysis
YANG · The Construction Theory of Denumerable Markov Processes
YOUNG, VALERO-MORA, and FRIENDLY · Visual Statistics: Seeing Data with Dynamic Interactive Graphics
ZELTERMAN · Discrete Distributions—Applications in the Health Sciences
* ZELLNER · An Introduction to Bayesian Inference in Econometrics
ZHOU, OBUCHOWSKI, and McCLISH · Statistical Methods in Diagnostic Medicine

*Now available in a lower priced paperback edition in the Wiley Classics Library.
†Now available in a lower priced paperback edition in the Wiley–Interscience Paperback Series.

CPSIA information can be obtained
at www.ICGtesting.com
Printed in the USA
BVOW06*0808251017
498193BV00012B/16/P